18 Springer Series in Solid-State Sciences

Edited by Manuel Cardona and Peter Fulde

Springer Series in Solid-State Sciences

Editors: M. Cardona P. Fulde H.-J. Queisser

1 **Principles of Magnetic Resonance**
2nd Edition 2nd Printing
By C. P. Slichter

2 **Introduction to Solid-State Theory**
By O. Madelung

3 **Dynamical Scattering of X-Rays in Crystals** By Z. G. Pinsker

4 **Inelastic Electron Tunneling Spectroscopy**
Editor: T. Wolfram

5 **Fundamentals of Crystal Growth I**
Macroscopic Equilibrium and Transport Concepts. By F. Rosenberger

6 **Magnetic Flux Structures in Superconductors** By R. P. Huebener

7 **Green's Functions in Quantum Physics**
By E. N. Economou

8 **Solitions and Condensed Matter Physics**
Editors: A. R. Bishop and T. Schneider

9 **Photoferroelectrics**
By. V. M. Fridkin

10 **Phonon Dispersion Relations in Insulators** By H. Bilz and W. Kress

11 **Electron Transport in Compound Semiconductors** By B. R. Nag

12 **The Physics of Elementary Excitations**
By S. Nakajima

13 **The Physics of Selenium and Tellurium**
Editors: E. Gerlach and P. Grosse

14 **Magnetic Bubble Technology**
By A. H. Eschenfelder

15 **Modern Crystallography I**
Symmetry of Crystals,
Methods of Structural Crystallography
By B. K. Vainshtein

16 **Electronic States in Organic Molecular Crystals** By E. Silinsh

17 **The Theory of Magnetism I**
Ground State and Elementary Excitations
By D. Mattis

18 **Relaxation of Elementary Excitations**
Editors: R. Kubo and E. Hanamura

19 **Solitons,** Mathematical Methods for Physicists
By G. Eilenberger

Relaxation of Elementary Excitations

Proceedings of the
Taniguchi International Symposium
Susono-shi, Japan, October 12–16, 1979

Editors:
R. Kubo and E. Hanamura

With 117 Figures

Springer-Verlag Berlin Heidelberg New York 1980

Professor Dr. *Ryogo Kubo*
Department of Physics, University of Tokyo
Tokyo, Japan

Professor Dr. *Eiichi Hanamura*
Department of Applied Physics, Faculty of Engineering,
University of Tokyo, Bunkyo-ku
Tokyo, Japan

Series Editors:
Professor Dr. Manuel Cardona
Professor Dr. Peter Fulde
Professor Dr. Hans-Joachim Queisser
Max-Planck-Institut für Festkörperforschung, Heisenbergstrasse 1
D-7000 Stuttgart 80, Fed. Rep. of Germany

ISBN-13:978-3-642-81484-6 e-ISBN-13:978-3-642-81482-2
DOI: 10.1007/978-3-642-81482-2

Softcover reprint of the hardcover 1st edition 1980

2153/3130-5 4 3 2 1 0

Preface

This is the Proceedings of the Taniguchi International Symposium on "Relaxation of Elementary Excitations" which was held October 12-16, 1979, at Susono-shi (at the foot of Mt. Fuji) in Japan. The pleasant atmosphere of the Symposium is evidenced in the picture of the participants shown on the next page.

The purpose of the symposium was to provide an opportunity for a limited number of active researchers to meet and to discuss relaxation processes and related phenomena not only of excitons and phonons in solids but also electronic and vibrational excitations in molecules and biological systems. First, the lattice relaxation, i.e., multi-phonon process, associated with electronic excitation, which plays important roles in self-trapping of an exciton and a particle (electron and hole) and also in degradation of semiconductor lasers, is discussed. Second, this lattice relaxation is studied as the intermediate state interaction in the second-order optical responses, i.e., in connection with the competitive behavior of Raman scattering and luminescence. Third, relaxation mechanisms and relaxation constants are determined by spectroscopic methods as well as by genuine nonlinear optical phenomena. Conversely the relaxation is decisive in coherent nonlinear optical phenomena such as laser, superradiance, and optical bistability. Fourth, the role played by relaxation processes is discussed for optical phenomena in macromolecules and biological system such as photosynthesis. Finally, it is shown that we have the concept of relaxation in infrared laser-induced molecular reactions and photodissociation of molecules which is common to that in solid state. This is interesting from the viewpoint of engineering because it may be used for isotope separation. Thus we hope that this Proceedings will give us a unified viewpoint on a great variety of relaxation phenomena in various materials.

We would like to explain the Taniguchi International Symposium. Mr. Toyosaburo Taniguchi, who is the former President of Toyobo Co., Ltd., and one of the leading figures in the textile industry in Japan, has supported many academic activities in natural and human sciences for a long time. He conceived the Taniguchi International Symposium which consists of about ten divisions, each covering a selected field. In order to understand his philosophy, his speech delivered at the reception is quoted in the Foreword.

As mentioned in his speech, the number of attendees is very limited in order to facilitate intense communication, and we feel obliged to share the information obtained in the symposium with scientists throughout the world. In this respect, we are grateful to Springer-Verlag for their cooperation in this purpose.

March 1980 *Ryogo Kubo and Eiichi Hanamura*

Foreword

Speech by Toyosaburo Taniguchi

I wish to extend a warm and sincere welcome to the many distinguished scholars who have come all these miles to attend the Second International Symposium, Division of the Theory of Condensed Matter, of the Taniguchi Foundation.

I sincerely feel that the need is greater than ever today for Japan to seriously consider how best we can strengthen and foster international understanding and friendship between nations, peoples and societies, and how best we can contribute towards the establishment of peace and prosperity in the world.

For more than twenty years, I have been supporting a symposium on mathematics in which distinguished scholars from all over the world such as the gentlemen gathered here today have engaged in free discussions which resulted, so I hear, in fruitful and satisfactory developments and growth on a global basis. The conception of this symposium is mostly unique, in that the participants, both Japanese and foreign, live together in community style, albeit for a limited period of time. I have heard from members of some of these study groups that this type of setup, as a background to the symposium, has helped to strengthen their ties of friendship and understanding with their colleagues on a most personal basis. It is no exaggeration to say that what developed in the mathematics groups led me to reorganize and strengthen the Taniguchi Foundation only a few years ago through additional funding.

In order to effectively translate the objectives of the Foundation into action with the funds available, it is necessary, in my belief, to select those fields which, because they are not necessarily in the limelight of popular interest, are low in funding. I would rather choose for the Foundation projects from modest unimpressive academic fields than those that stand out in gaudy, gorgeous popular acclaim.

I especially desire to see promising younger groups of scholars throughout the world assume an even more active role in our symposia, for they have greater potential to develop their abilities in adapting themselves to the changing international scene. With world-shaking events occurring daily, deep-rooted changes are taking place in every structure. It is really the younger generation who will lead us into the promise of tomorrow. I am convinced that their experiences in the symposia will be instrumental in forging the bonds of mutual understanding between peoples, between societies, and between nations. I share your hopes that those present at the symposia will enjoy themselves and come to exchange their views openly, frankly, and completely in a spirit of mutual trust and understanding leading to the establishment of lasting friendships for the benefit of all participants. I would like to add that the number of participants is limited to about 15 or 20--and no more--because of my desire to see a situation of living and studying matters together in a close-knit community setup. The more the number of participants increases, the more the substance of the symposium becomes vague and the perfume is gone; therefore I beg your understanding of our basic philosophies and thoughts.

The steady progress that the Taniguchi Foundation has made during such a short period of time has been made possible only by the continuing support and cooperation of all concerned. Today, the Foundation extends its activities to nearly ten divisions. In addition to the Theory of Condensed Matter, we hold symposia in Mathematics, Business History, Biophysics, Medical History, Brain Sciences, Ethnology and Neurobiology in Vision. I am convinced that were my late father, who founded this Foundation in 1929, to know this, he would be unbelievably happy. It is my fervent hope and prayer that these symposia will not be temporary in nature and will continue for ten years or more under the guidance of Mr. H. Shima, President of the Foundation, who is the former President of the National Space Development Agency of Japan. Planning and executing the symposia should be an exceedingly rewarding task, which I also deeply appreciate.

This reception should be a welcome relief from the tight symposium schedule, and, while I realize that your time in the Tokyo area is limited, I hope you will enjoy yourselves. In conclusion, I would like to end these words of welcome with our earnest prayer for the success of the symposium throughout all sessions, and also the sincere hope that all the participants, both Japanese and foreign, will return home bearing heartwarming memories of the symposium and our enjoyable time together this evening.

Thank you.

Participants in the Taniguchi International Symposium, October 12–16, 1979

(photographed with Mt. Fuji in the background)

Seated (left to right)
E. Hanamura Y. Toyozawa A. Laubereau R. Kubo C. H. Henry
Mrs. Knox R. S. Knox K. F. Freed E. Yablonovitch S. Mukamel

Standing (left to right)
H. Miyazaki T. Takagahara K. Nasu M. Inoue T. Shimizu
H. Hasegawa T. Yajima M. Aihara H. Sumi J. Tanaka S. Tsuchiya
Mrs. Sumi

Contents

Part I. *Lattice Relaxation*

Electrons, Holes and Excitons in Deformable Lattice. By Y. Toyozawa .. 3

Large Lattice Relaxation Processes in Semiconductors. By C.H. Henry .. 19

Nonradiative Transition During Lattice Relaxation in Localized Electron-Phonon System. By K. Nasu 34

Part II. *Intermediate State Interaction*

Resonant Raman Scattering and Luminescence. By T. Takagahara 45

First and Second Order Optical Responses in Exciton-Phonon System By H. Miyazaki and E. Hanamura 71

Vibronic Spectra of Molecular Crystals: Dynamic Theory and Comparison with Experiment. By E.I. Rashba 88

Ultrafast Vibrational Relaxation Processes of Polyatomic Molecules, Investigated by Picosecond Light Pulses. By A. Laubereau 101

Part III. *Nonlinear Optical Phenomena*

Relaxation Study in the Extremely Short Time Region by Stationary and Transient Nonlinear Optical Spectroscopy By T. Yajima, H. Souma, Y. Ishida and Y. Taira 121

Thermodynamics and Transient Dynamics of Simple Optical Systems with Instability. By H. Hasegawa 130

New Photon-Echo Phenomena in a Localized-Electron Phonon System By M. Aihara 156

Part IV. *Molecular Crystal and Biological System*

Relaxation of Electronic Excitation in Biology: The Initial Processing of Photon Energy and Information. By R.S. Knox 171

The Origin of the J Band. By J. Tanaka, M. Tanaka and N. Kanamaru 181

Part V. *Molecular System*

Infrared Laser Chemistry. By E. Yablonovitch 197

Multiphoton Vibrational Pumping of Optically Prepared NO_2 Molecules
By I. Burak, J. Tsao, Y. Prior and E. Yablonovitch 206

An Experimental Study on Unimolecular Reactions Induced by Infrared
Multiple Photon Absorption. By S. Tsuchiya 212

Theory of Collisionless Molecular Multiphoton Processes. By S. Mukamel 220

Unified Theory of Collisional Redistribution in Near-Resonance Light
Scattering. By S. Mukamel 237

Energetics of the Photodissociation of Polyatomic Molecules
By K.F. Freed 246

Determination of Molecular Relaxation Constants by Coherent Transient
Spectroscopy. By T. Shimizu 268

Index of Contributors 285

Part I

Lattice Relaxation

Electrons, Holes and Excitons in Deformable Lattice

Y. Toyozawa

Institute for Solid State Physics, The University of Tokyo
Roppongi, Minato-ku, Tokyo, Japan

Abstract

Theoretical and experimental studies on the behaviors of electron, hole and exciton interacting with phonon field and impurity field are reviewed, with particular attention to intrinsic and extrinsic self-trapping. The roles of short and long range components of the both fields are described with the help of the phase diagram for the stable and metastable states of the system. Various possible configurations of lattice distortion in the self-trapped state are considered, with the site off-diagonal (as well as diagonal) electron-phonon interaction, the symmetry and extension of atomic orbitals and the internal motion of exciton being taken into account. Nonradiative electron-hole recombination and exciton annihilation, with eventual defect reactions, are reviewed in the context of self-trapping and its generalization.

1. Localization Versus Delocalization

An electron, hole or exciton in the crystal lattice is subject to the competition of two opposing interactions: one for delocalization and the other for localization. The former is characterized by B, the maximum energy released when the electron (hole, or exciton) perfectly localized on a single site is allowed to extend throughout the crystal; within the "tightly bound electron" approximation with nearest neighbour transfer energy T, B is equal to $|T|$ times the number of nearest neighbours, and hence, to half the band width. The latter — the localizing interaction — is characterized by two kinds of energy — D, the fluctuation of site energy due to lattice vibrations, and E_{LR}, the relaxation energy of the lattice due to the localized electron [1].

The simplest representative system has Hamiltonian $H = H_e + H_{eL} + H_L$, given in the site (n) representation by

$$H_e = \sum_n |n\rangle E_a \langle n| + \sum_{n \neq m}\sum |n\rangle T_{nm} \langle m| \ , \qquad (1)$$

$$H_{eL} = -c\sum_n |n\rangle Q_n \langle n| \ , \qquad (2)$$

$$H_L = U_L + K_L = \sum_n (Q_n^2 + \omega^2 P_n^2)/2 \ . \qquad (3)$$

Q_n, the lattice distortion at the n-th site, is so normalized that ω, the lattice vibrational frequency (Einstein model), appears in the kinetic part

K_L, not in the potential U_L as usual. This normalization, though different from the conventional one, is more convenient for our present purpose since the adiabatic approximation corresponds to simply putting $\omega \to 0$. In the absence of transfer ($T_{nm} \to 0$), the electron would be localized on a single site, say n, with the adiabatic potential

$$W_e(Q) = E_a - cQ_n + Q_n^2/2 = (E_a - E_{LR}) + (Q_n - c)^2/2 \ , \tag{4}$$

$$E_{LR} \equiv c^2/2 \ ; \tag{5}$$

namely, the system would be stabilized by energy E_{LR} as the lattice is distorted to new equilibrium position $Q_n = c$ ($Q_{n'} = 0$ for $n' \neq n$), as shown schematically in Fig. 1(a). In the presence of transfer and the absence of lattice distortion, the electron would move freely throughout the lattice, with band energy $\varepsilon_K = E_a + \sum_n \exp(-iKR_n)T_{0n}$, as shown in Fig. 1(c). Referred to the perfectly localized state ($E = E_a$), the electron would be stabilized to energy $-B = \sum_n T_{0n}$ at the band bottom (it is assumed that $K = 0$ is at the band bottom). A simple minded conclusion from the above argument is that the electron will be localized by inducing lattice distortion around itself or delocalized without lattice distortion according as $E_{LR} \gtrless B$, as shown schematically in Fig. 1(b) where the two types of relaxed states are denoted by S (self-trapped state) and F (free state).

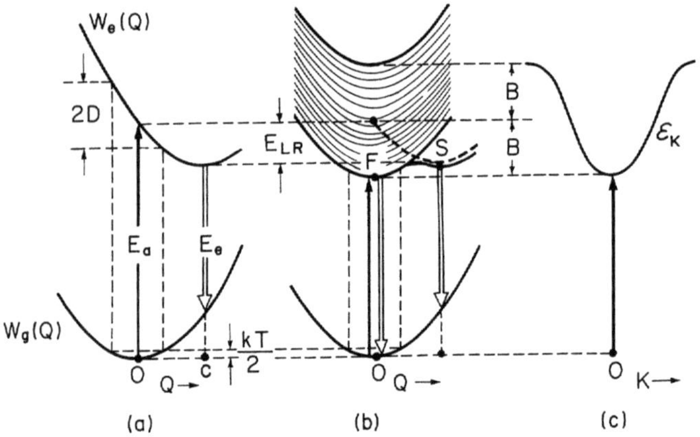

Fig. 1 Configuration coordinate models for localized excitation (a) and exciton (b), and excitonic band in rigid lattice (c). The upper half of the Figure applies also to an electron or a hole.

In the case of exciton, the upper and the lower parts of Fig. 1, representing the states with and without the exciton, respectively, are connected through the optical transitions (solid arrow for absorption, empty arrow for emission). With vanishing transfer, the configuration coordinate model for the localized excitation (Fig. 1(a)) predicts, within the classical Franck-Condon approximation, the absorption band of Gaussian shape: $F(E) = (2\pi D^2)^{-1/2}\exp[-(E - E_a)^2/2D^2]$ with

$$D^2 \equiv c^2 \langle\langle Q_n^2 \rangle\rangle = 2E_{LR}\,\kappa T \ , \tag{6}$$

since the vertical difference of the adiabatic potential, $W_e(Q) - W_g(Q) = E_a - cQ_n$, is a linear function of Q_n which obeys the Boltzmann distribution: $\exp[-Q_n^2/2\kappa T]$. D represents the fluctuation amplitude of the localized excitation energy. The lineshape will not be very different even in the presence of transfer so far as $B \ll D$. In the opposite limit: $B \gg D$, the spatially random potential $-cQ_n$ (temporal variation of $Q_n(t)$ can be neglected if $\hbar\omega \ll D$) will act as weak scatterer against the otherwise freely moving exciton. Since the scattering matrix element is given by $(H_{eL})_{K'K} = -N^{-1}\sum_n \exp[i(K-K')R_n]cQ_n$ according to (2), one can estimate the scattering lifetime τ_K of the exciton K by the golden rule:

$$\Gamma_K \equiv \frac{\hbar}{2\tau_K} = \pi\sum_{K'} \langle\langle|(H_{eL})_{K'K}|^2\rangle\rangle\delta(\varepsilon_{K'} - \varepsilon_K) \approx (\pi/2)\cdot(D^2/B) \ , \qquad (7)$$

where the state density is assumed to be constant (=N/2B) over the whole band. The absorption band is then of Lorentzian shape: $F(E) = \pi^{-1}\Gamma_0[(E-\varepsilon_0)^2 + \Gamma_0{}^2]^{-1}$ with lifetime broadening $2\Gamma_0$ of the optically created exciton with $K=0$. This width is much narrower than D, the fluctuation of local excitation energy, by a factor of order $(D/B) \ll 1$. This is an example of motional narrowing: the exciton moves to a neighbouring site in a time ($\sim \hbar/B$) much shorter than the time ($\hbar/D$) required (by the uncertainty principle) for the local fluctuation D to be perceptible in the spectra.

To sum up, the behavior of exciton immediately after the optical creation is governed by the ratio D/B: the absorption lineshape varies continuously from the Lorentzian with scattering lifetime broadening of free exciton (weak scattering: $D \ll B$) to the Gaussian reflecting the energy fluctuation of localized exciton (strong scattering: $D \gg B$) [2,3]. This is in contrast to the behavior of relaxed exciton which is governed by the ratio $E_{LR}/B \equiv g$: the stable state changes abruptly (see §2 and §3 for more exact description of the abruptness) from the free state F to the self-trapped state S as g exceeds a certain critical value $g_c \sim 1$ (weak or strong coupling according as $g \lessgtr g_c$), as shown schematically in Fig. 1(b). The luminescence (hollow arrows) from the relaxed exciton is expected to be sharp resonant line in the F-type but to be Stokes shifted (by $\sim 2E_{LR} - B$) broad band in the S-type, as is evident from Fig. 1(b). Since E_{LR}($\sim$eV) is usually much larger than κT and hence than D (see (6)), we may have the amphibious situation with weak scattering and strong coupling ($D < B < E_{LR}$): the exciton is almost free immediately after optical creation but will be self-trapped after relaxation. The excitons in alkali halides are typical example of this case.

2. Discontinuity in the Effective Mass of Phonon-Dressed Electron

We presented in §1 the very simple argument that the relaxed electron (hole, or exciton) is either in the free state (F) or the self-trapped state (S) according as $g = E_{LR}/B \lessgtr 1$. One may well suppose that the real situation would be intermediate between two such limiting situations; namely, one can lower the total energy by hybridizing the two ways of stabilization (delocalization and self-trapping). In fact, the electron can be self-trapped at any lattice site, thus forming the N-fold degenerate states because of translational symmetry. Non-vanishing $\hbar\omega$ will make the wave functions of the zero-point lattice vibration in different self-trapped states ($n \neq n'$) to overlap each other, thus lifting up the degeneracy to form a continuous band of phonon-dressed electron. Since the phonon-dress exists even in the F state, an appropriate way of answering the above question is to study the optimum strength of phonon-dressing as a function of coupling constant g.

The electron-phonon wave function of the self-trapped state described in §1 is given by

$$\Phi_n(c) = \chi_0(Q_1) \cdots \chi_0(Q_n - c) \cdots \chi_0(Q_N)|n\rangle \tag{8}$$

where $\chi_0(Q) \equiv (\pi\hbar\omega)^{-1/4}\exp[-Q^2/2\hbar\omega]$ is the wave function for the zero-point vibrational state of the harmonic oscillator in (3). In order to obtain the eigenstate of H with wave-vector K, we use the trial function

$$\Psi_K(\delta) = N^{1/2}\sum_n \exp(iKR_n)\Phi_n(\delta) \tag{9}$$

and minimize the energy expectation value $E_K(\delta)$ with respect to the variational parameter δ. The local distortion δ has the effect of reducing the transfer integral between $\Phi_n(\delta)$ and $\Phi_{n'}(\delta)$ $(n \neq n')$ by

$$\int\chi_0(Q_n - \delta)\chi_0(Q_n)dQ_n \cdot \int\chi_0(Q_{n'})\chi_0(Q_{n'} - \delta)dQ_{n'} = \exp[-S(\delta/c)^2] , \tag{10}$$

$$S \equiv E_{LR}/\hbar\omega \quad \text{(see (5))} . \tag{11}$$

Putting aside the zero-point vibrational energy $N\hbar\omega/2$ of the whole lattice, one obtains the energy of the phonon-dressed electron with $K = 0$ as

$$E_{K=0}(\delta) = E_a - B\exp[-S(\delta/c)^2] - E_{LR}[2(\delta/c) - (\delta/c)^2] . \tag{12}$$

In typical insulators and semiconductors, B and E_{LR} are of the order of eV while $\hbar\omega$ is a few hundredths of eV, so that S is usually as large as several tens. Then (12) has two well separated minima (see Fig. 2):

$$F : (\delta/c)_F \sim (\hbar\omega/B), \qquad E_F \sim E_a - B - (E_{LR}/B)\hbar\omega , \tag{13}$$

$$S : (\delta/c)_S \sim 1 - O(Se^{-S}), \quad E_S \sim E_a - E_{LR}(1 + O(e^{-S})) . \tag{14}$$

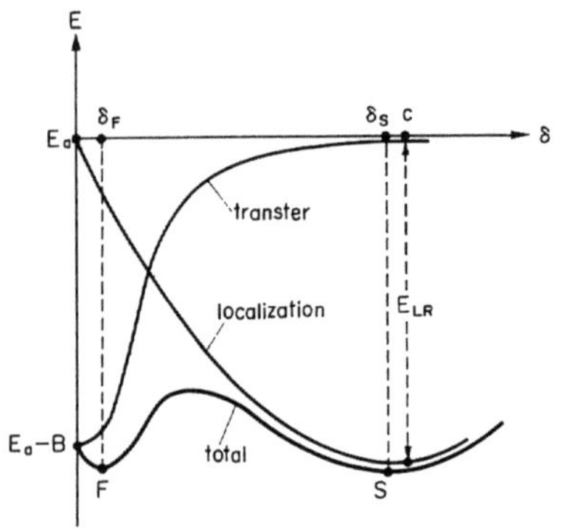

Fig. 2 Energy of an electron moving (with wave vector $K = 0$) in the deformable lattice as a function of the strength of associated lattice distortion: δ.

The local lattice distortion δ of the stable state changes discontinuously from (13) to (14) as $g = (E_{LR}/B)$ exceeds $g_c \sim 1$, and, correspondingly, the effective mass m^* of the phonon-dressed electron, proportional to $(d^2E_K/dK^2)^{-1} \propto \exp[S(\delta/c)^2]$ (see (10)) increases discontinuously from a value $m^* \sim m(1 + (\hbar\omega/B))$ (m is the band effective mass) to an enormous value of $\sim m\exp(S)$ which is practically infinite [4]. Namely, hybridizing the two ways of energy stabilization to minimize the total energy does not give rise to a significant modification of the statement in §1 as far as $\hbar\omega$ is much smaller than B and E_{LR}.

In molecular crystals, however, B can be as small as $\hbar\omega$; clearcut distinction between F and S states will be lost in this situation. More elaborate treatment of intramolecular vibrations with widely distributed frequencies is necessary for realistic description [5].

3. The Effect of Force Range for Self-Trapping

We have so far assumed that the electron is confined to the distorted site (see (8)). However, as we adiabatically switch on and increase the local distortion $Q_n = \delta$, a bound state will split off from the main band, with orbital radius decreasing from $R = \infty$ and binding energy increasing from zero (Fig. 1(b)). We have therefore to consider the spatial extension of the electron as a variable in order to appropriately describe the self-trapping process.

Secondly, we have considered the short ranged electron-phonon interaction (2) such that the electron is subject to the perturbing potential only at the distorted site. This is the case for the electron-acoustic phonon interaction ($Q_n \propto$ local dilation: $\Delta(R_n) \equiv \mathrm{div}\, u(R_n)$ with $u(R_n)$ being the displacement of the n-th atom; $c \propto E_d$: the deformation potential) and for the electron-intramolecular vibron interaction, but not for the electron-optical phonon interaction in polar crystal which is long-ranged. The localized charge $\rho(r) = -\mathrm{div}\, P(r)$ due to the ionic displacement polarization $P(r)$ will act on the electron with potential $e\Phi(r)$ which is coulombic at long distance and hence can always bind the electron however small ρ (assumed to be positive) may be. The last feature is in contrast to the case of short range interaction (2): the bound state appears only when the potential depth cQ_n exceeds a certain threshold, at the cost of a finite lattice distortion energy $Q_n^2/2$, and as the result, the system has to overcome a potential barrier in passing adiabatically from the F state to the S state [4,6]. Such a potential barrier does not exist in the case of long-ranged electron-phonon interaction.

The two points mentioned above can best be described with the use of continuum model. The adiabatic energy of the electron with (envelope) wave function $\Psi(r)$ in the field of elastic dilation $\Delta(r)$ (representing the acoustic mode) and electrostatic potential $\Phi(r)$ due to displacement polarization (representing the optical mode) is given by the functional

$$E[\Psi,\Delta,\Phi] = \frac{\hbar^2}{2m}\int (\nabla\Psi)^2 dr + E_d \int \Psi(r)^2 \Delta(r) dr + e\int \Psi(r)^2 \Phi(r) dr$$

$$+ (C/2)\int (\Delta(r))^2 dr + (\epsilon/8\pi)\int (\nabla\Phi(r))^2 dr\ , \qquad (15)$$

where C is the elastic constant and $\epsilon^{-1} \equiv \epsilon_\infty^{-1} - \epsilon_0^{-1}$ with static and high frequency dielectric constants $\epsilon_{0,\infty}$. The first, second and third lines on the right hand side correspond to (1), (2) and U_L of (3). Such a continuum model was introduced to study the lattice polarization effect in color centers [7], and was extended to include the elastic distortion for the study of self-trapping [1].

By minimizing (15) with respect to $\Psi(r)$, one would get the adiabatic potential $E[\Delta,\Phi]$ for the lowest electronic state as a functional of $\Delta(r)$ and $\Phi(r)$. As far as the extrema of the adiabatic potential are concerned, however, one can take an alternative but much simpler way by inverting the order of extremization: for fixed $\Psi(r)$, one minimizes (15) with respect to $\Delta(r)$ and $\Phi(r)$, to obtain $\Delta(r) = -(E_d/C)\Psi(r)^2$ and $\nabla^2\Phi = (4\pi e/\epsilon)\Psi(r)^2$ (from the latter equation follows immediately the above mentioned expression for ϵ: a real point charge $\rho_0(r) = e\delta(r)$ introduced into dielectric medium should give rise to the electronic and the total (electronic plus displacement) polarizations of the medium with respective electrostatic potentials $\Phi_e = e/\epsilon_\infty r$ and $\Phi_e + \Phi = e/\epsilon_0 r$). Inserting them into (15), one obtains the energy $E[\Psi]$. Let us take the trial function: $\Psi(r) = (2^{1/2}/a)^{3/2}\exp[-\pi(r/a)^2]$, in which we vary the parameter a in the range $a_0 < a < \infty$: the orbital radius a of the envelope function cannot be smaller than the lattice constant a_0. Then we obtain

$$E[\Psi] = E(a) = B\left(\frac{a_0}{a}\right)^2 - E_{LR}^{(ac)}\left(\frac{a_0}{a}\right)^3 - E_{LR}^{(op)}\left(\frac{a_0}{a}\right), \tag{16}$$

where $E_{LR}^{(ac)} \equiv E_d^2/2Ca_0^2$ and $E_{LR}^{(op)} \equiv e^2/\epsilon a_0$ are respectively the acoustic and optical mode contributions to the lattice relaxation energy and $B \equiv 3\pi\hbar^2/2ma_0^2$ the kinetic energy of the electron, all being defined for complete localization ($a = a_0$).

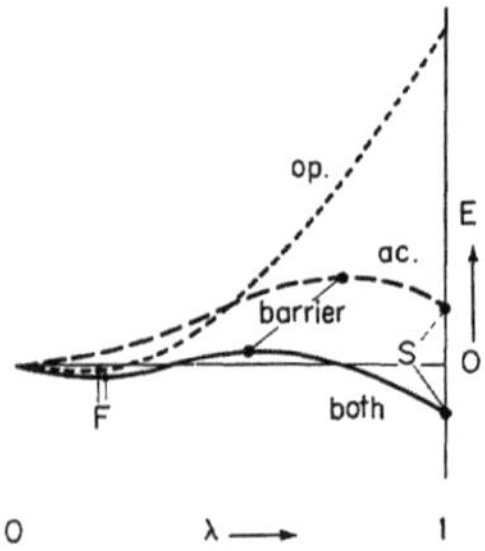

Fig. 3 Adiabatic energy of an electron in deformable lattice as a function of the degree of localization: λ.

To see the role of the force range, we plot (16) against $\lambda \equiv (a_0/a)$ for each of the short (ac) and long (op) range interactions by thin and dotted lines, respectively, in Fig. 3. The former has two minima at $\lambda_1 = 0$ and $\lambda_2 = 1$ corresponding to the F and S states, respectively, with potential barrier between them, provided that $g_s \equiv E_{LR}^{(ac)}/B > 2/3$. Whereas, the latter has only one minimum E_1 at $\lambda_1 = g_\ell/2$ (as far as $g_\ell \equiv E_{LR}^{(op)}/B < 2$), which represents the adiabatic situation of the polaron as first considered by PEKAR [8]. (When the polaron coupling constant α, introduced by FRÖHLICH and related with g_ℓ by $\alpha = (\sqrt{3\pi}/2)g_\ell(B/\hbar\omega)^{1/2}$, is smaller than 3π, one has to replace the above adiabatic solution by the dynamical solution $E_1 = -\alpha\hbar\omega$ where the finiteness of $\hbar\omega$ plays the essential role [9]. We shall ignore this elaboration in our qualitative description within the adiabatic approximation.)

In the presence of the both types of interaction, we may have two minima as shown by thick line in Fig. 3: the state F with $\lambda = \lambda_1$ which is accompanied by moderate P(r) but very weak $\Delta(r)$ and is characterized by large orbital radius of the electron (large polaron) as far as $\lambda_1 \ll 1$ (as is usually the case), and the state S ($\lambda = \lambda_2 = 1$) in which the electron is almost confined to a single site (small polaron or self-trapped state) due to the very strong P(r) and $\Delta(r)$ induced by itself. Fig. 4 is the phase diagram

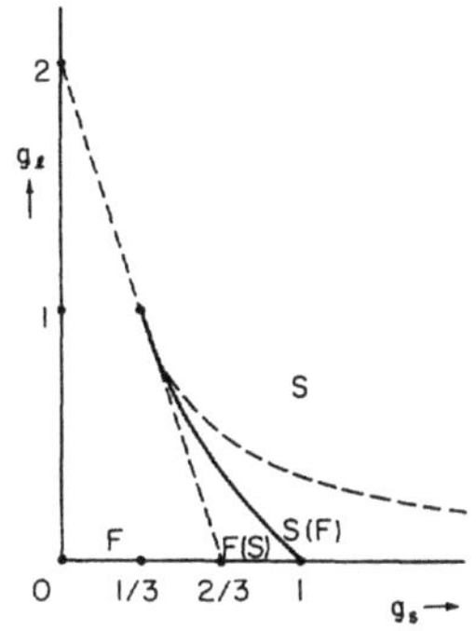

Fig.4 Phase diagram for an electron interacting with acoustic and optical phonon fields through respective coupling constants: g_s and g_ℓ.

in which each region of (g_s, g_ℓ) is characterized by the stable state, and also by the metastable state (shown in parentheses) if any. In the F(S) and S(F) regions, the stable and metastable states are separated by potential barrier, and the nature of the stable state changes discontinuously across the solid line. Evidently, the trigger for the discontinuity is the short range interaction: with this alone, the F state is stable or metastable (with finite potential barrier against self-trapping) but never unstable (the absence of the region S along the g_s axis).

We will now consider the observed behaviors of carriers in various substances in the light of this phase diagram. Simple estimation shows that $\hbar\omega/B < 0.05$ and $g_\ell < 1$ for electrons (in the conduction band), holes (in the valence band) and excitons (for which $g_\ell = 0$ because of electrical neutrality; see, however, §6) in most of typical inorganic insulators and semiconductors. Then the foregoing argument and the phase diagram predict that these carriers can rather distinctly be classified into the F type (nearly free state, large polaron) and the S type (self-trapped state, small polaron), with almost no intermediate situation. This is borne out by a variety of experiments — temperature dependence of mobility, cyclotron resonance, ESR, optical spectra and so on. In alkali halides, for instance, electrons belong to the F-type whereas holes and excitons to the S-type, mainly because of wide conduction band and relatively narrow ($B \lesssim 1$eV) valence (and hence, exciton) band. Holes and excitons belong to the S-type in AgCl but to the F-type in AgBr in spite of the similarity of the two substances in other respects. Low temperature luminescence of the mixed crystal $AgBr_{1-x}Cl_x$ of the F-type (resonant line) is taken over by the S-type (Stokes shifted broad band) rather abruptly as x exceeds 0.45 [10], which corresponds to the discontinuity line in the phase diagram. Even in alkali halides, very weak F-type emission have been observed in addition to the strong S-type luminescence, and the activation energy for the metastable F exciton to be self-trapped have been obtained from the temperature dependence of the intensity of the F-type emission [11]. Such an activation energy does not seem to exist for the holes in alkali halides, indicating that they belong to the S region in the phase diagram. Anomalous temperature dependence of the drift mobility of holes in AgBr [12] and TlBr [13] indicates the existence of the metastable S states slightly above the stable F state (F(S) region) [14]. Electrons, holes and excitons in II-VI, III-V and IV column semiconductors seem to mostly belong to the F region, presumably because of wide conduction and valence bands.

4. The Relaxation Modes of the Lattice

The Einstein model for the lattice vibrations assumed in §1 (see (3)) has a certain difficulty as regards the lattice relaxation. If $T_{nm} = 0$, the adiabatic potential for the electronic state localized at, say, the 0-th site is given by $E_a + (Q_0 - c)^2/2 + \sum_{n(\neq 0)} Q_n^2$, and hence, Q_0 simply vibrates around c, there being no interaction which makes Q_0 to relax towards c. This difficulty is removed by considering the off-diagonal terms $Q_n Q_{n'}$ $(n \neq n')$ in U_L (associated with the dispersion of ω) which in fact dissipate the energy stored on the 0-th site into all the other sites so that Q_0 finally relaxes to c. The wave-front of the lattice vibrational energy thus released should propagate with sound velocity if the acoustic mode is considered.

In the presence of T_{nm}, the free electron will be self-trapped through the gradual decrease of the spatial extension a of the electron wave function and the lattice distortion from $a = \infty$ to $a = a_0$ in (16). So to speak, we have to consider the "variable range" interaction mode for the abscissa of the configuration coordinate model in Fig. 1(b), in contrast to the fixed range interaction mode in the case of localized electron at lattice defect [15]. Putting $E_{LR}^{(op)} = 0$ in (16) for simplicity, one finds that the lattice has to be thermally activated to the potential barrier of $E_a = (4/27)Bg_s^{-2}$ at $\lambda_a = (2/3)g_s^{-1}$, or to tunnel quantum-mechanically to $\lambda_t \sim g_s^{-1}$ (the latter is in fact not unique; (16) applies to the extremum points only). If $g_s \gg 1$, only the long wavelength modes of lattice vibrations will participate the activation and tunneling processes, and the use of the continuum model is well justified for this early stage of self-trapping [16]. As for the relaxed or self-trapped state $(\lambda = 1)$, however, the use of discrete lattice model is essential as will be seen below.

According to (1) with nearest neighbour transfer T only, the energy of the band bottom is given by $E(a_0) = E_a(a_0) - \nu|T(a_0)|$, which varies with the uniform dilation $\Delta = 3\delta a_0/a_0$ with the linear coefficient

$$E_d = \frac{d}{d\Delta} E(a_0) = (a_0/3)(dE_a/da_0) - (\nu a_0/3)(d|T|/da_0) = E_d' + E_d'' . \quad (17)$$

The first and second terms should contribute respectively to the site-diagonal and -off-diagonal parts of the electron-phonon interaction, of which only the former was taken into account in (2). Although they are incorporated into a single deformation potential E_d in the continuum model (15), the two parts play different rôles for the microscopic configuration of self-trapped state. Let us assume E_d'' to be positive since the transfer $|T|$ is usually a decreasing function of the interatomic distance a_0. Self-trapping does take place when $|E_d' + E_d''|$ is large enough as in the continuum model, but the microscopic configuration is such that the electron is self-trapped symmetrically around one site (S_1) for E_d' dominant, and symmetrically around two (S_2) or even more sites (S_n) for E_d'' dominant, as shown by SONG [17] and UMEHARA [18]. The effect of $E_d''(>0)$ making the particular pair(s) of atoms sharing the electron to come closer, and that of $E_d'(\gtrless 0)$ making the lattice around the particular atom keeping the electron to contract or expand, act constructively in the case of $E_d' > 0$ which they studied, and the transition between S_1 and S_2 or S_n in the first quadrant of (E_d', E_d'') phase diagram is expected to be relatively mild. When $E_d' < 0$, however, the two effects act destructively, and S_1 or S_2 self-trapped electron will push away the surrounding atoms, making the atomic or the molecular nature of the center more distinguished. The transition between S_1 and S_2 will also be more distinct, the stable and metastable states being separated by a significant potential barrier.

The preference among S_1, S_2 and S_n also depends upon the symmetry of the atomic orbital from which the band is formed. S_2 may be favored by the p-orbital which is strongly directional and ready to form a molecular bond between a particular pair of atoms. The s-orbital, on the other hand, may favor S_1 or S_n since it has no directionality.

The magnitude of E_d'' depends sensitively on the radius of the atomic orbital from which the band is formed. With the tight binding approximation, one can put $|T(a_0)| \sim T_0 \exp(-\alpha a_0)$, where α is of the order of reciprocal radius of the atomic orbital. T_0 is of the order of atomic binding energy and hence given by $\sim \hbar^2\alpha^2/2m$. From (17), one obtains $E_d'' \propto \alpha^3 \exp(-\alpha a_0)$. As a function of α, it takes sharp maximum at $\alpha a_0 = 3$, which usually corresponds to the valence band. It is smaller for the conduction (α smaller) and the inner core (α larger) bands.

In contrast, E_d' is insensitive against α. The a_0-dependence of the atomic level $E_a(a_0)$ in the crystal comes mainly from the Madelung energy: $+Me^2/a_0$ in ionic crystals, and from the (smaller) electronic polarization energy: $-(1-\epsilon_\infty^{-1})e^2/2a_0$ which exists also in non-polar crystals. As the result, E_d' for the valence or core band hole is weakly positive in non-polar crystals and definitely negative in ionic crystals.

The conduction band is usually too wide to apply such argument based on the assumption of well localized atomic orbital; even the decomposition of E_d into E_d' and E_d'' may not be very meaningful. On the other hand, there seems to be a correlation in sign between the deformation potential E_d and the work function W (electron affinity) of the conduction band since the lattice dilation will make the conduction band bottom closer to the vacuum zero level. This is most clearly seen in liquid [19] and solid [20] He^4: in conformity with the negative sign of W($\sim$-1eV) [21], the electron is considered to form a bubble ($\Delta > 0$) which is favored by $E_d < 0$. This is in contrast to the behavior of positive hole: it is considered to form an iceberg (S_n), which is in consistence with the the above argument: the valence band in He is formed from the 1s orbital, and both E_d' and E_d'' are positive favoring the contraction. The relaxed exciton in liquid He^4 is considered to be in the excimer state He_2^* (S_2 in our terminology) with a bubble around it, presumably because of the p-like nature of exciton (s-like hole and p-like electron) and of negative E_d contributed mainly by the electron.

In other rare gas solids, the electrons are considered to be in the F state while the situation is subtle for the holes: the hole mobility [22] increases with temperature in Ne and Ar (indicating S) almost constant in Kr, and decreases in Xe (indicating F), although the absolute value is rather small in any case presumably due to narrow valence bands. The microscopic configuration is not clear. In contrast, the excitons in Ar, Kr and Xe are in the S_2 state [23,24], presumably due to the p-like nature of the exciton. The S-type exciton in spite of the F-type hole in Xe may be due to the constructive contribution of electron and hole to the total E_d (E_d is positive for the electron in view of positive W [21], E_d' and E_d'' for the hole are positive as mentioned before). The exciton in Ne has two relaxed states: S_1 and S_2 as are observed in luminescence [23] and transient absorption spectra [24], and the bubble formation around S_1 has recently been observed by time-resolving the latter spectra [24].

It is well established that self-trapped holes and excitons in alkali halides are of the S_2 type, and moreover, behave almost like a di-halogen molecule ion. It is partly due to the p-like nature of the hole and the exciton, and partly due to the negative E_d' of the hole in ionic crystals.

The positive hole in AgCl is self-trapped around a silver ion in the d-like state, with tetragonal lattice distortion around it (S_1 of Jahn-Teller type) [25]. It is ascribed to the strong admixture of Ag 4d orbitals to the Cl 3p orbitals in the valence band, particularly around the band top at the L-point of the Brillouin zone.

5. Extrinsic Self-Trapping

Let us consider an isoelectronic impurity or any other neutral defect. An electron in the conduction band, being subject to its perturbing potential which is naturally short-ranged, can be bound by it only when the potential is deep enough. One can then conceive of the following situation which is called the extrinsic self-trapping [26]: the electron-phonon interaction is not strong enough for self-trapping to take place in the host lattice, the impurity potential itself is not deep enough to bind the electron in the rigid lattice, but the cooperation of the two effects is enough to deeply bind the electron at the impurity with strong lattice distortion, as shown schematically in Fig. 5(a) where the adiabatic potentials similar to Fig. 1(b) are shown for various depths of the short-ranged impurity potential but with the electron-phonon coupling constant g_S fixed below the self-trapping threshold. The extrinsic self-trapped state is separated from the free state by a potential barrier so that the discontinuous transition takes place between them across the solid line in the phase diagram Fig. 5 (b) (except in the limit of $g_S \to 0$).

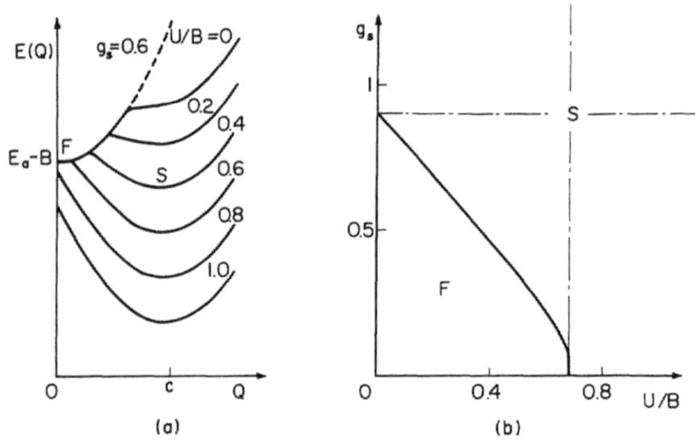

Fig. 5 Configuration coordinate model for extrinsic self-trapping (a), and the stable state as a function of short-ranged impurity potential: U and electron-phonon coupling: g_S.

One can extend the above argument so as to include the charged impurity and the long-ranged electron-phonon interaction, with the use of (15). We have only to add the short- and long-range parts of the impurity potentials, -u(r) and -v(r) (both assumed to be attractive), to the corresponding parts of the phonon fields in the second line of (15). Since the fraction of electronic charge found at the impurity site is proportional to $(a_0/a)^3$, the contribution of the short-range part to the electronic energy can be written as $-U(a_0/a)^3$. The contribution of the long-range part can be written as $-V(a_0/a)$ as is always the case for coulombic potential. Note that they are of the same form as the second and the third terms, respectively, of (16): the exponent of the a-dependence is determined by the force range.

One can then apply the phase diagram of Fig. 4 to the electron in the coexistent impurity potential and phonon fields by simply redefining the short- and long-range coupling constants as $g_s \equiv (E_{LR}^{(ac)} + U)/B$ and $g_\ell \equiv (E_{LR}^{(op)} + V)/B$, except that "F" is to be interpreted as the state shallowly bound by the charged impurity in the case of $V \neq 0$.

LANGER et al. [27] found that the electron trapped by In impurity in CdF_2 (monovalent donor since In^{+++} replaces Cd^{++}) has two relaxed states: the shallow coulombic state (metastable), and the deep state with strong lattice distortion (stable) which can be converted to the former through optical ionization. This system corresponds to S(F) region of our phase diagram. Even in III-V semiconductors with smaller electron-phonon coupling, some of the deep impurity states are known to have strong lattice distortion caused by bound electron or hole [28] as have been studied in great detail with the use of photocapacitance technique. Such a strong coupling situation is essential for the multiphonon recombination of electron and hole as will be described in §7. The phenomenon of persistent photocurrent found in many III-V semiconductors was explained by a configuration coordinate model [29] which exactly corresponds to the extrinsic self-trapping: namely the free carrier can be trapped by the impurity only with the help of lattice distortion which necessarily requires a finite activation energy.

A typical example of extrinsic self-trapping of exciton is TlCl containing isoelectronic Br impurities [30]. In the rigid lattice, a Br impurity cannot bind the exciton (indirect) as is evident from the amalgamation type absorption edge of mixed crystals $TlCl_{1-x}Br_x$. On the other hand, the luminescence from the relaxed exciton is of the S-type in TlCl:Br although it is of the F-type in pure TlCl and pure TlBr. This indicates the necessity of cooperation of impurity potential and lattice relaxation for the bound state to be realized.

6. Self-Trapping of Electron-Hole Pair

We have so far considered the exciton along the same line as the electron or the hole, with the only difference that the former is not subject to long-ranged phonon interaction because of its electrical neutrality. However, the internal motion of exciton, ignored so far, can also undergo a drastic change if the lattice relaxation energy of the electron or the hole overcomes the coulombic binding energy R of the electron-hole pair.

The system of an electron and a hole interacting with the acoustic phonon field through their respective deformation potentials E_d^e and E_d^h was studied by SUMI [31]. The phase diagram of the relaxed state in a typical case with band-effective-mass ratio $m_h/m_e = 4$ and relatively weak coulombic attraction is reproduced schematically in Fig. 6 with rescaling of the abscissa and the ordinate. The lattice distortions induced by the electron and the hole act constructively or destructively according as E_d^e and E_d^h have the same or the different signs. In the former case, the exciton can be self-trapped even when neither a single electron nor a single hole is self-trapped, as shown by the tilted borderline between F_{ex} and S_{ex} regions. If the hole is self-trapped for itself, the electron can be shallowly bound in its coulombic field — the region S_h:e in Fig. 6; as one increases E_d^e, however, one gets into the region S_{ex}, with abrupt contraction of electronic orbital radius across the borderline (self-contraction). This discontinuity corresponds to the F(S)-S(F) transition in the "extrinsic" version of Fig. 4 if the self-trapped hole is replaced by a charged impurity. It would be interesting to study which of the situations is realized in the

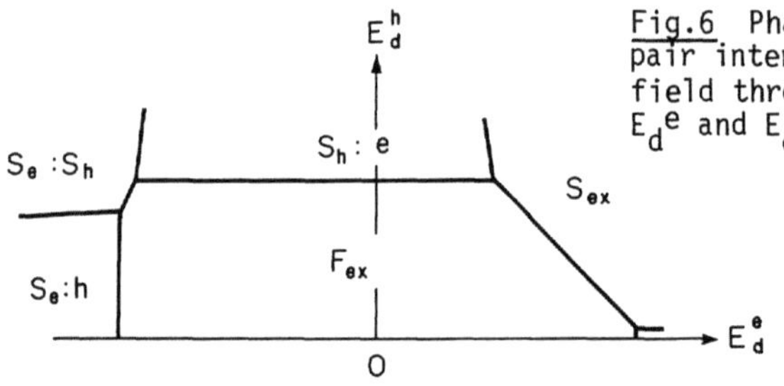

Fig.6 Phase diagram for electron-hole pair interacting with acoustic phonon field through deformation potentials: $E_d{}^e$ and $E_d{}^h$.

ground and excited electronic states of the self-trapped excitons in alkali halides and rare gas solids.

Another interesting feature is found in the second quadrant of Fig. 6. In the S_e:S_h region, the electron and the hole are self-trapped individually rather than from an exciton: the stabilizing configuration of the lattice is opposite for the electron (dilation) and the hole (contraction), resulting in the short-ranged repulsive force between the two particles, which counterbalances the long-ranged coulombic attraction to keep them off at moderate distance — at most a few times of lattice constant. Although such a situation has so far not been found in crystals, there is an argument [32] that a similar situation is realized in amorphous semiconductors. On the other hand, a generalized version of this situation will have much to do with the defect formation to be described in the next section.

The activation energy required for a large radius Wannier exciton to be self-trapped is a subtle matter. It should be finite for any type of exciton due to the electrical neutrality, as mentioned in §3. On the other hand, a hole (or an electron) can be self-trapped without activation energy if it belongs to the region S of Fig. 4, as is the case for alkali halides. The Wannier exciton consisting of such a hole and a light mass electron ($m_e/m_h \ll 1$ is assumed for the conduction and valence bands) will behave almost like a single hole especially when the electron orbital radius of the exciton is much larger than the lattice constants. A simple consideration [33] shows that the activation energy for self-trapping of such an exciton is a rapidly decreasing function of the radius a_n (and hence, of the principal quantum number n) of the electron-hole relative motion, in agreement with recent observation on alkali halides [34].

7. Multiphonon Recombination and Defect Reaction

As was mentioned in §5, the deep impurity level with strong lattice distortion plays the important rôle of the intermediate state in nonradiative recombination of electrons and holes in semiconductors. A great number of phonons corresponding to the band gap energy are released at one coup of this process: such is possible only with strong electron-phonon coupling, the situation which is appropriately described by the configuration coordinate model. This was borne out by HENRY and LANG [35], who performed extensive study of carrier capture cross-sections of various deep levels with detailed theoretical analysis.

The whole process of recombination, consisting of successive electron and hole captures (or vice versa), can be described by the configuration coordinate model of Fig. 7, where the uppermost, the middle and the lowest

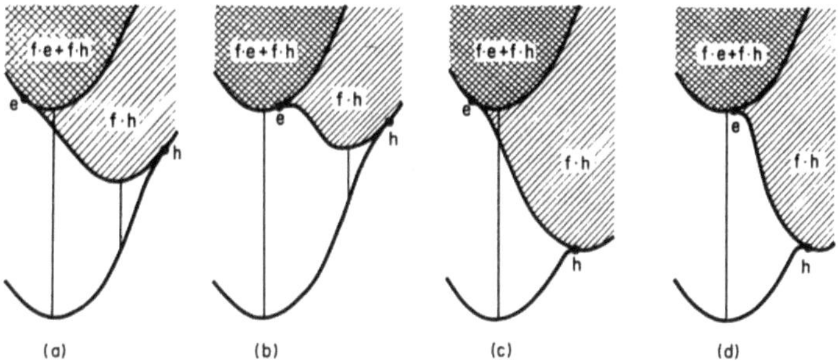

Fig. 7 Configuration coordinate models for nonradiative recombination of electron and hole via a deep impurity state.

curves represent the energies of the state with free electron and free hole (double continuum), the state with trapped electron and free hole (single continuum) and the ground state without electron-hole, respectively, as functions of local distortion around the impurity [36]. We classify the situation into four cases: one needs extrinsic self-trapping (the carrier can be bound by the impurity only with the help of lattice distortion, see §5) for the first capture in (b), for the second capture (which means annihilation) in (c), and for both in (d). The case (d), however, is impossible since the impurity is charged at least in one of the initial and the intermediate states, giving rise to bound states even without the help of lattice distortion. It is to be noted that the lower one of any two adjacent energy states in Fig. 7 represents a discrete bound state split off from the continuum of the upper state, and that always the lower curve is bent down as it approaches and merges in the upper curve at e- or h-point (the upper curve representing the extended state of the relevant carrier should not be influenced by the local distortion). The impurity behaves as electron or hole capture according as the system is in the neighbourhood of e- or h-point. The electron captured at e is already tightly bound and well incorporated into the impurity (different charge state) when the lattice is relaxed or overshoots (towards the right) along the middle curve. Once the hole is captured at h, the lattice relaxes (towards the left) along the lowest curve; beyond the bending region, the hole is bound tightly enough so that it has annihilated the pre-captured electron — completion of recombination. Consistent description of the whole process is possible only with the many-body representation [37]. The deep impurity states observed optically or thermally correspond to the minimum point of the middle curve, and it is essential for the calculation of deep impurity levels to take into account the lattice distortion, especially in those states responsible for nonradiative recombination. Most favorable for recombination is the situation where e is close to the minimum of the uppermost curve and h to that of the middle curve. In the case (c), the optical depth of the electron binding is larger than the band gap so that a finite thermal energy is required for the hole to annihilate this electron.

In addition to those states described in Fig. 7, we have to consider the exciton states below the ionization continuum. Confining ourselves to perfect crystal, we can conceive two possible situations of self-trapping as shown in Fig. 8. The self-trapped exciton will be radiatively annihilated in case (a) as is observed in rare gas solids and alkali halides. In case (b), the self-trapped exciton can be annihilated only through thermal activation (or tunneling) of the lattice [38].

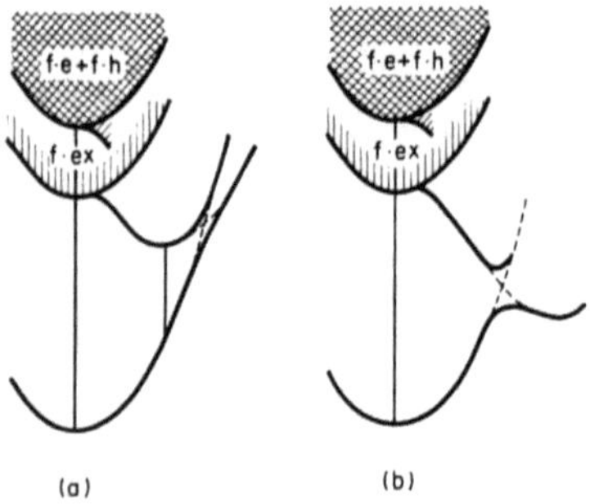

Fig.8 Configuration coordinate model for self-trapping and annihilation of an exciton.

It is interesting to note that the situation (b) is also realized in alkali halides if the concept of self-trapping is generalized in the way described below. In view of the importance of anion displacements for self-trapping of exciton in alkali halides, let us define the deformation potentials for the electron and the hole in the anion sub-lattice. They have different signs since the contraction of the sub-lattice results in increase of electron energy and decrease of hole energy. If both the electron and the hole had strong enough phonon coupling constants g, they would be self-trapped individually with a certain distance apart (S_e:S_h) as described in §6. This is not the case for the electron (due to the wide conduction band). However, if an anion is allowed to displace to a neighbouring site (resulting in a vacancy and a doubly occupied site), the total system will be stabilized a great deal since the electron and the hole will be deeply bound respectively around the vacancy and the doubly occupied site. This is nothing but the F- and H-centers formation which was established [39] to be the primary stage of defects formation during the electron-hole recombination in alkali halides. Since the pair in this relaxed state does not luminesce but annihilates only through thermal activation, this state corresponds to case (b) of Fig. 8, where the abscissa is to be extended to the configuration of one atomic distance displacement of the anion. In this way, we can interpret this state as a modification of the S_e:S_h state: the electron and the hole are extrinsically self-trapped around their respective defect centers produced by themselves. Such a viewpoint is consistent with the excitonic mechanism for defects formation proposed before [40,41]. To be more exact, a variety of experimental studies [39,42~45] indicate that the higher electronic states, rather than the ground state, of the self-trapped exciton are efficiently relaxed to the configuration (b) with defects: it is quite reasonable that less tightly bound electron-hole pair is more readily subject to dichotomy in the phonon field.

Even in semiconductors, the recombination through deep states gives rise to defects reaction of various types, causing degradation of light emitting diodes, as has been studied extensively for these years [46]. The relaxed excited state in case (c) of Fig. 7 might then include the already defect-reacted state. It should also be mentioned that similar processes for defects formation have been considered in amorphous semiconductors [47].

In conclusion, we have presented a unified description of a variety of electronic-atomic processes in insulators and semiconductors by generalizing the concept of self-trapping.

References

1 Y. Toyozawa: in Proc. 4th Intern. Conf. Vacuum Ultraviolet Radiation Physics (Vieweg-Pergamon, Braunschweig,1974) p. 317
2 Y. Toyozawa: Prog. Theor. Phys. 20, 53 (1958)
3 H. Sumi: J. Phys. Soc. Jpn. 32, 616 (1972)
4 Y. Toyozawa: Prog. Theor. Phys. 26, 29 (1961)
5 V.M. Kenkre and R.S. Knox: Phys. Rev. B 9, 5279 (1974)
6 E.I. Rashba: Optika i Spektrosk. 2, 75; 88 (1957); Izv. Akad. Nauk SSSR, Ser. fiz. 40, 1793 (1976)
7 R. Kubo: J. Phys. Soc. Jpn. 3, 254 (1948); 4, 322 (1949); J.H. Simpson: Proc. Roy. Soc. A 197, 269 (1949); See also, N.F. Mott and R.W. Gurney: Electronic Processes in Ionic Crystals (Oxford, 1940) Chapt. 3
8 S. Pekar: Journ. Physics U.S.S.R. 10, 341 (1946)
9 H. Fröhlich: Advances in Phys. 3, 325 (1954)
10 H. Kanzaki, S. Sakuragi and K. Sakamoto: Solid State Commun. 9, 999 (1971)
11 Ch. Lushchik, I. Kuusman and V. Plekhanov: J. Luminescence 18/19, 11 (1979), and the references thereof.
12 R.K. Ahrenkiel and R. Van Heyningen: Phys. Rev. 144, 576 (1966)
13 T. Kawai, K. Kobayashi, M. Kurita and Y. Makita: J. Phys. Soc. Jpn. 30, 1101 (1971)
14 Y. Toyozawa and A. Sumi: in Proc. 12th Intern. Conf. on Physics of Semiconductors (B.G. Teubner, Stuttgart 1974) p. 179
15 Y. Toyozawa and M. Inoue: J. Phys. Soc. Jpn. 21, 1663 (1966)
16 S.V. Iordansky and E.I. Rashba: Zh. Eksp. Teor. Fiz. 74, 1872 (1978)
17 K.S. Song: J. Phys. Soc. Jpn. 26, 1131 (1969)
18 M. Umehara: J. Phys. Soc. Jpn. 47, 852 (1979)
19 J. Poitrenaud and F.I.B. Williams: Phys. Rev. Lett. 29, 1230 (1972)
20 G.A. Sai-Halasz and A.F. Dahm: Phys. Rev. Lett. 28, 1244 (1972)
21 N. Schwentner, F.-J. Himpsel, V. Saile, M. Skibowski and W. Steinmann: Phys. Rev. Lett. 34, 528 (1975)
22 P.G. Lecomber, R.L. Loveland and W.E. Spear: Phys. Rev. B 11, 3124 (1975)
23 I.Ya. Fugol': Advances in Physics 27, 1 (1978)
24 T. Suemoto and H. Kanzaki: J. Phys. Soc. Jpn. 46, 1554 (1979), and private communication on more recent work
25 M. Höhne and M. Stasiw: Phys. Stat. Solidi 28, 247 (1968); M. Fukai, Y. Hayashi, H. Yoshioka, H. Kanzaki and S. Sakuragi: J. Phys. Soc. Jpn. 30, 1510 (1971)
26 Y. Shinozuka and Y. Toyozawa: J. Phys. Soc. Jpn. 46, 505 (1979)
27 U. Piekara, J.M. Langer and B. Krukowska-Fulde: Solid State Commun. 23, 583 (1977)
28 H. Kukimoto, C.H. Henry and F.R. Merritt: Phys. Rev. B 7, 2486 (1973); C.H. Henry, H. Kukimoto, G.L. Miller and F.R. Merritt: ibid. B 7, 2499 (1973)
29 D.V. Lang and R.A. Logan: Phys. Rev. Lett. 39, 635 (1977); D.V. Lang, R.A. Logan and M. Jaros: Phys. Rev. B 19, 1015 (1979)
30 J. Nakahara and K. Kobayashi: J. Phys. Soc. Jpn. 40, 180 (1976); K. Takahei and K. Kobayashi: ibid. 44, 1850 (1978)
31 A. Sumi: J. Phys. Soc. Jpn. 43, 1286 (1977)
32 D. Emin: private communication
33 Y. Toyozawa and Y. Shinozuka: J. Phys. Soc. Jpn. 48, 472 (1980)
34 H. Nishimura, H. Miyazaki, Y. Tanaka, K. Uchida and M. Tomura: J. Phys. Soc. Jpn. 47, 1829 (1979)
35 C.H. Henry and D.V. Lang: Phys. Rev. B 15, 989 (1977)
36 Y. Toyozawa: Solid State Electronics 21, 1313 (1978)

37 Y. Shinozuka: Thesis, University of Tokyo (1980)
38 Similar situation has been considered by P.W. Anderson in the context of anomalies in CuCl: Informal Communication
39 Y. Kondo, M. Hirai and M. Ueta: J. Phys. Soc. Jpn. 33, 151 (1972)
40 Y. Toyozawa: J. Phys. Soc. Jpn. 44, 482 (1978)
41 C.H. Leung and K.S. Song: Phys. Rev. B 18, 922 (1978)
42 R.T. Williams: Phys. Rev. Letters 36, 529 (1976)
43 Y. Suzuki, M. Okumura and M. Hirai: J. Phys. Soc. Jpn. 47, 184 (1979)
44 K. Kan'no, M. Itoh and Y. Nakai: J. de Phys. 41, Colloque (1980), to be published
45 K. Soda and N. Itoh: J. Phys. Soc. Jpn. to be published
46 L.C. Kimerling: Solid State Electronics 21, 1391 (1978)
47 R.A. Street: Solid State Commun. 24, 363 (1977); Phys. Rev. B 17, 3984 (1978)

Large Lattice Relaxation Processes in Semiconductors

C.H. Henry

Bell Laboratories, Murray Hill, NJ, 07974 USA

ABSTRACT

Large lattice relaxation was first revealed in electron spin resonance studies of vacancies and vacancy-impurity complexes in Si. These centers were found to have large Jahn-Teller distortions. More recent studies, mainly capacitance measurements, have revealed a number of other large lattice relaxation phenomena. These are large Stokes shifts in optical transitions, recombination enhanced defect motion, self-trapping at an impurity and capture by multiphonon emission. All of the processes mentioned above are briefly reviewed, and for capture by multiphonon emission, a detailed model is presented.

1. Introduction

In semiconductors, the strength of the coupling between electronic energy levels of defects and impurities and the surrounding lattice has an enormous range. On one extreme are delocalized shallow donors, acceptors and bound excitons, which show strong zero phonon optical transitions and weak phonon side bands. On the other extreme are highly localized nonradiative recombination centers, which allow the electron to drop from the conduction band to the valence band by means of a trapped electron level which crosses the energy gap as the lattice coordinate changes.

TOYOZAWA [1] has explained that these two extremes arise because of the competition between the delocalizing effect of the broad energy bands and the localizing effect of the short range impurity potential and lattice relaxation. Until recently, most experimental attention in semiconductors was given to study of the delocalized states by optical spectroscopy and a commonly held view was that large lattice relaxation effects were not important. An exception was the beautiful spin resonance studies of vacancies and vacancy-impurity complexes in Si by WATKINS and CORBETT [2], which revealed large Jahn-Teller distortions. In the last eight years a resurgence in the study of deep levels by electrical methods, particularly by junction capacitance measurements, has uncovered a number of other important lattice relaxation phenomena. These are the observation of large Stokes shifts in optical transitions, recombination enhanced defect motion, evidence for an electron trapped by lattice relaxation at an impurity and nonradiative recombination by multiphonon emission. All of these effects, except multiphonon emission, will be reviewed briefly in Section 2. In Section 3, we will discuss in detail a model of capture of carriers by multiphonon emission.

2. Review of Large Relaxation Phenomena in Semiconductors

A. Jahn-Teller distortion of the lattice vacancy in Si

In a study of the electron spin resonance (ESR) of the isolated vacancy in Si, WATKINS [2] found that the ESR data could be interpreted with the simple model shown in Fig. 1. Formation of the vacancy produces four broken bond orbitals a, b, c and d. From these orbitals, four orthogonal states can be constructed: a_1, a symmetric singlet state; and t_2, a three-fold degenerate triplet state. This degeneracy is removed by a Jahn-Teller distortion in which the four atoms surrounding the vacancy pull together in pairs. This simple analysis has been confirmed by recent self-consistent calculations of the electronic structure of the vacancy by BARAFF et. al. [3] and LIPARI et. al. [4]. Analysis of the available experimental data by BARAFF et. al. [5] indicates that the neutral vacancy, which has two electrons in the t_x state, is stabilized by a Jahn-Teller energy about 0.8 eV.

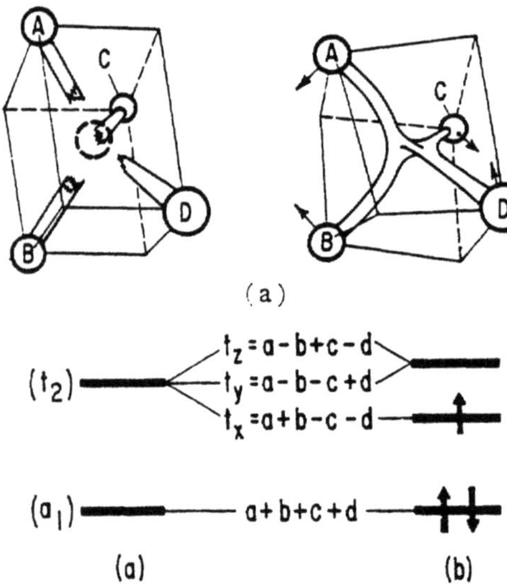

Fig.1 Lattice vacancy and associated molecular orbitals: a) before Jahn-Teller distortion b) after.

B. Stokes shifts and thermal broadening of optical transitions

The traditional evidence for large lattice relaxation of electronic levels in insulators is a large Stokes shift between the peak energies of absorption and emission. An analogous shift can be seen in photocapacitance measurements. In these measurements, the optical cross sections σ_p^0 and σ_n^0 for exciting an electron from the valence band to the unoccupied level and from the occupied level to the conduction band can be measured. An example is

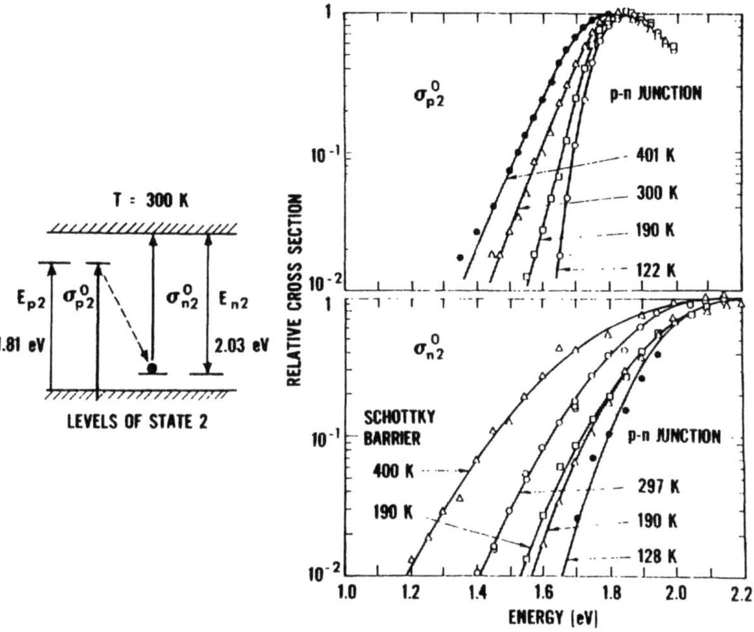

Fig.2 Optical cross sections for the second electron state of O in GaP. The diagram indicates the energy levels, before and after capture with the lattice at equilibrium, deduced by determining the threshold energies E_{p2} and E_{n2} from an analysis of this data. After HENRY and LANG [6].

the cross sections of the second electron associated with substitutional O in GaP shown in Fig. 2.

In the absence of lattice relaxation, the thresholds for the two transitions should add up approximately to the energy gap E_g. For large lattice relaxation, the sum of the threshold energies will be substantially more than E_g. The electronic structure of the O center is still controversial [7], [8]. It is possible that some of the energy attributed here to lattice relaxation is due to an internal electronic transition. However, the substantial broadening of the absorption edges with increasing temperature is additional evidence for the lattice relaxation model. HENRY and LANG [6] point out that the thermal broadening is consistent with a configuration coordinate diagram fitting the optical energies shown in Fig.2 as well as the thermal ionization energy of 0.89 eV.

C. Recombination enhanced defect motion

A baffling problem in light emitting diode and laser technology is understanding how devices degrade and why they only degrade during forward biased operation, when minority carriers are injected into the light generating region of the device. Perhaps the most spectacular example of this type of degradation is the formation of "dark line defects", nonradiative regions

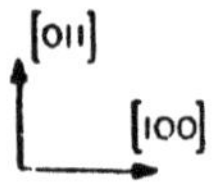

Fig.3 Schematic of the growth process of the dislocation network forming the dark line defect. The successive growth stages are illustrated in (a), (b), (c), (d). After PETROFF and HARTMAN [9].

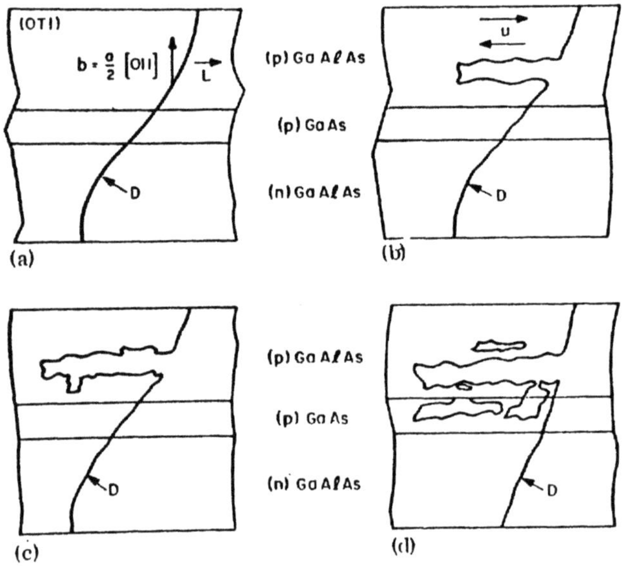

oriented approximately in the <100> direction that form during the first tens of hours of laser and LED operation.

PETROFF and HARTMAN [9] showed that these defects are a network of dislocation dipoles originating from a single dislocation threading through the active layer of the laser as shown in Fig.3. PETROFF and HARTMAN and HUTCHINSON et. al. [10] concluded that the dislocation dipoles form by a climb mechanism and that the dislocation dipole encloses an additional plane of Ga and As atoms. At present, the source of these atoms is a mystery.

Dark line defect formation and other types of degradation require transport of atoms within the crystal at or near room temperature. The mechanism of recombination enhanced defect motion, discovered by LANG and KIMERLING [11], provides a qualitative explanation of how this motion can take place in the presence of minority carriers. LANG and KIMERLING studied the annealing of radiation damage centers in GaAs and GaP using the method of deep level transient spectroscopy, DLTS, previously invented by LANG [12]. DLTS is a form of capacitance detection of deep levels in which, under favorable circumstances, each deep level in the energy gap shows up as a distinct peak when the sample temperature is varied. The deep levels detected in GaAs irradiated at room temperature are shown in Fig.4. Each peak is proportional to the concentration of level within the space charge region of the p-n junction.

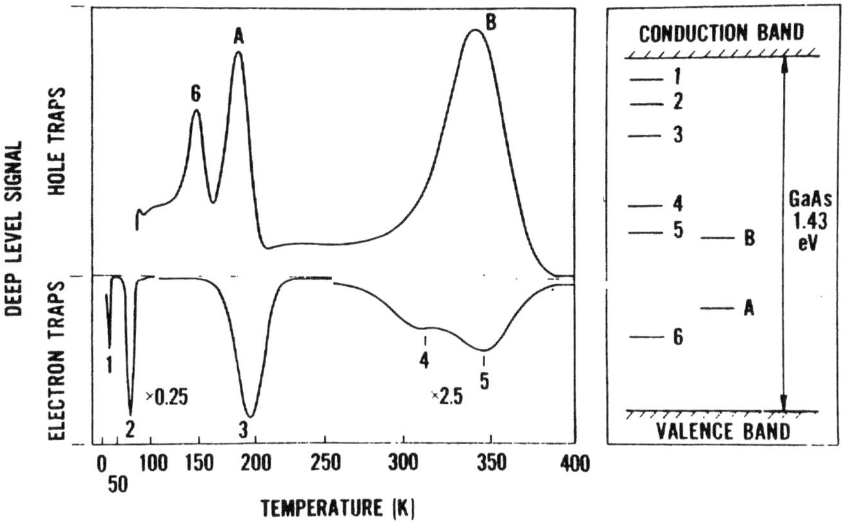

Fig.4

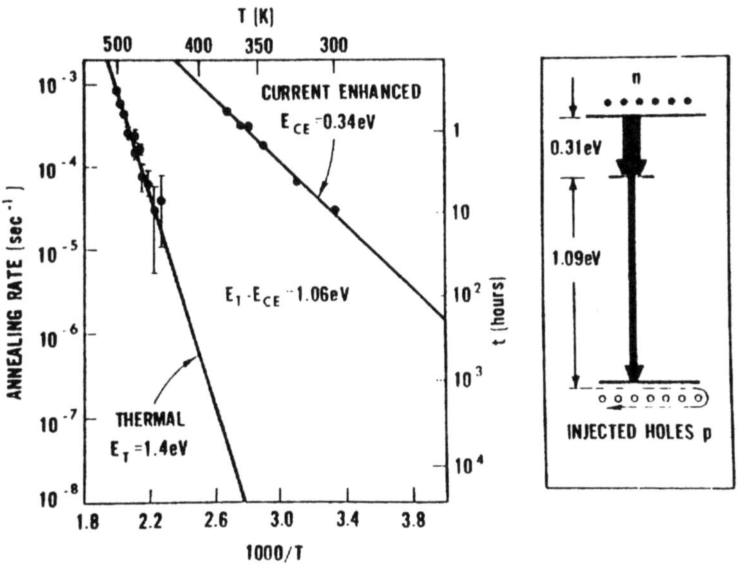

Fig.5

Fig.4 DLTS spectrum for a GaAs p^+n junction containing eight levels. Levels A and B are always present in LPE GaAs. Levels 1-6 are due to radiation damage. The levels in the upper and lower halves of the energy gap are observed by applying reduced bias pulses and injection pulses. After LANG and KIMERLING [11].

Fig.5 Thermal and current enhanced annealing rates versus 1/T and the energy level diagram for a radiation damage center in GaAs exhibiting recombination enhanced defect motion. After LANG and KIMERLING [11].

The annealing of the radiation damage centers could be easily followed by measuring the decrease in the peak heights with time. Fig.5 shows the results for the level three. The level thermally annealed at temperatures above 150°C with an activation energy of about 1.4 eV. When current of $1 A/cm^2$ was passed through the p-n junction, the annealing rate increased by 4-6 orders of magnitude. A detailed study showed that the annealing rate was directly proportional to the rate of electron-hole recombination through level three.

Figure 5 shows that the activation energy for recombination enhanced annealing is approximately 1.06 eV less than the thermal activation energy. The reduction in the activation energy is slightly less than the energy released during electron-hole recombination (1.09 eV). A study by LANG and KIMERLING [11,13] of 12 radiation damage centers in both GaAs and GaP showed that all centers exhibited recombination enhanced annealing and in all cases, the energy lost during recombination was somewhat greater than the decrease in activation energy. It appears that much of the energy released during recombination can take the place of thermal energy in causing diffusion jumps. In four cases the activation energy went to zero and athermal recombination enhanced annealing was observed [13].

In nonradiative recombination by multiphonon emission, discussed below, the electronic energy lost is initially channeled into a single vibrational coordinate. This energy will rapidly propagate away from the recombination center. The efficiency with which this energy contributes to defect motion makes it likely that the same lattice coordinate involved in recombination is also involved in defect motion. A configuration diagram illustrating a possible one coordinate mechanism is shown in Fig.6.

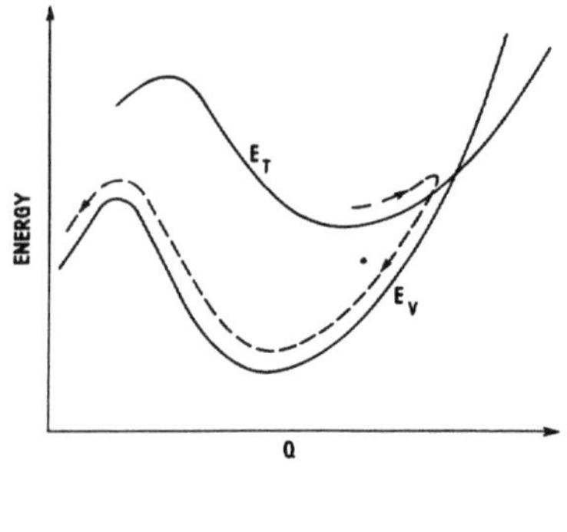

Fig.6 Configuration coordinate diagram showing how recombination enhanced defect motion might occur. E_T and E_V represent the total energy (elastic and electronic) when the electron is in the trap level and when it is at the top of the valence band, respectively. Recombination of the trapped electron with a free hole at the crossing leaves coordinate Q highly excited, enabling the lattice (possibly after the addition of some thermal energy) to cross the barrier.

Recently, TROXELL et. al. [14] showed that the Al interstitial in Si moves by recombination enhanced defect motion. The study of recombination enhanced defect motion in Si should be very fruitful because many of the defects in Si are reasonably well understood. Hence, a more detailed picture of recombination enhanced motion is likely to emerge in Si than in GaAs or GaP, where the basic defects are only beginning to be identified.

D. Persistent photoconductivity and extrinsic self-trapping

In alkali halide and other ionic crystals, lattice relaxation leads to self-trapping of free holes (V_k center formation). Self-trapping does not occur in semiconductors, where both electrons and holes are highly mobile. TOYOZAWA [1] has pointed out that in semiconductors, deep levels arise from the combined effects of attraction by a short range impurtiy potential and lattice relaxation. One possibility he considers is that, prior to relaxation, the impurity potential alone cannot form a bound state. He refers to this model as extrinsic self-trapping. Extrinsic self-trapping leads to an unusual configuration coordinate diagram in which the barrier for capture into the bound state lies between the lattice equilibrium positions. Such a diagram, shown in Fig.7, was developed by LANG and LOGAN [15] to explain the properties of the D-X center, a trap responsible for persistent photoconductivity in the alloy $Al_xGa_{1-x}As$.

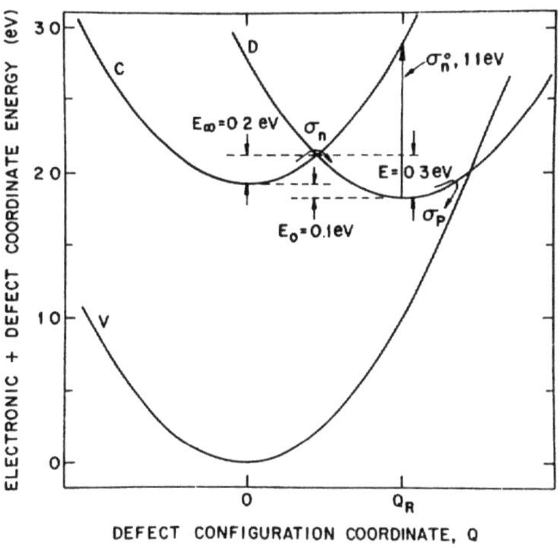

Fig.7 Configuration-coordinate model for the donor-defect center in $Al_xGa_{1-x}As$ (Te). After LANG and LOGAN [15].

At Q=0, the unoccupied trap level is far above the bottom of the conduction band C, whereas at Q_R the lattice equilibrium position of the occupied state, the trap level, is 1.1 eV below the conduction band. The 0.2 eV barrier energy was established from the thermal activation energy for electron capture, the 0.30 eV barrier energy was determined from the activation energy for thermal ionization of the trapped electron and the 1.1 eV energy was determined from analysis of the photoionization cross section.

LANG and LOGAN'S measurements show that the activated behavior of the capture cross section persists to temperatures below 70 K, where the capture cross section is so small that if carriers are photoionized out of the bound state, many hours are required for recapture. Consequently, if the sample is cooled in the dark to below 60 K from room temperature and then exposed briefly to light, the sample becomes conducting and the conductivity lasts for many hours. This effect was discovered and studied by NELSON [16].

NELSON and LANG both found that the D-X center concentration correlates with donor doping. The D-X center has been observed with concentrations comparable to the doped electron carrier density, as high as $10^{18}/cm^3$. LANG et. al. [17] have identified this center as a donor-arsenic vacancy complex. Such a complex should have axial symmetry along the <111> direction for group 6 donors and lower symmetry for group 4 donors, in agreement with recent measurements of NARAYANAMURTI et. al. [18].

3. Capture of Carriers by Multiphonon Emission (MPE)

An important problem in semiconductor device technology understanding is how regions of rapid nonradiative recombination are able to develop in relatively pure materials. A major aspect of this is to understand how impurities can be rapid nonradiative recombination centers. Such impurities must have large capture cross sections for both electrons and holes. A trap of this

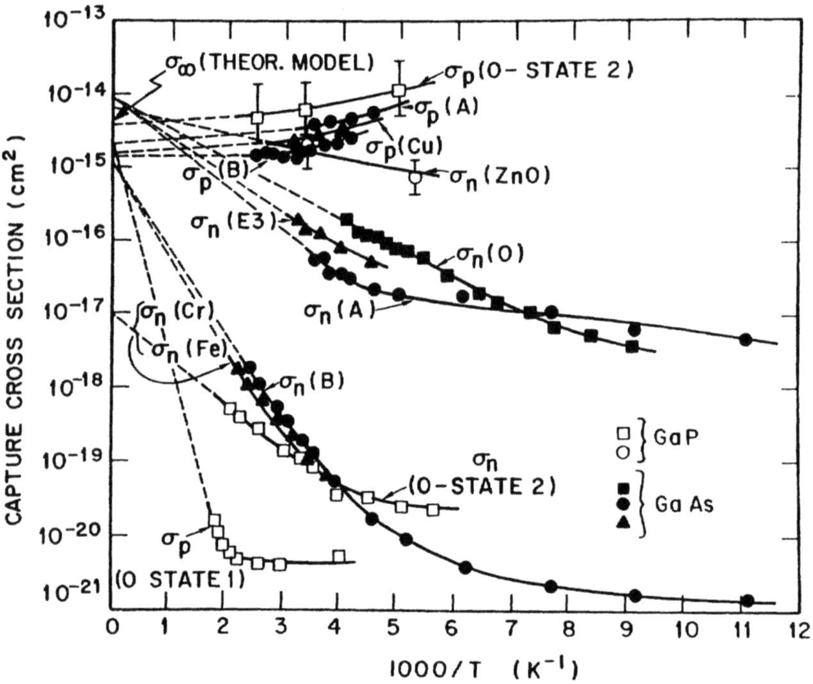

Fig.8 Carrier-capture cross sections vs inverse temperature for various deep levels in GaAs and GaP. An n subscript denotes electron capture while a p subscript denotes hole capture. Levels shown for GaAs are due to Cu, Fe, Cr, O, the E3 radiation damage defect, and two unidentified by commonly occurring levels A and B. For GaP the cross sections are associated with the Zn-O center and the two states of the oxygen center. The dashed lines show the temperature dependence of the cross sections extrapolated to $T = \infty$. The value of σ_∞ predicted by the theoretical model is indicated on the ordinate. After HENRY and LANG [6].

type controlling the minority carrier lifetime in GaP,has been found by HAMILTON et. al. [19]. HENRY and LANG [6] have shown that many impurities in GaAs and GaP have thermally activated cross sections that approach 10^{-15}-10^{-14} cm^2 at high temperature and that this behavior is expected for centers capturing carriers by multiphonon emission as a consequence of a strong coupling between the bound electron level and the lattice. Their data are shown in Fig.8.

This section will be devoted to presenting the model for electron capture by MPE of HENRY and LANG [6]. We will find that this model can account for the features of Fig.8: 1) a limiting cross section at high temperatures of between 10^{-14} and 10^{-15} cm^2, 2) thermally activated cross sections at high temperature which become independent of temperature at low temperatures. The model is relatively simple and intuitive, but it is only an approximate calculation. A rigorous model of capture by MPE from a continuum state into a localized level has not yet been made by anyone.

A. Impurity model

We will calculate capture for a model in which the impurity is represented by a neutral spherical square well potential coupled to a single lattice coordinate Q. The coupling to the lattice is brought about by the depth of the well changing with Q, as shown in Fig.9. The changes in well depth

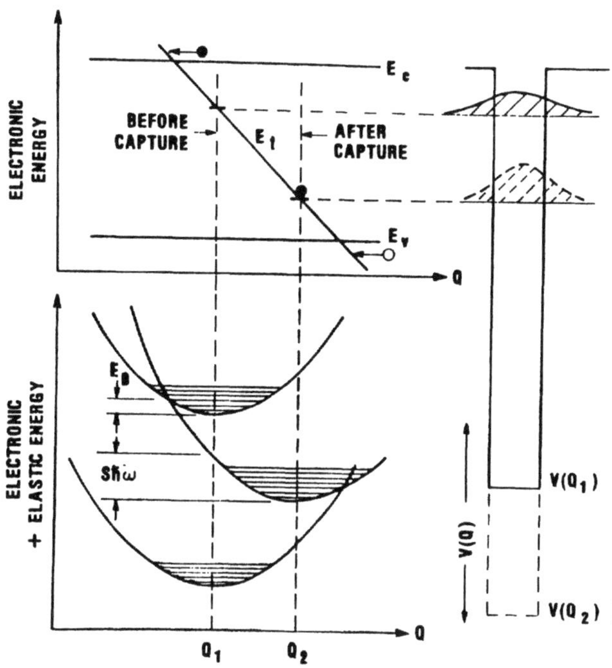

Fig.9 Model showing the electronic energies of the bottom of the conduction band, trap level, and top of the valence band (upper left); the configuration coordinate diagram found by adding the elastic energy to the electronic energies (lower left); and the potential well of depth V, that changes with Q (right).

do not alter the energies of the free carrier levels E_c and E_v. The configuration coordinate diagram is found by adding the elastic energy $k/2\ Q^2$ to the energy level diagram shown in the top of Fig.9.

Prior to electron capture the lattice is vibrating about Q_1. For sufficiently strong vibrations the bound level crosses into the conduction band, where capture of an electron can take place. After capture, the lattice relaxes to a new equilibrium position Q_2. It does so because the total energy (electronic and elastic) can be lowered by the lattice deforming so as to lower the energy of the trapped electron. A new equilibrium position is reached because this deformation increases the elastic energy of the crystal. Vibrations about Q_2 enable the bound electron to cross into the valence band, where hole recombination can take place. Therefore, the electron drops from the conduction band to the valence band in two steps as it rides across the forbidden energy gap in a trap level. For this process to be rapid, the probability of the bound state to cross into the bands during thermal vibrations must be high for both transitions.

B. Breakdown of the adiabatic approximation and an estimate of the electronic matrix element

In this section, we will consider a transition from a continuum state to the bound state of a spherical square well, treating the configuration coordinate Q as a classical variable. Even if the lattice deforms so as to bring the unoccupied bound level nearly degenerate with the occupied free electron state, no transition will take place unless there is an off diagonal part of the Hamiltonian coupling these states. With the aid of Fig.10, let us consider how such an off diagonal term can develop.

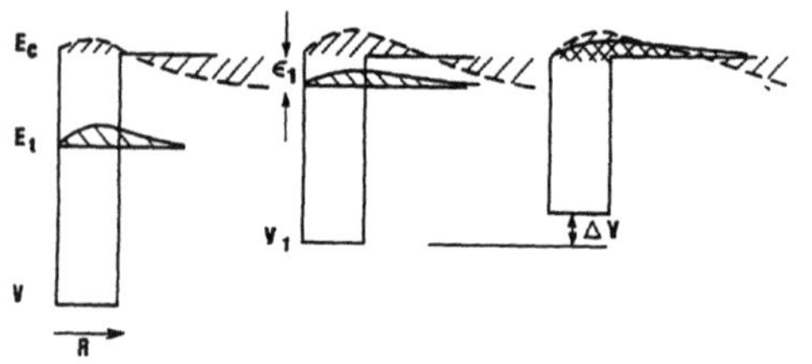

Fig.10 Bound and continuum wave functions of the spherical square well. It is assumed that the wave functions adiabatically follow changes in the well depth as long as $E_c-E_t \geq \varepsilon_1$ and that the wave functions remain rigid (nonadiabatic) for $E_c-E_t < \varepsilon_1$.

Assume the bound and continuum states are diagonal when the states are far apart and the potential has depth V. Suppose the lattice moves to a smaller value of Q, decreasing the well depth to V_1. In the adiabatic approximation, the bound and continuum states will adjust to the change and continue to diagonalize the Hamiltonian. Thus, as E_t approaches E_c, the bound state will spread out, decreasing its amplitude within the well and the continuum state will become more of a resonant state, increasing in amplitude within the well. The lattice moves through the crossing in a small fraction of a vibrational period. As the crossing is approached, the adiabatic solutions must change more and more rapidly. Eventually, therefore, the adiabatic approximation breaks down. We will assume for simplicity that the wave functions remain adiabatic until bound electron level is ε_1 below the crossing. For the remainder of the crossing, we assume adiabatic approximation breaks down and the change in the wave function is negligible.

With these rather crude assumptions, the states are diagonal when E_t is ε_1 below the continuum. An additional decrease in well depth ΔV will bring the rigid bound state to the crossing. Therefore ΔV is the operator of the off-diagonal part in the Hamiltonian, that can couple the states. The matrix element is $\langle t|\Delta V|c\rangle$, where $|c\rangle$ and $|t\rangle$ are the electronic wave functions which are the adiabatic solutions at ε_1 below the crossing. Using the bound and continuum states of the spherical square well HENRY and LANG [6] showed that the squared matrix element can be written approximately as

$$|\langle c|\Delta V|t\rangle|^2 = \frac{2\pi}{\hbar\Omega}\left(\frac{\hbar^2}{2m^*}\right)\varepsilon_1^{1/2} \qquad (1)$$

where m^* is the effective mass of the electron and Ω is the volume of the crystal. Remarkably, the matrix element is independent of the range and depth of the potential well and depends only slowly on ε_1. The lack of dependence on well parameters may imply that the results reached are not very sensitive to the details of the impurity.

C. Estimate of how far below the crossing the adiabatic approximation breaks down

To estimate ε_1, HENRY and LANG [6] calculated the increase in amplitude of an initially empty bound level as it crosses an initially occupied free state at a uniform rate, using first order time dependent perturbation theory and then compared this result with the adiabatic solution. The results are shown in Fig.11. The results are normalized by dividing $|a_t|^2$ and $|a_c|^2$ by

$$P_{LZ} = \frac{2\pi}{\hbar}\,\frac{|\langle c|\Delta V|t\rangle|^2}{\dot{E}_t} \qquad (2)$$

the probability of a transition during a single crossing predicted by the Landau-Zener theory [20] in the limit of weak coupling (nonadiabatic limit). We see that the adiabatic and time dependent solutions agree far below the crossing, but begin to rapidly diverge at $x \approx -1/2$. Therefore, ε_1 was set equal to $\Delta E/2$. HENRY and LANG [6] estimated ΔE by averaging over different vibrational amplitudes contributing to level crossing and capture and concluded $\varepsilon_1 \approx 60$ meV. They also showed that by averaging the Landau-Zener transition probability with a classical distribution of amplitudes of vibration, they could calculate a capture cross section that agrees exactly with the high temperature limit of the capture cross section derived using quantized lattice vibrations. The derivation of the latter will be presented in the next section.

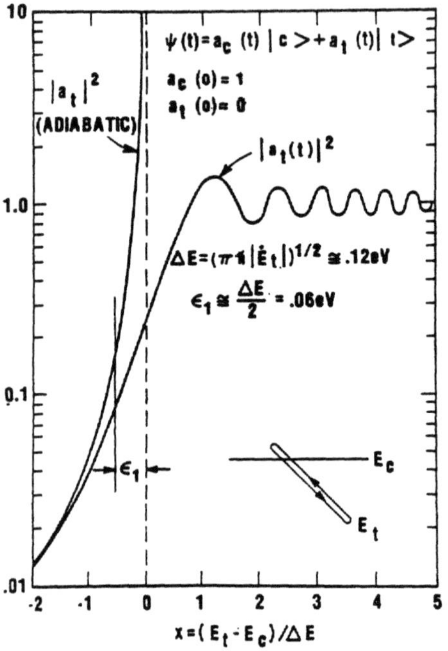

Fig.11 Squared amplitude of the bound state divided by P_{LZ}, the single pass transition probability of the Landau-Zener theory [20]. As the crossing is approached, the amplitude of the adiabatic solution approaches 1 and this amplitude divided by P_{LZ} approaches infinity. The squared amplitude of the time dependent solution approaches P_{LZ} above the crossing. After HENRY and LANG [6].

D. Calculation of capture cross section

Carrier capture involves a transition from an initially occupied vibronic state of the free electron to a vibronic state of the trapped electron at the same energy. We assume that the vibrational wave functions only overlap strongly near the crossing, where the electronic part of the matrix element is a constant given by $\langle t|\Delta V|c\rangle$, calculated in Section B. The capture cross section is defined as

$$\sigma_c = (\text{transition probability per sec})\ \frac{\Omega}{\langle v\rangle} \tag{3}$$

where Ω is the volume of the crystal and $\langle v\rangle$ is the incident velocity of the free electron.

If we denote the quantized vibrational states as $|n_c\rangle$ and $|n_t\rangle$ and use Fermi's golden rule to calculate the transition probability per second we have

$$\sigma_c = \frac{2\pi}{\hbar}\frac{\Omega}{\langle v\rangle}\,|\langle t|\Delta V|c\rangle|^2 \underset{n_c}{\text{Ave}} \sum_{n_t} |\langle n_t|n_c\rangle|^2\,\delta(E_c-E_t) \tag{4}$$

The last term is a thermally weighted average of the overlap of vibrational wave functions at the same energy. This is the same as the normalized emission line shape $f(h\nu)$

$$f(h\nu) \equiv \underset{n_t}{\text{Ave}} \sum_{n_c} |\langle n_t|n_c\rangle|^2 \, \delta(E_t+h\nu-E_c) \quad , \tag{5}$$

that would be calculated in a theory of radiative capture, evaluated at $h\nu = 0$. Hence the capture cross section is given by

$$\sigma_c = \frac{2\pi}{\hbar} \frac{\Omega}{\langle v\rangle} |\langle t|\Delta V|c\rangle|^2 \, f(0) \tag{6}$$

where the matrix element has been evaluated in (1).

The function f(0) has been evaluated in closed form by HUANG and RHYSE [21], KUBO and TOYOZAWA [22] and LAX [23]. It is an expression involving modified Bessel functions. When the lattice relaxation is large, f(0) reduced to simple asymtotic forms valid at high and low temperature

$$f(0) = \begin{cases} (4\pi S\hbar\omega kT)^{-1/2} \exp(-E_B/kT) \\ (4\pi S\hbar\omega \frac{\hbar\omega}{2})^{-1/2} \exp(-2E_B/\hbar\omega) \end{cases} \tag{7}$$

where E_B and $S\hbar\omega$ is defined in Fig.9. Thus the cross section should be thermally activated at high temperature and become constant for $T < \hbar\omega/2k$. A typical phonon energy in GaAs is $\hbar\omega$ = 25 meV, corresponding to constant behavior for T < 150 K. The curves in Fig.8 tend to flatten out between 150 and 250 K. The persistence of thermal activation of capture into the D-X center to below 70 K (Section 2D) implies a very low frequency vibration is involved in the relaxation of this center.

E. Probability of reemission before lattice relaxation

It is possible that after capture, but before the lattice relaxes, the electron will be reemitted into another free electron state. This possibility is illustrated in Fig.12. The probability of reemission from t into a single state c′ is infinitesimal. However, because there are a continuum of states c′, the probability of reemission after capture is finite. HENRY and LANG [6] crudely estimated this probability using the Landau-Zener theory [20]. According to this theory, the probability of no transition when state t crosses level c′ is

$$\exp\left(-\frac{2\pi}{\hbar} \frac{|\langle t|\Delta V|c\rangle|^2}{\dot{E}_t}\right) \quad .$$

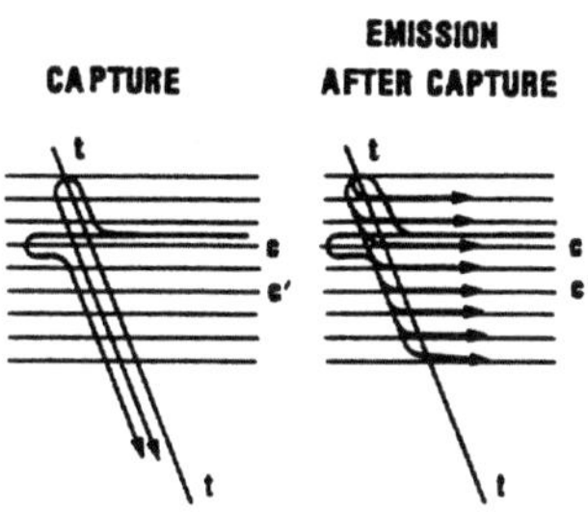

Fig.12 Illustration of capture and reemission after capture as the bound state moves in and out of the free electron continuum during a vibration of the lattice. After HENRY and LANG [6].

The matrix element will be assumed to have the constant value given by (1) and valid for states near the bottom of the conduction band. The probability that the electron will remain in t is the product of these probabilities given by

$$P_{tt} = \exp\left(- \frac{2\pi}{\hbar} \frac{|\langle t|\Delta V|c\rangle|^2}{\mathring{E}_t} \int dN_{c'}\right) \qquad (8)$$

and the net capture cross section is

$$\sigma = \sigma_c \, P_{tt} \qquad (9)$$

where σ_c is given by (6). Eqs. (6) and (8) show that as $|\langle c|\Delta V|t\rangle|^2$ increases, σ will increase, reach a maximum and then decrease. This feature is also present in the Landau-Zener 2 level theory [20].

HENRY and LANG [6] evaluate P_{tt} to be approximately 0.6. They find that at high temperatures σ $\sigma_\infty \exp(-E_B/kT)$. The exponential term comes from f(0) given by (7). HENRY and LANG calculate σ_∞, the high temperature limit of σ, to be $\approx 6\times10^{-15}$ cm^2 in agreement with the data in Fig.8.

4. Discussion

We have reviewed five different phenomena that can be interpreted as consequences of large lattice relaxation of trapped electron states in semiconductors. These are the Jahn-Teller distortion of the Si vacancy, large Stokes shifts in the optical transitions of the second electron bound to O in GaP, the formation of a bound state by the D-X center only after relaxation, recombination enhanced defect motion of radiation damage centers, and nonradiative capture and recombination by multiphonon emission. These examples illustrate the following:

1) lattice relaxation largely determines the energies of deep levels;
2) lattice relaxation provides a mechanism for rapid nonradiative recombination;
3) lattice relaxation provides a mechanism for defect motion at low temperature.

The above phenomena were described in terms of extremely simple models, which rely heavily on configuration coordinate diagrams. For example, in discussion of capture by multiphonon emission, the impurity potential was approximated as a spherical square well of a depth determined by the lattice coordinate. Such simple models of complex phenomena are appropriate because they clarify the physical mechanisms and because, in many cases, detailed experimental knowledge of defect structure and lattice distortions is lacking. However, this information is available for a number of defects, especially for vacancy related centers in Si. Recently, realistic calculations have been made of the electronic structure and lattice relaxation of the isolated Si vacancy. It is likely that a more detailed understanding of nonradiative recombination and recombination enhanced defect motion will also emerge from studies of these phenomena in well characterized defects.

References

1 See Y. Toyozawa's paper in this volume.

2 G.D. Watkins, Proceedings of the International Conference on Crystal Defects, Conference Journal of the Physical Society of Japan 18, 22 (1963).

3 G.A. Baraff, E.O. Kane and M. Schlüter, Phys. Rev. Lett. 43, 956 (1979).

4 N.O. Lipari, J. Bernholc and S.T. Pantelides, Phys. Rev. Lett. 43, 1354 (1979).

5 G.A. Baraff, E.O. Kane and M. Schlüter, Phys. Rev. (to be published).

6 C.H. Henry and D.V. Lang, Phys. Rev. B 15, 989 (1977).

7 T.N. Morgan, Phys. Rev. Lett. 40, 190 (1978).

8 M. Gal, B.C. Covenett and P. Smith, Phys. Rev. Lett. 43, 1611 (1979).

9 P. Petroff and R.L. Hartman, J. Appl. Phys. 45, 3001 (1974).

10 P.W. Hutchinson, P.S. Dobson, S. O'Hara and D.H. Newmann, Appl. Phys. Lett. 26, 250 (1975).

11 D.V. Lang and L.C. Kimerling, Phys. Rev. Lett. 33, 489 (1974).

12 D.V. Lang, J. Appl. Phys. 45, 3023 (1974).

13 D.V. Lang and L.C. Kimerling, Appl. Phys. Lett. 28, 248 (1976).

14 J.R. Troxell, A.P. Chatterjee, G.D. Watkins and L.C. Kimerling, Phys. Rev. B 19, 5336 (1979).

15 D.V. Lang and R.A. Logan, Phys. Rev. Lett. 39, 635 (1977).

16 R. J. Nelson, Appl. Phys. Lett. 31, 351 (1977).

17 D.V. Lang, R.A. Logan and M. Jaros, Phys. Rev. B 19, 1015 (1979).

18 V. Narayanamurti, R.A. Logan and M.A. Chin, Phys. Rev. Lett. 43, 1536 (1979).

19 B. Hamilton, A.R. Peaker, S. Bromwell, W. Harding and D.R. Wight, Appl. Phys. Lett. 26, 702 (1975).

20 L.D. Landau and E.M. Lifshitz, Quantum Mechanics (Addeson-Wesley, Massachusetts, 1958), p. 310.

21 K. Huang and Rhys, Proc. R. Soc. 204, 406 (1950).

22 R. Kubo and Y. Toyozawa, Prog. Theor. Phys. 13, 160 (1955).

23 M. Lax, J. Chem. Phys. 20, 1752 (1952).

Nonradiative Transition During Lattice Relaxation in Localized Electron-Phonon System

K. Nasu

Institute for Solid State Physics, University of Tokyo
7-22-1, Roppongi, Minato-ku, Tokyo, Japan

1. Introduction

It was about 25 years ago that a theory for the multi-phonon nonradiative transition(MNT) in a localized electron-phonon system was derived by KUBO and TOYOZAWA(KT)[1], and it has been widely used to explain various multi-phonon nonradiative processes in crystals and molecules ever since. This theory, however, should be understood to give a static transition rate, because it is derived under the hypothesis that the phonon system at the initial state of the transition is in a thermal equilibrium. Although many theoretical studies subsequent to KT have so far been worked out for the MNT problem, they are also mainly concerned with the static MNT.

There are, however, many experimental evidences that tell us the occurrence of a dynamical MNT, that is, the nonradiative transition not after the thermal equilibrium, but during the lattice relaxation. This is closely related with the resonant second order optical process in a

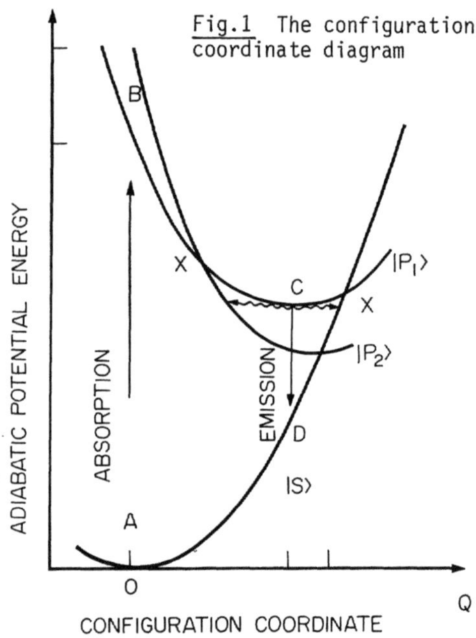

Fig.1 The configuration coordinate diagram

localized electron-phonon system[2,3]. Let us consider the process according to the schematic configuration coordinate diagram model as shown in Fig. 1. When the electron is excited from its ground state S to one of its excited state P_1, the phonon wave packet is vertically transferred on to the adiabatic potential energy surface(APES) of P_1(:from the point A to the point B in Fig. 1). After the excitation, the wave packet relaxes to the equilibrium position associated with P_1 state(the point C in Fig. 1) within a time of the order of 10^{-13} second. According to the conventional concept of the MNT, it is considered to occur after the relaxation and its final state is another excited state P_2(or the ground state S). It occurs very slowly as compared with the period of the phonon vibration, because the phonon must tunnel or overcome the barrier of the APES between P_1 and P_2 as schematically denoted by a wavy line in Fig. 1.

When the APES of P_1 has an intersection with that of P_2, however, the dynamical MNT is expected to occur more probably than the static one, because during the relaxation the phonon wave packet passes the intersection(the point X in Fig.1). We can find many experimental examples for the dynamical MNT: the nonradiative luminescence quenching of the F center[4], polarization correlation between the light absorption and the luminescence in Tl^+ type centers[5], the defect formation from the exciton in alkali halide crystals[6], nonradiative recombination of the electron with the hole in semiconductors[7] and so on. The photoexcitation energy transfer from one impurity center to the other is also considered to be the same phenomenon.

When we attempt to analyze these experiments quantitatively, we must have the knowledge about the transition probability around the intersection. In this respect, the LANDAU-ZENER-STUECKELBERG(LZS) formula[8] widely used in the molecular collision theory is very interesting. It gives a transition probability when a particle is compelled to pass the intersection at a uniform velocity V by an external force. According to this formula the transition probability P becomes

$$P=1-\exp(-2\pi T^2/V), \tag{1.1}$$

where T is the transfer matrix element between the initial and the final states. Although the LZS formula has a very simple form as shown in eq. (1.1), it covers the whole region from the adiabatic limit($T^2/V\rightarrow\infty$) to the diabatic limit($T^2/V\rightarrow 0$). For this reason, very recently, several studies have been worked out for the MNT, in which the potential crossing problem is managed by making use of the LZS formula phenomenologically. When we apply the formula to the second order optical process, however, we should be careful of its applicability conditions, because we are not concerning with the motion of the classical particle , but concerning with the relaxational motion of the phonon wave packet created by the light absorption.

2. Lattice Relaxation and Time Resolved Polarization of Emission

In order to clarify the dynamics of the MNT during the relaxation and the applicability conditions of the LZS formula, let us consider the resonant second order optical process(light absorption→ nonradiative transition→ emission process) in a three level system shown in Fig.2. The ground state has an s-like symmetry, and the excited states P_1 and P_2 have x- and y-like symmetries respectively. The excited states are nearly degenerate and interacting with each other. The energy matrix element of the total Hamiltonian($\equiv \mathcal{H}$) of the system with respect to the ground state can be written as (with $\hbar = 1$),

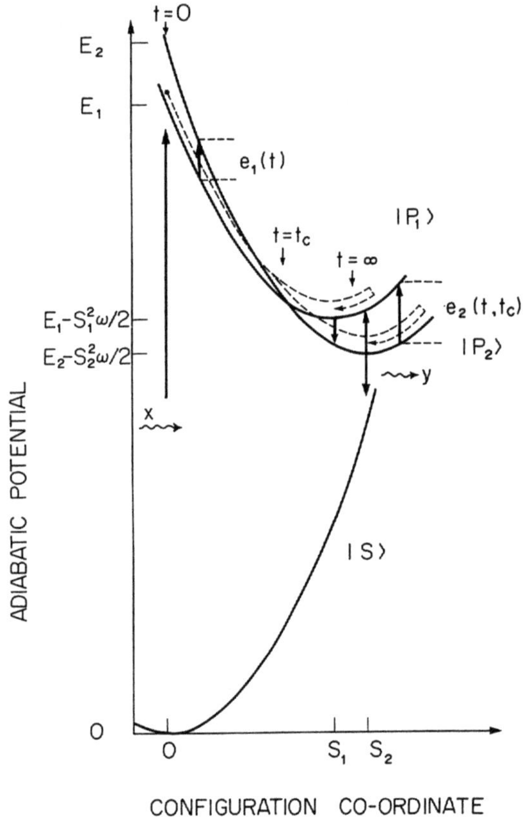

Fig.2 Schematic configuration coordinate diagram for the three level system

$$H_g \equiv \langle S|\mathcal{H}|S\rangle = \sum_k \omega_k a_k^+ a_k . \quad (2.1)$$

Here, ω_k is the frequency of the phonon, a_k^+ and a_k are its creation and annihiration operators and k is the normal mode index. We can represent the energy matrix in the subspace of the excited states by the 2 x 2 matrix form:

$$H_e = \begin{pmatrix} H_1 & 0 \\ 0 & H_2 \end{pmatrix} + \begin{pmatrix} 0 & T \\ T & 0 \end{pmatrix}, \tag{2.2}$$

taking $|P_j\rangle (j=1,2)$ as a set of base functions. Here H_j $(j=1,2)$ is

$$H_j = E_j + H_g + \sum_k S_{jk}\omega_k(a_k^+ + a_k)/\sqrt{2}, \quad (j=1,2). \tag{2.3}$$

E_j is the electronic transition energy and S_{jk} is the dimensionless coupling constant. T is the matrix element of the interaction which triggers the MNT. Practically, this interaction arises from various origins. However, in order to simplify the problem we assume it is a constant number.

We calculate the total intensity of the x and y polarized emissions($\equiv$ $L_1(t)$ and $L_2(t)$) at time t after the pulse excitation of the ground state electron by the x polarized white light at time t=0, at 0 K. The theory of the time resolved emission spectrum by TOYOZAWA[2] tells us that $L_j(t)(j=1,2)$ is given by

$$L_j(t) = e^{-2\gamma t}\langle\{\exp(itH_e)\}_{1j}\{\exp(-itH_e)\}_{j1}\rangle, \tag{2.4}$$

where γ is the radiative damping time constant assumed to common to the excited states, and $\langle\ldots\rangle \equiv Tr(\ldots\exp(-H_g/kT))/Tr(\exp(-H_g/kT))$. Since $L_j(t)$ satisfies the sum rule: $L_1(t)+L_2(t)=\exp(-2\gamma t)$, we discuss hereafter only $L_2(t)$, which is rewritten as

$$L_2(t) = e^{-2\gamma t}\langle\{\exp_-[i\int_0^t dt_1 \mathcal{T}(t_1)]\}_{12}\{\exp_+[-i\int_0^t dt_1' \mathcal{T}(t_1')]\}_{21}\rangle, \tag{2.5}$$

where

$$\mathcal{T}(t) = \exp[it\begin{pmatrix} H_1 & 0 \\ 0 & H_2 \end{pmatrix}]\begin{pmatrix} 0 & T \\ T & 0 \end{pmatrix}\exp[-it\begin{pmatrix} H_1 & 0 \\ 0 & H_2 \end{pmatrix}]. \tag{2.6}$$

It is convenient to cast $L_2(t)$ into a form of power series expansion with respect to T^2:

$$L_2(t) = e^{-2\gamma t}\sum_{n=o}(-1)^{n-1}\, T^{2n}\, L_2^n(t). \tag{2.7}$$

Here $L_2^n(t)$ is the expanding coefficient, which will be discussed in the later sections.

3. Second Order Perturbation

In this section , we calculate $L_2^1(t)$. From eq.(2.5) we get

$$L_2^1(t)=\int_0^t dt_1\int_0^t dt_1'\langle\exp[it_1H_1]\exp[-i(t_1-t_1')H_2]\exp[-it_1'H_1]\rangle. \tag{3.1}$$

By using the displacement operator $\exp(ip_j)$ of the phonon we can represent H_j as

$$H_j=\exp(ip_j)H_g\exp(-ip_j)+E_j-\sum_k S_{jk}^2\omega_k/2, \tag{3.2}$$

where $p_j\equiv i\sum_k S_{jk}(a_k^+-a_k)/\sqrt{2}$. Then, eq.(3.1) is rewritten as

$$L_2^1(t)=\int_0^t dt_1\int_0^t dt_1'\langle e^{ip_1}e^{-ip_{12}(t_1)}e^{ip_{12}(t_1')}e^{-ip_1}\rangle$$
$$\times\exp[i(E_1-E_2-\sum_k(S_{1k}^2-S_{2k}^2)\omega_k/2)(t_1-t_1')], \tag{3.3}$$

where, $p_{12}\equiv p_1-p_2$ and $p_{12}(t)\equiv\exp(iH_gt)p_{12}\exp(-iH_gt)$. The bracket in eq(3.3) can be calculated by the cumulant expansion method as

$$\ln\langle\ldots\rangle=[\sum_k S_{12k}^2(e^{-i\omega_k(t_1-t_1')}-1)+\sum_k S_{1k}S_{12k}(e^{i\omega_kt_1}-e^{-i\omega_kt_1}-e^{i\omega_kt_1'}+e^{-i\omega_kt_1'})]/2, \tag{3.4}$$

where $S_{12k}=S_{1k}-S_{2k}$.

When the electron-phonon coupling related with the MNT($P_1\to P_2$) is strong: $\langle P_{12}^2\rangle \gg 1$, $\langle\ldots\rangle$given by eq.(3.4) becomes a damping function of $\beta(\equiv t_1-t_1')$ with the damping rate much greater than the phonon frequency. In this case we can approximate $L_2^1(t)$ as

$$L_2^1(t)=\int_0^t dt_1\int_0^t dt_1'\exp[-ie_1(\alpha)-D^2\beta^2/4], \tag{3.5}$$

where $\alpha\equiv(t_1+t_1')/2$, $\qquad D\equiv[\sum_k S_{1k}^2\omega_k^2]^{1/2}$, (3.6)

and

$$e_1(t)\equiv E_2-E_1-\sum_k S_{1k}S_{12k}\omega_k(\cos\omega_kt-1). \tag{3.7}$$

D is the damping rate and is much greater than the averaged phonon frequency($\equiv\bar{\omega}$),

$$\bar{\omega}\equiv[\sum_k S_{12}^2\omega_k^2/(\sum_k S_{12k}^2)]^{1/2}\ll D. \tag{3.8}$$

It also can be interpreted as the energy width in which the MNT becomes possible. This width originates from the zero point vibration of the phonon and is proportional to the distance between the two equilibrium points of the excited state.

The quantity $e_1(t)$ denotes the FRANCK-CONDON energy(FCE) of the MNT from the APES of P_1 to that of P_2. As shown in Figs.2 and 3, it starts from E_2-E_1, and when the phonon has a continuous frequency spectrum it converges to the FCE at the equilibrium point of P_1:

$$e_1(\infty)=(E_2-\sum_k S_{2k}^2\omega_k/2)-(E_1-\sum_k S_{1k}^2\omega_k/2)+\sum_k S_{12k}^2\omega_k/2, \tag{3.9}$$

because of the phase radomization effect. Thus $e_1(t)$ reflects the relaxational motion of the phonon wave packet that is created on the APES of the P_1 state by the white excitation. If the wave packet passes the intersection of the two APES's of the excited states only once at time t_c during the relaxation, $L_2^1(t)$ increases appreciably near t_c, and this time is given by the equation $e_1(t_c)=0$.

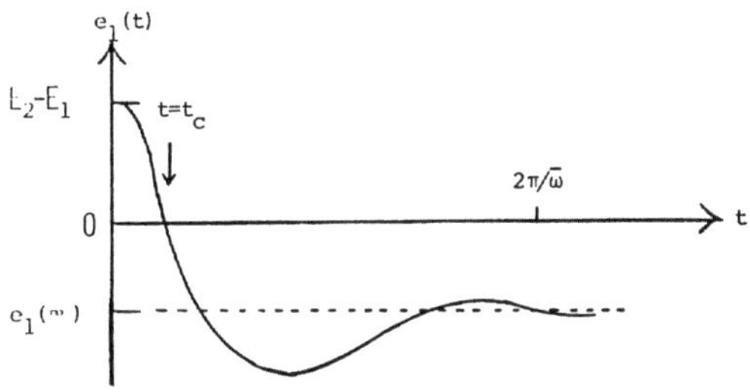

Fig.3 Time dependence of $e_1(t)$

In order to estimate the time integral in eq.(3.5), we expand $e_1(t)$ with respect $(t-t_c)$ and rewrite $L_2^1(t)$ as

$$L_2^1(t)=\int_0^t dt_1\int_0^t dt_1' \exp[-i\{e_1'(t_c)(\alpha-t_c)+e_1''(t_c)(\alpha-t_c)^2/2+\ldots\}\beta-D^2\beta^2/4], \tag{3.10}$$

where $e_1'(t_c)$ and $e_1''(t_c)$ is the velocity and the acceleration of the wave packet at the intersection:

$$e_1'(t)=\frac{\partial e_1(t)}{\partial t}=\sum_k S_{1k}S_{12k}\omega_k^2\sin\omega_k t, \tag{3.11}$$

$$e_1''(t)=\frac{\partial^2 e_1(t)}{\partial t^2}=\sum_k S_{1k}S_{12k}\omega_k^3\cos\omega_k t. \tag{3.12}$$

If the starting point of the relaxation is far removed from the intersection, the wave packet passes the intersection momentarily as compared with the period of the phonon vibration. Then we can neglect the acceleration as compared with the velocity. In this case $L_2(t)$ can be calculated easily from eq.(3.10), and it turns out to increase abruptly from zero to $2\pi T^2/|e_1'(t_c)|$ around t_c as shown in Fig.4. This value conforms to the lowest order term of the LZS formula in eq.(1.1). This is the dynamical MNT.

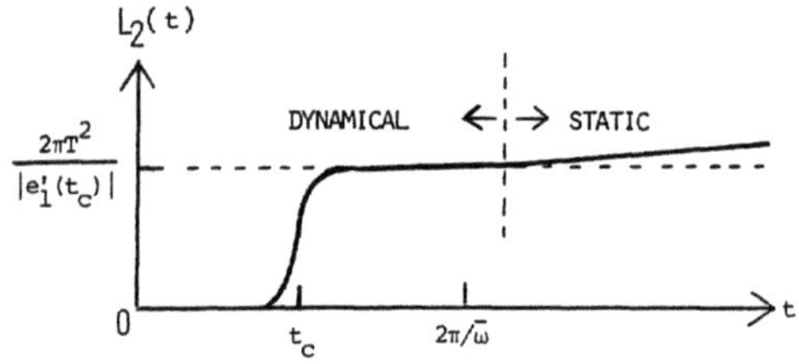

Fig.4 $L_2(t)$ as a function of t

For sufficiently large time $t(\gg 2\pi/\bar{\omega})$, the relaxation completes and we can replace $e_1(t)$ by $e_1(\infty)$ in eq.(3.5). Then we can easily calculate $L_2(t)$ in this time region and it turns out to increase gradually as compared with $2\pi/\bar{\omega}$ as shown in Fig.4, and its time derivative is given as

$$\frac{\partial}{\partial t}L_2(t)=2\pi^{1/2}D^{-1/2}\,T^2\exp[-e_1(\infty)^2/D^2]. \tag{3.13}$$

This is the transition rate of the static MNT mentioned before. Thus the MNT can be divided into two parts: the static and the dynamical parts. When the passage of the intersection occurs only once during the relaxation, the condition that determines which part of the two is more dominant becomes as

$$|e_1'(t_c)|^{-1} \gtrless \gamma^{-1}\pi^{-1/2}D^{-1}\exp[-e_1(\infty)^2/D^2]. \tag{3.14}$$

Incidentally, let us consider the multi-dimensional effect of the lattice relaxation. Only to simplify our illustration, we have described in Fig.2 as if the phonon mode associated with the $S\to P_1$ transition were identical with that of the $P_1\to P_2$ transition. However, it is not always so. In order to see this, we must describe the APES at least in a two dimensional configuration coordinate space as shown in Fig.5. It is a bird's-eye view of the APES. The points B,C and E are the equilibrium points of the S,P_1 and P_2 states respectively. The line FG denotes the equi-potential line between P_1 and P_2 states. Q_{sx} is the phonon mode associated with the $S\to P_1$ transition and Q_{xy} is the phonon mode associated with the $P_1\to P_2$ transition.

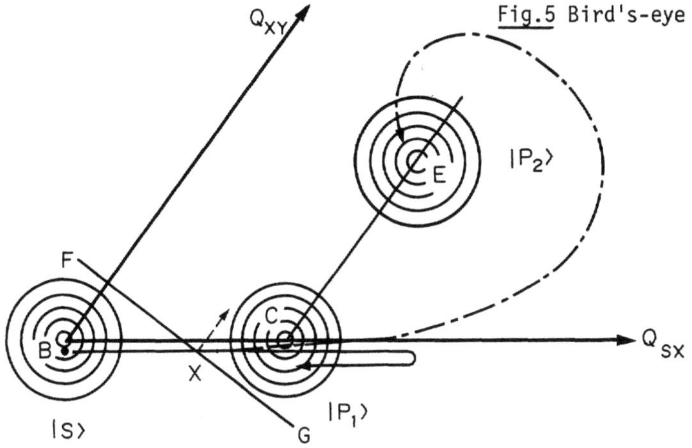

Fig.5 Bird's-eye view of the APES.

After the optical excitation, the phonon wave packet at the point B relaxes toward the point C. During the relaxation it goes across the equi-potential line FG around the point X. Consequently, when we determine the velocity (derivative of the FCE for the $P_1 \to P_2$ transition) around the point X, we must project the motion of Q_{sx} mode on to Q_{xy} mode. One can easily see that the summation of $S_{1k}S_{12k}$ over k in eq.(3.11) denotes this projection. Thus our theory includes the multi-dimensional effect exactly.

4. Higher Order Process

The calculation of higher order terms of T^2 is somewhat complicated, however, as proved by NASU and KAYANUMA[9,10] the dynamical part of the MNT conforms to the LZS formula at every order of T^2, under the condition

$$D \ll |e_1'(t_c)|^{1/2}. \tag{4.1}$$

This means that the distance between the equilibrium points of the excited states is very short as compared with the distance between the P_1 and the ground states.

Thus we can clarify the dynamical properties of the MNT during the relaxation according to the theory of the resonant second order optical process.

References

1 R.Kubo and Y. Toyozawa:Prog. Theor. Phys. 13,160(1955)

2 Y. Toyozawa: J. Phys. Soc. Japan 41,400(1976)

3 T. Takagawara, E. Hanamura and R. Kubo: J. Phys. Soc. Japan 43,802;811; 1522(1978)

4 R. Bartram and M.Stoneham:Solid State Comm. 17,1595(1975)

5 A. Fukuda, S. Makishima, T. Mabuchi and R. Onaka:J. Phys. Chem. Solids 28,1763(1967)

6 M. Kabler and R. Williams: Phys. Rev. B18,1949(1978)

7 C. Henry and D. Lang: Phys. Rev. B15,989(1975)

8 C. Zener: Proc. Roy. Soc. A137,696(1932)

9 K. Nasu and Y. Kayanuma: J.Phys. Soc . Japan 45,1341(1978).

10 Y. Kayanuma and K.Nasu: Solid State Comm. 27,138(1978)

Part II

Intermediate State Interaction

Resonant Raman Scattering and Luminescence

T. Takagahara

Department of Applied Physics, Faculty of Engineering, University of Tokyo
Hongo, Bunkyoku, Tokyo, Japan

Abstract

A general theory of second order optical process is developed on the basis of the damping theory. It is clarified that the competitive behavior between the Raman scattering and luminescence depends not only on the relaxation phenomena in the material system but also on the properties of excitation light and observation method. A general theory of time-resolved emission spectrum is formulated on the basis of microscopic treatment of the photo-detection process. The time-dependent behaviors of the Raman scattering and luminescence under pulsed excitation are clarified. As an illustration light scattering from a localized electron-phonon system is studied both on the stationary and time-resolved emission spectra. Then the theory is applied to the analysis of the recent experiments on the emission spectra from excitonic molecules in CuCl crystal and qualitatively satisfactory agreement between the theory and experiments is obtained. Finally a new formulation of resonant Raman scattering is presented in terms of the Keldysh-Schwinger formalism. This is a complete generalization of our stochastic theory to include the frequency and wavevector dependence of the relaxation phenomena.

1. Introduction

In recent years there has been much controversy on the fundamental problem of light scattering under resonant excitation [1-12]. The most important point of this problem is to clarify how the relaxation phenomena in the material system are reflected in the resonant secondary emission. As is well known, the secondary emission spectrum is composed of three components in general. The first is the resonance Raman scattering which is correlated with the incident light with respect to energy and polarization and the second of them is referred to as hot luminescence which is the emission after the phase relaxation is completed but while the energetic relaxation is going on. The third of them is known as ordinary luminescence which is the emission after the thermal equilibrium is established in the excited state. The composition ratio of these three components in the resonant secondary emission is determined by the relaxation processes in the material system. We have developed a general theory of resonant light scattering paying special attention to the role of the relaxation processes in the material system [11]. As noted above, there are two kinds of relaxation processes— one is the transverse or phase relaxation and the other the longitudinal or population relaxation. On the basis of the detailed study presented in [11], we may roughly say that the intensity ratio of the Raman scattering to the luminescence is determined by the ratio of the longitudinal relaxation rate to the transverse one. It is to be remarked that this statement is valid only in the ideal case of a three-level system in the narrowing limit. In general, however, we cannot always define the longitudinal and transverse relaxation rates unambiguously. Bearing these reservations in mind, we can discuss some aspects of the competition between the Raman scattering and luminescence in terms of the ratio between these relaxation rates.

Moreover the competition between the Raman scattering and luminescence depends sensitively on the properties of incident light and observation method. This fact must be taken into account in the analysis of experimental results as will be done in §5. In order to look into the competition between the two components and to understand the dynamics of relaxation in the intermediate state, the study of transient behaviors of these components under pulsed excitation is very important. In §3 we present a general theory of time-resolved emission spectrum and elucidate the different behaviors of the Raman scattering and luminescence. The transient behavior of the Raman component is sensitively affected by the experimental conditions, i.e., slit width , pulse shape and so on. On the contrary, the luminescence has an intrinsic component whose lifetime is identical to that of the intermediate state.

As an illustration of our theory light scattering from a localized electron-phonon system is studied on the stationary emission spectrum in §2 and on the time-resolved emission spectrum in §4. In §5 our theory is applied to the analysis of the recent experiments on resonant Raman scattering from excitonic molecules in heavily excited CuCl crystal. The agreement between our theory and experiments is qualitatively satisfactory. We have so far developed the theory on the basis of the damping theory in the Markovian approximation. In this formalism, however, the frequency and wavevector dependence of the relaxation phenomena cannot be suitably incorporated. In §6 we present a new formulation in terms of the Keldysh-Schwinger perturbation formalism. It will be shown that the new theory is a complete generalization of our stochastic theory with inclusion of the frequency and wavevector dependence of the relaxation phenomena.

2. Secondary Emission Spectrum from Localized Electron-Phonon System

In order to see the general features of resonant second order optical process, we have performed a model calculation on a localized electron-phonon system [12]. A localized electron interacts with a localized phonon mode and the latter is coupled to crystal modes. As the electronic states we consider only two levels, i.e., the ground and excited levels. The vibrational frequency of the local mode is in general different in the two electronic states. The Hamiltonian of the system is written explicitly as

$$H_g = \Omega_g b^+ b + \sum_k \omega_k R_k^+ R_k + \sum_k g_k (b^+ R_k + b R_k^+), \tag{2.1}$$

$$H_e = \varepsilon + \Omega_e (b^+ - \Delta)(b - \Delta) + \sum_k \omega_k R_k^+ R_k + \sum_k g_k \{(b^+ - \Delta) R_k + (b - \Delta) R_k^+\}, \tag{2.2}$$

where b and $\{R_k\}$ represent the annihilation operators of the local mode and crystal modes, respectively, Ω_g and Ω_e the local mode frequencies in the ground and excited electronic states and Δ the shift of the equilibrium position of the local mode. The energy of the local mode dissipates into the crystal modes through their interaction. When the local mode frequency is within the energy spectrum of the crystal modes, i.e., in the case of pseudolocal mode, the interaction between them is of Fermi resonance type and is given in the form of linear coupling as in (2.1). On the other hand, in the case of high frequency local mode like in U-center, the interaction is necessarily anharmonic due to the energy conservation and for example given by

$$\sum_{k,k'} g_{kk'}\{(b^{+}-\Delta)R_k R_{k'}+(b-\Delta)R_k^{+}R_{k'}^{+}\} .$$

The theoretical framework, however, is all the same in both cases except for the temperature dependence of the relaxation operator. In any case the irreversible process of energy dissipation of the local mode into the crystal modes is treated in terms of the damping theory by means of the NAKAJIMA [13]-ZWANZIG [14] projection technique. Namely eliminating the crystal modes, we introduce the stochastic motion of the local mode. The resulting relaxation operator of the local mode is given by

$$\Gamma_{\alpha\beta}\rho=(A_\alpha+A_\beta)b_\alpha^{+}\rho b_\beta+(B_\alpha+B_\beta)b_\alpha\rho b_\beta^{+}-(A_\alpha+B_\alpha)b_\alpha^{+}b_\alpha\rho-(A_\beta+B_\beta)\rho b_\beta^{+}b_\beta-(A_\alpha+A_\beta)\rho \quad , \quad (2.3)$$

with

$$b_g=b \text{ and } b_e=b-\Delta \; , \qquad (2.4)$$

where the indices α and β denote the electronic states. The explicit expressions of A and B are given by

$$A_\lambda = \sum_k |g_k|^2 \langle R_k^{+}R_k\rangle \delta(\omega_k-\Omega_\lambda), \qquad (2.5)$$

$$B_\lambda = \sum_k |g_k|^2 \langle R_k R_k^{+}\rangle \delta(\omega_k-\Omega_\lambda). \qquad (2.6)$$

Γ_{gg} and Γ_{ee} describe the vibrational relaxation in the ground and excited electronic states, respectively. The approach to the thermal equilibrium is guaranteed by the relation of detailed balance:

$$B_g=A_g\, e^{\beta\Omega_g/k_BT} \quad , \quad B_e=A_e\, e^{\beta\Omega_e/k_BT} . \qquad (2.7)$$

Then the secondary emission spectrum is expressed as [12]:

$$I(\omega_1,\omega_2) = 2\mathrm{Re}[\,(I) + (II) + (III)\,], \qquad (2.8)$$

$$(I)=\mathrm{Tr}\frac{1}{\gamma_e/2-i\Delta\omega_2-\Gamma_{eg}+iH_{eg}^{\times}}\;\frac{1}{i(\Delta\omega_1-\Delta\omega_2)-\Gamma_{gg}+iH_g^{\times}}\;\frac{1}{\gamma_e/2+i\Delta\omega_1-\Gamma_{ge}+iH_{ge}^{\times}}\,\rho_0, \quad (2.9)$$

$$(II)=\mathrm{Tr}\frac{1}{\gamma_e/2-i\Delta\omega_2-\Gamma_{eg}+iH_{eg}^{\times}}\;\frac{1}{\gamma_e-\Gamma_{ee}+iH_e^{\times}}\;\frac{1}{\gamma_e/2+i\Delta\omega_1-\Gamma_{ge}+iH_{ge}^{\times}}\,\rho_0 \; , \qquad (2.10)$$

$$(III)=\mathrm{Tr}\frac{1}{\gamma_e/2-i\Delta\omega_2-\Gamma_{eg}+iH_{eg}^{\times}}\;\frac{1}{\gamma_e-\Gamma_{ee}+iH_e^{\times}}\;\frac{1}{\gamma_e/2-i\Delta\omega_1-\Gamma_{eg}+iH_{eg}^{\times}}\,\rho_0 \; , \qquad (2.11)$$

where $\Delta\omega_1$ and $\Delta\omega_2$ are the incident and scattered photon energies relative to ε , γ_e the radiative or longitudinal decay rate of the excited electronic state and ρ_0 the initial equilibrium state of the local mode.

Before proceeding to the numerical calculation we look into the general aspects of competition between the Raman scattering and luminescence on the basis of expressions (2.8)-(2.11). The first term (I) contains the factor $\Delta\omega_1-\Delta\omega_2$ in its denominator and gives rise to the Raman scattering and Rayleigh scattering. On the other hand, the second and third terms (II) and (III) contain the factors $\Delta\omega_1$ and $\Delta\omega_2$ in a factorized form and therefore correspond to the luminescence. The relative intensity ratio of the Raman scattering to the luminescence can be roughly estimated by the middle denominators of the corresponding expressions because the denominators at the right- and left-ends are common to all three terms and represent the absorption and emission processes, respectively. Then the intensity ratio of the Raman scattering to the luminescence is roughly estimated as

$$\text{Raman:luminescence} \cong \frac{1}{\Gamma_{gg}} : \frac{1}{\gamma_e-\Gamma_{ee}} \cong \frac{1}{B} : \frac{1}{\gamma_e} . \tag{2.12}$$

Namely the relative intensity ratio of the Raman scattering to the luminescence is determined by the relative magnitude of γ_e to B.

This expectation is realized in Fig.1. In this case the local mode frequency is assumed not to change in the electronic transition, i.e., $\Omega_g=\Omega_e=\Omega_0$. In the numerical calculation the following parameters are chosen:

$$\Delta=1.0\ ,\quad \gamma_e=0.2,\quad \Delta\omega_1=2.5,\quad e^{-\hbar\Omega_0/k_BT}=0.2,\quad B=1.0,\ 0.1,\ 0.01,$$

where frequencies and decay rates are scaled by Ω_0. When B is one order of magnitude smaller than γ_e, the Raman lines at half-integer points of $\Delta\omega_2$ are dominant. When B is increased to be comparable to γ_e, the luminescence lines at $\Delta\omega_2=1.0$ and 2.0 are rather strong and may be considered to be hot luminescence as compared with the ordinary luminescence shown by dashed line. When B is one order of magnitude larger than γ_e, the luminescence line at $\Delta\omega_2=0.0$ is predominant and the Raman structures disappear.

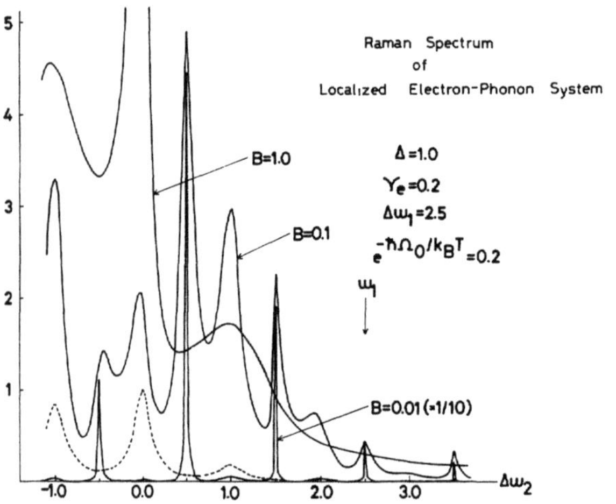

Fig.1 Secondary emission spectrum from a localized electron-phonon system. The intensity of the Rayleigh line $\delta(\Delta\omega_2-2.5)$ is given as 2.06, 2.15 and 1.48 for B=0.01, 0.1 and 1.0, respectively. The dashed line represents the ordinary luminescence for B=0.1.

As another example we study a case with the frequency change of the local mode in the electronic transition. The parameters are chosen as follows:

$$\Delta=1.0,\quad \gamma_e=0.02,\quad \Delta\omega_1=2.5,\quad \Omega_g=1.0,\quad \Omega_e=0.7,\quad k_BT=0.2,$$

$$B_g=A_g\, e^{\hbar\Omega_g/k_BT}=\frac{\gamma}{1-e^{-\hbar\Omega_g/k_BT}}\ ,\quad B_e=A_e\, e^{\hbar\Omega_e/k_BT}=\frac{\gamma}{1-e^{-\hbar\Omega_e/k_BT}}\ ,$$

$$\gamma=0.04,\ 0.005,\ 0.001,$$

where the frequencies and decay rates are scaled by Ω_g. In this case the Raman lines appear at $\Delta\omega_2=\Delta\omega_1\pm\ell\Omega_g$ while the luminescence lines at $\Delta\omega_2=m\Omega_e-n\Omega_g$ (ℓ, m, n:integers) and they can be distinguished from each other even under resonant excitation. In Fig.2, the Raman lines are denoted by $R_{\pm\ell}$ and the luminescence lines by m'-n corresponding to the above positions. The overall features of competition between the Raman scattering and the luminescence are almost the same as in Fig.1. The absence of 4'-3 and 1'-1 luminescence lines in the case of γ=0.005 may be explained as being hidden into the wings of strong Raman or luminescence lines in the neighborhood.

We mention briefly on the possibility to observe these changes of the secondary emission spectrum. As for the local mode associated with an impurity or a defect in solids, the vibrational relaxation is much faster than the radiative decay, i.e., $B\gg\gamma_e$ and the ordinary luminescence is the dominant contribution. On the other hand, in the gas phase the relaxation of molecular vibrations is caused mainly by inter-molecule collisions and can be controlled continuously by changing the gas pressure. The molecular gas will be the most suitable sample to observe the changes of secondary emission spectra as

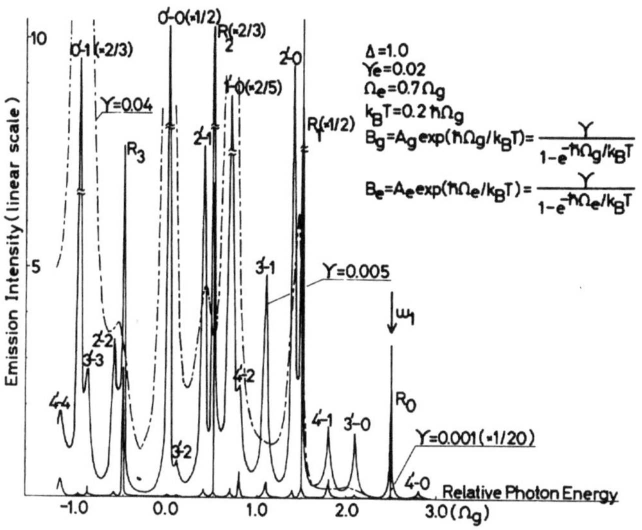

Fig.2 Secondary emission spectrum from a localized electron-phonon system with frequency change of the local mode. The intensity of the Rayleigh line $\delta(\Delta\omega_2-2.5)$ is given as 2.40, 2.38 and 2.39 for γ=0.04, 0.005 and 0.001, respectively.

shown in Figs.1 and 2. A demonstrative experiment was already reported on the change of secondary emission spectra from molecular iodine gas by varying the gas pressure [15].

3. General Theory of Time-Resolved Emission Spectrum

In this section we will derive the general formula of the time-resolved emission spectrum on the basis of the microscopic treatment of the photo-detection process. The photo-detection process is shown schematically in Fig.3 corresponding to the experimental block diagram in [16,17]. The incident laser light is scattered by the sample and only some fraction of the scattered light passes through the Kerr shutter while it is open, say from T to T+ΔT and enters the photo-detector. The time T can be controlled by the optical delay circuit whereas the duration time of observation ΔT is determined by the time constant of the Kerr cell. The scattered light which passed through the Kerr shutter goes into the filter. The filter is a kind of frequency selective device, e.g. prism or Fabry-Perot interferometer which allows photons within definite energy range to pass through to the photo-detector.

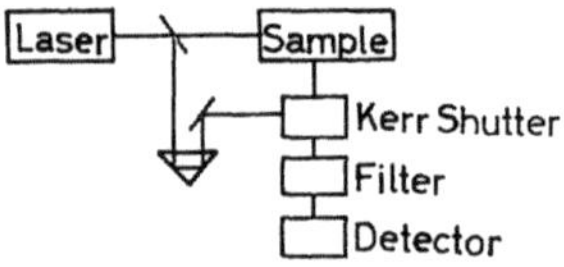

Fig.3 Schematic block diagram of experimental setup for the time-resolved measurement

Now with the above formulation of the photo-detection process we derive the general expression of time-resolved emission spectrum by means of the perturbation theory with respect to the electromagnetic interaction. The probability amplitude corresponding to the process in which an incident photon of frequency ω_1 is absorbed, a photon of frequency ω_2 is emitted and the latter is absorbed by the photo-detector during the time interval (T, T+ΔT) is given by

$$\Delta\psi_{fg}(T)=\int_T^{T+\Delta T} dt_0 \int_{-\infty}^{t_0} dt_1 \int_{-\infty}^{t_1} dt_2 \sum_{\omega_2} \langle f| e^{-iH(T+\Delta T-t_1)} V_2^+ e^{-iH(t_1-t_2)} V_1^- e^{-iHt_2}|g\rangle$$
$$\cdot {}_D\langle f| e^{-iH_D(T+\Delta T-t_0)} V_D^- e^{-iH_D t_0}|g\rangle_D \, e^{-i\omega_1 t_2+i\omega_2(t_1-t_0)} \varepsilon(t_2), \quad (3.1)$$

where H denotes the Hamiltonian of the material system, V the interaction between the radiation field and the material system, H_D and V_D the corresponding quantities for the photo-detector, $\varepsilon(t)$ the envelope of the incident pulse and $|g\rangle$, $|f\rangle$, $|g\rangle_D$ and $|f\rangle_D$ denote the initial and final states of the material system and of the photo-detector, respectively. Summation with respect to ω_2 is to be carried out over the frequency range of the filter. Factoring out the average frequency, we may write as

$$\cdot \sum_{\omega_2} e^{-i\omega_2(t_0-t_1)} = e^{-i\bar{\omega}_2(t_0-t_1)} \sum_{\omega_2} e^{-i(\omega_2-\bar{\omega}_2)(t_0-t_1)}$$
$$= e^{-i\bar{\omega}_2(t_0-t_1)} C(t_0-t_1) \ . \quad (3.2)$$

This function C is referred to as the response function of the filter in [18].

The integration range of t_0 is limited because the photo-detection time t_0 is restricted within the duration time of observation, i.e., from T to T+ΔT. In our treatment the time required for the scattered photon to traverse from the sample to the detector is neglected.

Then the photon counting rate per unit time is obtained as

$$I(\omega_1,\omega_2;T) \equiv \frac{1}{\Delta T}|\Delta\psi_{fg}(T)|^2$$
$$= \frac{1}{\Delta T}\int_T^{T+\Delta T} dt_0 \int_T^{T+\Delta T} dt_0' \int_{-\infty}^{t_0} dt_1 \int_{-\infty}^{t_1} dt_2 \int_{-\infty}^{t_0'} dt_1' \int_{-\infty}^{t_1'} dt_2' \, \varepsilon(t_2)\varepsilon(t_2')$$
$$\cdot e^{-i\omega_1(t_2-t_2')-i\bar{\omega}_2(t_0-t_1)+i\bar{\omega}_2(t_0'-t_1')} \, C(t_0-t_1)C^*(t_0'-t_1')$$
$$\cdot F(t_1,t_2;t_1',t_2')F_D(t_0,t_0'), \tag{3.3}$$

with

$$F(t_1,t_2;t_1',t_2') = \sum_f \mathrm{Av}_g \langle f| e^{-iH(T+\Delta T-t_1)} V_2^+ e^{-iH(t_1-t_2)} V_1^- e^{-iHt_2} |g\rangle\langle g| e^{iHt_2'}$$
$$\cdot V_1^+ e^{iH(t_1'-t_2')} V_2^- e^{iH(T+\Delta T-t_1')} |f\rangle , \tag{3.4}$$

$$F_D(t_0,t_0') = {}_D\langle g| e^{iH_D t_0'} V_D^+ e^{iH_D(t_0-t_0')} V_D^- e^{-iH_D t_0} |g\rangle_D$$
$$= \int_{-\infty}^{\infty} d\omega \, e^{i\omega(t_0-t_0')} F_D^{abs}(\omega) , \tag{3.5}$$

where $F_D^{abs}(\omega)$ is nothing but the absorption spectrum of the photo-detector. The expression (3.3) is quite general but rather complicated. Hereafter for the sake of simplicity we assume that the spectral width of $F_D^{abs}(\omega)$ is much broader than any intrinsic linewidth and the spectral width of the filter so that $F_D(t_0,t_0')$ can be replaced by $\delta(t_0-t_0')$, namely any photon impinging on the detector is instantaneously absorbed. Under this assumption the general expression (3.3) is simplified to

$$I(\omega_1,\omega_2;T) = \frac{1}{\Delta T}\int_T^{T+\Delta T} dt_0 \int_{-\infty}^{t_0} dt_1 \int_{-\infty}^{t_1} dt_2 \int_{-\infty}^{t_0} dt_1' \int_{-\infty}^{t_1'} dt_2' \, \varepsilon(t_2)\varepsilon(t_2')$$
$$\cdot e^{-i\omega_1(t_2-t_2')+i\bar{\omega}_2(t_1-t_1')} C(t_0-t_1)C^*(t_0-t_1')F(t_1,t_2;t_1',t_2'). \tag{3.6}$$

This is the expression of the time-resolved emission spectrum to be used in the following. Now there remains a task to determine the functional form of C(t). This is a rather cumbersome task because we must go into the details of each experimental apparatus. In the case of a Fabry-Perot interferometer the spectral transmission function identical with the Fourier transform of C(t) is given by the Airy formula [19]. This formula reduces to the simple form:

$$C(t) \cong \theta(t)\, e^{-\gamma_0 t}, \tag{3.7}$$

when specified to the neighborhood of any one spectral component as shown by EBERLY and WODKIEWICZ [18]. In the following we will use this form for the sake of simplicity.

Now we would like to examine the relationship between the present theory and the previous ones. In our previous paper [20] we defined provisionally the time-resolved emission spectrum by

$$\Upsilon(\omega_1,\omega_2;t) \equiv \frac{d}{dt} W(t) , \tag{3.8}$$

with

$$W(t)=\int_{-\infty}^{t} dt_1 \int_{-\infty}^{t_1} dt_2 \int_{-\infty}^{t} dt_1' \int_{-\infty}^{t_1'} dt_2' \; \varepsilon(t_2)\varepsilon(t_2') \; e^{-i\omega_1(t_2-t_2')+i\omega_2(t_1-t_1')} \cdot F(t_1,t_2;t_1',t_2') . \tag{3.9}$$

As remarked in [21], however, $W(t)$ is not always a monotonously increasing function and therefore $\Upsilon(\omega_1,\omega_2;t)$ may become negative. On the other hand, it is obvious from its definition that the new formula (3.3) or (3.6) is positive definite. We can show that the new formula (3.6) may be roughly approximated by

$$I(\omega_1,\omega_2;T)= \frac{1}{\Delta T} \int_{T}^{T+\Delta T} dt_0 \int_{-\infty}^{t_0} dt_1 \; |C(t_0-t_1)|^2 \; \Upsilon(\omega_1,\omega_2;t_1) . \tag{3.10}$$

Therefore the new definition is given by a convolution of the previous definition with the characteristic function of the filter.

Next we examine the time-integrated emission spectrum under pulsed excitation which is usually identified with the stationary emission spectrum in experiments. In this case we can replace $(T, T+\Delta T)$ by $(-\infty, \infty)$ and obtain the expression of the time-integrated emission spectrum apart from a constant factor as

$$I(\omega_1,\omega_2) = \int d\omega_1 \; E(\omega_1) F_S(\omega_1,\omega_2) , \tag{3.11}$$

with

$$E(\omega)= \sum_{\omega_1} \int_{-\infty}^{\infty} d\left(\frac{t_2+t_2'}{2}\right) \int_{-\infty}^{\infty} d(t_2-t_2') \; e^{i(\omega-\omega_1)(t_2-t_2')} \varepsilon(t_2)\varepsilon(t_2') , \tag{3.12}$$

$$F_S(\omega_1,\omega_2)=\lim_{t\to\infty} \frac{d}{dt} \int_{-\infty}^{t} dt_1 \int_{-\infty}^{t_1} dt_2 \int_{-\infty}^{t} dt_1' \int_{-\infty}^{t_1'} dt_2' \; e^{-i\omega_1(t_2-t_2')+i\omega_2(t_1-t_1')} \cdot F(t_1,t_2;t_1',t_2') , \tag{3.13}$$

where $F(t_1,t_2\;;t_1',t_2')$ is given by (3.4) and $E(\omega)$ represents the Fourier transform of the incident pulse and $F_S(\omega_1,\omega_2)$ the stationary emission spectrum. In the case of nano-second laser pulse the contribution to $E(\omega)$ from the pulse envelope is negligible so that the integration in (3.11) needs to be carried out only over the frequency distribution of the incident light.

In the rest of this section some model calculations will be carried out on the basis of (3.6) in order to see the competition between the Raman scattering and luminescence in their time-dependent behaviors. First let us consider the time-resolved emission spectrum from a three-level system which was discussed previously in order to describe the resonant secondary emission from an excitonic molecule [21]. The energy level scheme of this system is shown in Fig.4. As will be seen later, the time-resolved emission spectrum depends on the shape of the incident pulse. As a physically feasible and mathematically convenient pulse shape we assume $\varepsilon(t)$ to be given by

$$\varepsilon(t) = \begin{cases} 1- e^{-g_1 t} \,, & (0 \leqq t < T_p) \\ (1- e^{-g_1 T_p})\, e^{-g_2(t-T_p)} \,. & (T_p \leqq t) \end{cases} \qquad (3.14)$$

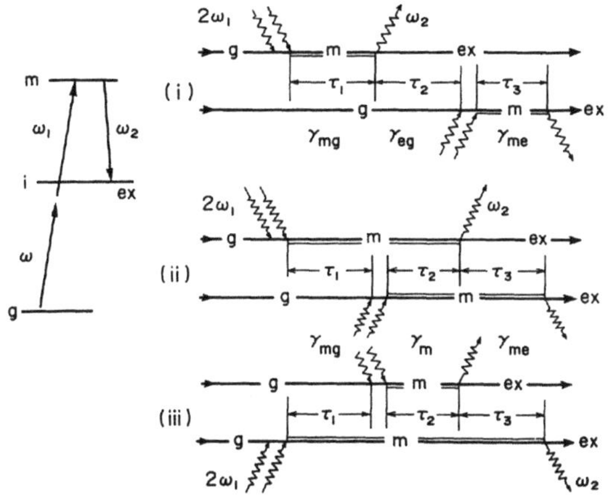

Fig.4 Energy level scheme and decomposed diagrams of the third order optical process via an excitonic molecule

With this pulse shape and the response function of the filter C(t) given by (3.7) we can calculate the expression (3.6). Since the explicit expression is rather lengthy, only the characteristic time constants of the Raman scattering and the luminescence will be written down. The same notations as in [21] will be used. The ground, intermediate and final states are denoted by $|g\rangle$, $|m\rangle$ and $|e\rangle$, respectively and the parameter $\gamma_{\alpha\beta}$ represents the relaxation constant of the tetradic Green function $\langle\langle\alpha\beta|G|\alpha\beta\rangle\rangle$. The energies of the incident and scattered photons are scaled respectively as $\Delta\omega_1=\omega_1-\omega_{mg}$ and $\Delta\omega_2=\omega_2-\omega_{me}$. Then the characteristic time constants for the two time regions before and after T_p are given as follows:

$0 \leqq T < T+\Delta T < T_p$

Raman: $1,\ g_1,\ 2g_1,\ \gamma_{eg}+\gamma_0+i(\Delta\omega_1-\Delta\omega_2),\ \gamma_{eg}+\gamma_0+g_1+i(\Delta\omega_1-\Delta\omega_2),\ \gamma_{me}+\gamma_0-i\Delta\omega_2,$

$2\gamma_0+\gamma_e,\ \gamma_{mg}-i\Delta\omega_1,\ \gamma_{mg}+g_1-i\Delta\omega_1\ ,$ (3.15)

Luminescence: $1,\ g_1,\ 2g_1,\ \gamma_{eg}+\gamma_0+i(\Delta\omega_1-\Delta\omega_2),\ \gamma_{eg}+\gamma_0+g_1+i(\Delta\omega_1-\Delta\omega_2),$

$\gamma_{me}+\gamma_0-i\Delta\omega_2,\ 2\gamma_0+\gamma_e,\ \gamma_{mg}-i\Delta\omega_1,\ \gamma_{mg}+g_1-i\Delta\omega_1,\ \gamma_m\ ,$ (3.16)

$T_p \leqq T < T+\Delta T$

Raman: $2g_2,\ \gamma_{eg}+\gamma_0+g_2+i(\Delta\omega_1-\Delta\omega_2),\ \gamma_{me}+\gamma_0-i\Delta\omega_2,\ 2\gamma_0+\gamma_e,\ \gamma_{mg}+g_2-i\Delta\omega_1,$ (3.17)

Luminescence: $2g_2$, $\gamma_{eg}+\gamma_0+i(\Delta\omega_1-\Delta\omega_2)$, $\gamma_{me}+\gamma_0-i\Delta\omega_2$, $2\gamma_0+\gamma_e$, $\gamma_{mg}+g_2-i\Delta\omega_1$,

$$\gamma_m \ , \tag{3.18}$$

where 1 represents a time independent term. In the above calculation we identified the Raman scattering and luminescence respectively with the terms whose denominators contain the factor $\Delta\omega 1-\Delta\omega_2$ and $\Delta\omega_2$. As seen from (3.15)-(3.18) the characteristic time constants of these emission components are affected by those of the incident laser pulse g_1 and g_2 and that of the response function of the filter γ_0. It is to be noted, however, that there is a luminescence component with the intrinsic time constant γ_m^{-1} of the intermediate state.

Now let us show some numerical examples of the overall behavior of the time-resolved emission spectrum. Figure 5a corresponds to the case of square pulse excitation. The parameters are chosen arbitrarily but with appropriate relevance to the excitonic molecule discussed in [21]:

$\Delta\omega_1=-1.0$ [meV], $\gamma_{mg}=0.09$, $\gamma_{eg}=0.06$, $\gamma_{me}=0.15$, $\gamma_m=0.18$, $\gamma_e=0.12$,

$\gamma_0=0.01$, $\Delta T=25$ [ps], $T_p=100$.

The emission intensity observed during the time interval $(T, T+\Delta T)$ is plotted at time $t=T$. At the very early stage of pulse excitation there is seen a very broad spectrum which reflects the indefiniteness of the incident light frequency of order t^{-1}. As time goes on, there appears an oscillating structure with respect to the emission frequency at a given time. As can be seen from the analytic expression (3.6), although omitted, this oscillating spectral structure is composed of superposition of many components with oscillation period given by t^{-1} or $(\Delta T)^{-1}$. Of course the spectral intensity at any given emission frequency is temporally oscillating. During the pulse excitation the Raman line at $\Delta\omega_2=-1$ [meV] is stronger than the luminescence line at $\Delta\omega_2=0$ [meV]. After the pulse is turned off, there can be seen that the peak position of the spectral intensity is removing from the Raman line to the luminescence line and finally the latter becomes stronger than the former.

In order to see the pulse shape dependence we varied the steepness parameter $g=g_1=g_2$ as $g^{-1}=1.0$ [ps] (Fig.5b) and 10 [ps] (Fig.5c). The square pulse excitation corresponds to $g^{-1}=0$. The overall spectral features do not show so much change with variation of g from 1.0 to ∞. It is to be noted, however, that the intensity ratio of the luminescence line to the Raman line is increased with the increase of the steepness parameter g. The ratio of peak intensity of the luminescence to that of the Raman component at time t=200 [ps] is given as 3.2, 3.0 and 0.36 for $g=\infty$, 1.0 and 0.1 [$(ps)^{-1}$], respectively. It can be understood as follows. The intensity of the luminescence line may be roughly proportional to the spectral intensity of the incident light within the resonance width of the intermediate state. The spectral intensity of the incident light is given by the Fourier transform of its pulse envelope since in the present case monochromatic excitation is considered. The Fourier transform has a broader spectrum for larger value of g. Hence the shortening of the pulse rise time g^{-1} leads to the enhancement of the luminescence relative to the Raman component.

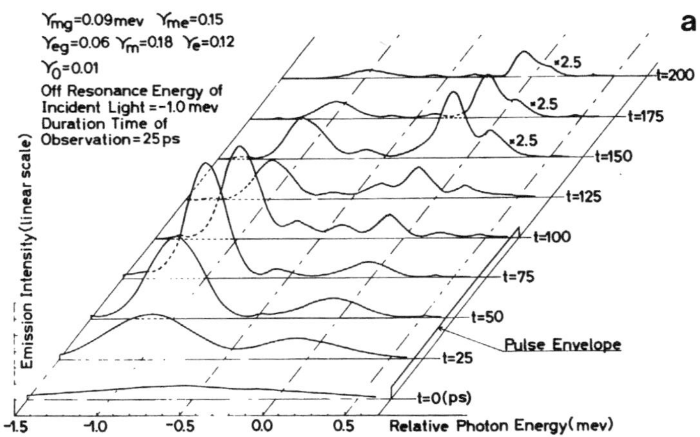

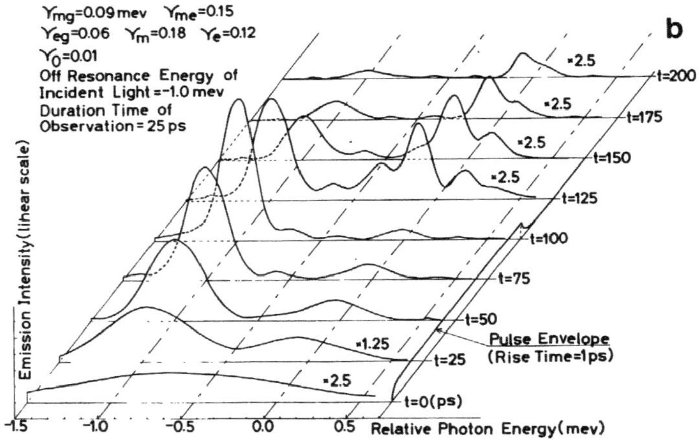

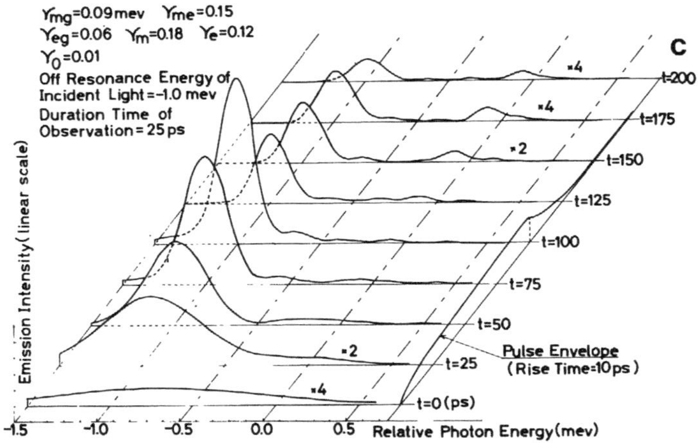

Fig.5 Time-resolved emission spectrum from a three-level system under a pulsed excitation, of the form of (3.14) with $g_1=g_2=g$. a) $g^{-1}=0$ [ps] b) $g^{-1}=1.0$ c) $g^{-1}=10$

4. Time-Resolved Emission Spectrum from Localized Electron-Phonon System

In this section we discuss the time-resolved emission spectrum from a localized electron-phonon system. The same problem was discussed by a few authors [10,22,23]. Their discussions are, however, restricted only to limiting situations, namely strong electron-phonon coupling, white light excitation or oversimplified model of relaxation of the vibronic levels. Here we present a quite general theory [12], making use of the model discussed in §2. Combining the theories presented in §2 and §3, we obtain the expression of the time-resolved emission spectrum from a localized electron-phonon system as

$$I(\omega_1,\omega_2;T)=\frac{1}{\Delta T}\int_T^{T+\Delta T} dt\, L^{-1}\{(I)+(II)+(III)\}+\text{c.c.}\,, \tag{4.1}$$

$$(I)=\frac{1}{2\pi i}\int du\frac{\phi(u)\phi(s-u)}{(s+2\gamma_0)(s+\gamma_e/2+\gamma_0-i\Delta\omega_2-\Gamma_{eg}+iH^\times_{eg})(s-u+\gamma_0+i(\Delta\omega_1-\Delta\omega_2)-\Gamma_{gg}+iH^\times_g)}$$
$$\cdot\frac{}{(s-u+\gamma_e/2+i\Delta\omega_1-\Gamma_{ge}+iH^\times_{ge})}\,, \tag{4.2}$$

$$(II)=\frac{1}{2\pi i}\int du\frac{\phi(u)\phi(s-u)}{(s+2\gamma_0)(s+\gamma_e/2+\gamma_0-i\Delta\omega_2-\Gamma_{eg}+iH^\times_{eg})(s+\gamma_e-\Gamma_{ee}+iH^\times_e)}$$
$$\cdot\frac{}{(s-u+\gamma_e/2+i\Delta\omega_1-\Gamma_{ge}+iH^\times_{ge})}\,, \tag{4.3}$$

$$(III)=\frac{1}{2\pi i}\int du\frac{\phi(u)\phi(s-u)}{(s+2\gamma_0)(s+\gamma_e/2+\gamma_0-i\Delta\omega_2-\Gamma_{eg}+iH^\times_{eg})(s+\gamma_e-\Gamma_{ee}+iH^\times_e)}$$
$$\cdot\frac{}{(s-u+\gamma_e/2-i\Delta\omega_1-\Gamma_{eg}+iH^\times_{eg})}\,, \tag{4.4}$$

where L^{-1} denotes the inverse Laplace transform with respect to s , $\phi(s)$ the Laplace transform of the pulse envelope $\varepsilon(t)$ and $\Gamma_{\alpha\beta}$ and $H^\times_{\alpha\beta}$ are defined in §2. The response function of the filter C(t) is assumed to be given by (3.7).

In the numerical calculation we are necessarily obliged to take the representation which diagonalizes the operators $\Gamma_{\alpha\beta}$ and $H^\times_{\alpha\beta}$ in order to transform the denominators in (4.2)-(4.4) into c-numbers. The eigenvalue problem of the operators $\Gamma_{\alpha\beta}$ and $H^\times_{\alpha\beta}$ was solved by the author, HANAMURA and KUBO [12] making use of the method due to LEE, HUANG and YANG [24]. In the following we introduce a more convenient representation due to LOUISELL [25]. The relaxation operator given in (2.4) can be brought into self-adjoint form by an appropriate transformation:

$$\tilde{\Gamma}_{\alpha\beta}\rho = e^{-\chi_{\alpha\beta}b^+_\alpha b_\alpha}\,\Gamma_{\alpha\beta}\, e^{\chi_{\alpha\beta}b^+ b}\rho$$
$$=-(A_\alpha+A_\beta)\rho+\sqrt{(A_\alpha+A_\beta)(B_\alpha+B_\beta)}\,(b^+_\alpha\rho b_\beta+b_\alpha\rho b^+_\beta)-(A_\alpha+B_\alpha)b^+_\alpha b_\alpha\rho-(A_\beta+B_\beta)\rho b^+_\beta b_\beta\,, \tag{4.5}$$

with

$$\chi_{\alpha\beta} = -\frac{1}{2}\log\frac{B_\alpha+B_\beta}{A_\alpha+A_\beta}\,. \tag{4.6}$$

Now we introduce new operators B_1, B_2, B_1^+ and B_2^+ defined by

$$B_1(\alpha\beta)\rho=\frac{1}{\sqrt{d_{\alpha\beta}^2-1}}(\rho b_\beta-d_{\alpha\beta}b_\alpha\rho) \ , \quad B_1^+(\alpha\beta)\rho=\frac{1}{\sqrt{d_{\alpha\beta}^2-1}}(\rho b_\beta^+-d_{\alpha\beta}b_\alpha^+\rho) \ ,$$
$$B_2(\alpha\beta)\rho=\frac{1}{\sqrt{d_{\alpha\beta}^2-1}}(b_\alpha^+\rho-d_{\alpha\beta}\rho b_\beta^+) \ , \quad B_2^+(\alpha\beta)\rho=\frac{1}{\sqrt{d_{\alpha\beta}^2-1}}(b_\alpha\rho-d_{\alpha\beta}\rho b_\beta) \ , \tag{4.7}$$

with $d_{\alpha\beta}=e^{-\chi_{\alpha\beta}}$. These operators satisfy the Bose commutation relations:

$[B_i, B_j^+]=\delta_{ij}$, $[B_i, B_j]=0$ and $[B_i^+, B_j^+]=0$.

In terms of these operators we obtain a diagonal representation of $\tilde{\Gamma}_{\alpha\beta}$ as

$$\tilde{\Gamma}_{\alpha\beta}=-(B_\alpha-A_\beta)B_1^+(\alpha\beta)B_1(\alpha\beta)-(B_\beta-A_\alpha)B_2^+(\alpha\beta)B_2(\alpha\beta) \ . \tag{4.8}$$

The eigenvalue and eigenvector of $\Gamma_{\alpha\beta}$ are given by

$$\Gamma_{\alpha\beta}(n_1,n_2)=-(B_\alpha-A_\beta)n_1-(B_\beta-A_\alpha)n_2 \ , \quad (n_1, n_2=0, 1, 2, \ldots) \tag{4.9}$$

$$|n_1,n_2>_{\alpha\beta}= e^{\chi_{\alpha\beta}b^+b}\ \frac{(B_1^+(\alpha\beta))^{n_1}(B_2^+(\alpha\beta))^{n_2}}{\sqrt{n_1!\ n_2!}}\ |0,0>_{\alpha\beta} \ , \tag{4.10}$$

with

$$|0,0>_{\alpha\beta}=\sqrt{1-d_{\alpha\beta}^{-2}}\ \sum_{n=0}^{\infty} e^{n\chi_{\alpha\beta}}|n>_{\alpha\beta}<n| \quad , \tag{4.11}$$

where $|n>_\alpha$ denotes the n-th vibrational state of the electronic level α. It can be easily checked that these eigenvectors are also eigenvectors of $H_{\alpha\beta}^\times$ when $\Omega_\alpha=\Omega_\beta$:

$$H_{\alpha\beta}^\times|n_1,n_2>_{\alpha\beta}=(n_1\Omega_\alpha-n_2\Omega_\beta)|n_1,n_2>_{\alpha\beta}\equiv h_{\alpha\beta}(n_1,n_2)|n_1,n_2>_{\alpha\beta} \ . \tag{4.12}$$

Of course there is one to one correspondence between these eigensolutions and those found in [12].

In terms of these eigenvectors of $\Gamma_{\alpha\beta}$ and $H_{\alpha\beta}^\times$ we can rewrite (4.2)-(4.4) in the form of c-numbers. Here for simplicity the expression for square pulse excitation of duration time T is written down:

$$(I)=\frac{1}{2\pi i}\int du\frac{e^{uT}}{u(u+s)}\ \Sigma\ U_{eg}^{gg}(0,0|m,n)U_{gg}^{eg}(m,n|k,l)U_{ge}^{gg}(k,l|i,j)U_{gg}^{ge}(i,j|0,0)$$
$$\cdot\frac{1}{(s+\gamma_4)(s+\gamma_1)(s+u+\gamma_2)(s+u+\gamma_3)} \ , \tag{4.13}$$

with

$\gamma_1=\frac{\gamma_e}{2}+\gamma_0-\Gamma_{eg}(m,n)+i(h_{eg}(m,n)-\Delta\omega_2)$, $\gamma_3=\frac{\gamma_e}{2}-\Gamma_{ge}(i,j)+i(\Delta\omega_1+h_{ge}(i,j))$,

$\gamma_2=\gamma_0-\Gamma_{gg}(k,l)+i(\Delta\omega_1-\Delta\omega_2+h_{gg}(k,l))$, $\gamma_4=2\gamma_0$,

$$(II)=\frac{1}{2\pi i}\int du \frac{e^{uT}}{u(u+s)} \Sigma\, U^{gg}_{eg}(0,0|m,n)U^{eg}_{ee}(m,n|k,l)U^{ee}_{ge}(k,l|i,j)U^{ge}_{gg}(i,j|0,0)$$

$$\cdot\frac{1}{(s+\gamma_4)(s+\gamma_1)(s+\gamma_2)(s+u+\gamma_3)} , \qquad (4.14)$$

with

$$\gamma_1=\frac{\gamma_e}{2}+\gamma_0-\Gamma_{eg}(m,n)+i(h_{eg}(m,n)-\Delta\omega_2) , \quad \gamma_3=\frac{\gamma_e}{2}-\Gamma_{ge}(i,j)+i(\Delta\omega_1+h_{ge}(i,j)),$$

$$\gamma_2=\gamma_e-\Gamma_{ee}(k,l)+ih_{ee}(k,l) , \quad \gamma_4=2\gamma_0 ,$$

$$(III)=\frac{1}{2\pi i}\int du \frac{e^{uT}}{u(u+s)} \Sigma\, U^{gg}_{eg}(0,0|m,n)U^{eg}_{ee}(m,n|k,l)U^{ee}_{eg}(k,l|i,j)U^{eg}_{gg}(i,j|0,0)$$

$$\cdot\frac{1}{(s+\gamma_4)(s+\gamma_1)(s+\gamma_2)(s+u+\gamma_3)} , \qquad (4.15)$$

with

$$\gamma_1=\frac{\gamma_e}{2}+\gamma_0-\Gamma_{eg}(m,n)+i(h_{eg}(m,n)-\Delta\omega_2) , \quad \gamma_3=\frac{\gamma_e}{2}-\Gamma_{eg}(i,j)+i(h_{eg}(i,j)-\Delta\omega_1),$$

$$\gamma_2=\gamma_e-\Gamma_{ee}(k,l)+ih_{ee}(k,l) , \quad \gamma_4=2\gamma_0 ,$$

where the summation should be taken over the indices i, j, k, l, m and n and the local mode frequency is assumed to be equal in both electronic states. The U-function is defined as the expansion coefficient of the eigenvector $|i,j\rangle_{\alpha\beta}$ in terms of other eigenvectors $|i',j'\rangle_{\alpha'\beta'}$:

$$|i,j\rangle_{\alpha\beta}=\sum_{i',j'} U^{\alpha'\beta'}_{\alpha\beta}(i',j'|i,j)|i',j'\rangle_{\alpha'\beta'} . \qquad (4.16)$$

The various U-functions can be written in terms of the following function V with various arguments:

$$V(p+s,\ p,\ q+t,\ q;x,\ y)=\sum_{k=Max(0,-s)}^{p} \frac{\sqrt{p!(p+s)!}}{k!(k+s)!}\ \frac{x^k(-x)^{k+s}}{(p-k)!} \cdot \sum_{\ell=max(0,-t)}^{q}$$

$$\cdot\frac{\sqrt{q!(q+t)!}}{\ell!(\ell+t)!}\ \frac{(-y)^{\ell}y^{\ell+t}}{(q-\ell)!} \cdot , \qquad (4.17)$$

where $-p\leqq s$ and $-q\leqq t$.

From (4.2)-(4.4) or (4.13)-(4.15) we can see that the characteristic decay constants of the Raman and luminescence components during the pulse excitation are given by

$$\text{Raman:}\frac{\gamma_e}{2}+\gamma_0-\Gamma_{eg},\ \gamma_0-\Gamma_{gg},\ \frac{\gamma_e}{2}-\Gamma_{ge},\ 2\gamma_0,\ 1 ,$$

$$\text{luminescence:}\frac{\gamma_e}{2}+\gamma_0-\Gamma_{eg},\ \gamma_0-\Gamma_{gg},\ \gamma_e-\Gamma_{ee},\ \frac{\gamma_e}{2}-\Gamma_{ge},\ 2\gamma_0,\ 1 , \qquad (4.18)$$

where 1 stands for the time independent term. On the other hand, the characteristic decay constants after the pulse is switched off are given by

$$\begin{aligned}&\text{Raman:} \frac{\gamma_e}{2}+\gamma_0-\Gamma_{eg},\ 2\gamma_0\ ,\\ &\text{Luminescence:} \frac{\gamma_e}{2}+\gamma_0-\Gamma_{eg},\ 2\gamma_0,\ \gamma_e-\Gamma_{ee}\ .\end{aligned} \tag{4.19}$$

In the above $\gamma_e/2+\gamma_0-\Gamma_{eg}$ represents the spectral width of emission lines plus γ_0 and $\gamma_e-\Gamma_{ee}$, the vibrational relaxation plus the longitudinal or radiative decay of the excited electronic state. It is important to note that the latter time constant is contained only in the luminescence component and is not shared by the Raman component. The luminescence from the higher-lying eigenstate of the operator Γ_{ee} can be referred to as hot luminescence and that from the lowest eigenstate is nothing but the ordinary luminescence.

In the numerical calculation of (4.13)-(4.15) the summation over the infinite series was truncated by about one hundred thousand terms in order to keep the relative error within 10^{-3} at most. Figure 6 shows the result under a square pulse excitation for the case without frequency change of the local mode, i.e., $\Omega_g=\Omega_e=\Omega_0$. The relevant parameters are chosen as:

$$\Delta=1.0,\ \Delta\omega_1=2.5,\ B=0.1,\ \gamma_e=0.2,\ \gamma_0=0.04,\ e^{-\hbar\Omega_0/k_BT}=0.1,$$

$$T=100,\ T_p=500,$$

where the frequencies and relaxation rates are scaled in units of Ω_0 and the unit of time is taken as Ω_0^{-1}. The emission intensity measured during the time interval $(T,T+\Delta T)$ is plotted at t=T. The emission spectrum observed during the time interval (0,100) is not so broad as in Fig.5 and shows already well-defined structures. This is because ΔT is much longer than the intrinsic time constants γ_0^{-1}, γ_e^{-1} and B^{-1}. At t=300 the emission spectrum seems to have reached the stationary value. After the pulse is switched off, it can be seen that the relative intensity ratio of the Rayleigh and Raman components to the luminescence decreases. Even in the asymptotic time region of t=1000 the

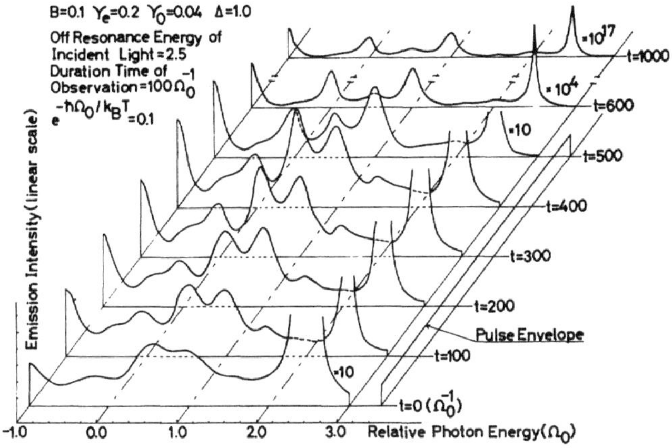

Fig.6 Time-resolved emission spectrum from a localized electron-phonon system under a square pulse excitation

emission spectrum does not seem to approach to the ordinary luminescence and there remain the Rayleigh and Raman components appreciably. This is because the longest lifetime is $(2\gamma_0)^{-1}$ in this case and is shared by both the Raman and luminescence components as seen from (4.19). The emission spectrum in the asymptotic time region is determined by the components with decay rate $2\gamma_0$ and this does not necessarily coincide with the ordinary luminescence. On the contrary if γ_0 is comparable to γ_e, the longest lifetime is given by γ_e^{-1} corresponding to the zero eigenvalue of Γ_{ee}. Then in this case the emission spectrum approaches to the ordinary luminescence asymptotically. In conclusion the time-resolved emission spectrum reflects the dynamics of relaxation in the intermediate state but at the same time depends sensitively on the details of observation conditions.

5. Resonant Raman Scattering and Luminescence due to Excitonic Molecule

Now we apply the preceding theory to the interpretation of resonant secondary emission spectra from excitonic molecules in CuCl observed recently by Japanese research groups [15,16,26]. We can create the excitonic molecule directly and very effectively by means of the giant two-photon absorption [27]. The absorption coefficient is as large as that of the single exciton and is by an order of 10^6 larger than that of the ordinary two-photon absorption. This extremely strong absorption comes from the mutual enhancement due to the giant oscillator strength and the resonance effect. The relevant emission process of the excitonic molecule is as follows: one of the two excitons composing an excitonic molecule is annihilated radiatively and the other exciton is left behind as a polariton or a longitudinal exciton in the crystal. When the incident light is resonant with the giant two-photon absorption of the excitonic molecule, there arises a question whether the third order optical process of two-photon absorption and subsequent photo-emission can be regarded as two separable processes of absorption and emission or as an inseparable process of light scattering. This question will be discussed in this section.

The resonant secondary emission spectrum can be calculated by means of the damping theory [21]. The energy level scheme of the system and three decomposed diagrams are shown in Fig.4. $|g\rangle$, $|i\rangle$, $|m\rangle$ and $|e\rangle$ denote the initial ground state, the intermediate exciton state, the excitonic molecule state and the final exciton state, respectively. The transition probability per unit time apart from a constant factor is given by [21]

$$I(2\omega_1,\omega_2)=\frac{\gamma_{mg}}{(2\Delta\omega_1)^2+\gamma_{mg}^2}\,\frac{\gamma_{me}}{(\Delta\omega_2)^2+\gamma_{me}^2}\left\{\frac{2}{\gamma_m}+\frac{\gamma_{eg}[1+2\Delta\omega_1\Delta\omega_2/\gamma_{mg}\gamma_{me}]}{(2\Delta\omega_1-\Delta\omega_2)^2+\gamma_{eg}^2}\right.$$
$$\left.-\frac{(2\Delta\omega_1-\Delta\omega_2)[2\Delta\omega_1/\gamma_{mg}-\Delta\omega_1/\gamma_{me}]}{(2\Delta\omega_1-\Delta\omega_2)^2+\gamma_{eg}^2}\right\}, \quad (5.1)$$

where $\gamma_{\alpha\beta}$ denotes the decay constant of the tetradic Green function $\langle\langle\alpha\beta|G|\alpha\beta\rangle\rangle$. The energies of the incident and scattered photons are scaled respectively as $2\Delta\omega_1=2\omega_1-\omega_{mg}$ and $\Delta\omega_2=\omega_2-\omega_{me}$. This is the expression for the monochromatic and stationary excitation. In usual experiments, however, the stationary emission spectrum is identified with the time-integrated emission spectrum under pulsed excitation of nano-second duration. Taking account of this fact and the spectral width of the incident light and using the results obtained in §3, we finally obtain the expression of the emission spectrum to be compared with experiments as

$$I(2\omega_1^0,\omega_2)=\int d(2\omega_1)\frac{1}{\sqrt{\pi}\ \sigma}\exp[-\frac{(2\omega_1-2\omega_1^0)^2}{\sigma^2}]\ I(2\omega_1,\omega_2)\quad , \tag{5.2}$$

with

$$\sigma=\sqrt{2\sigma_f^2+\sigma_p^2}\quad , \tag{5.3}$$

where the pulse envelope and the spectral distribution of the incident light are assumed to be given by

$$\varepsilon(t)=\exp[-(t\sigma_p/2)^2]\quad , \tag{5.4}$$

$$f(\Delta\omega_1)=\frac{1}{\sqrt{\pi}\ \sigma_f}\exp[-(\frac{\Delta\omega_1}{\sigma_f})^2]\quad . \tag{5.5}$$

Recently MITA and UETA [26] observed the coexistence of resonant Raman scattering and luminescence in the emission spectrum from excitonic molecules in CuCl excited by a nano-second laser pulse with very narrow spectral width of 0.053 meV as shown in Fig.7a. In this case the contribution of pulse width σ_p to σ in (5.3) is one or two orders of magnitude smaller than σ_f and can be neglected. The relevant relaxation rates appearing in (5.1) can be determined from experiments as follows:

γ_{eg} = HWHM of Raman line ,

γ_{me} = HWHM of luminescence line ,

γ_{mg} = FWHM of giant two-photon absorption ,

γ_m = Decay rate of luminescence line in time-resolved experiments .

The calculated emission spectra making use of these parameters are shown in Fig.7b. These should be compared with the spectra from f to ℓ in Fig.7a. The agreement between the two is qualitatively satisfactory.

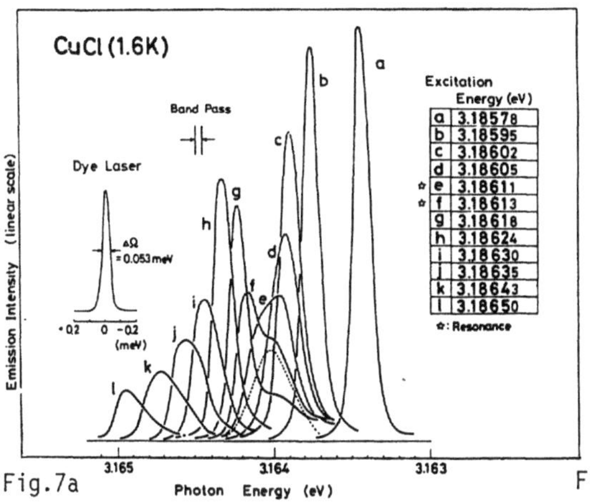

Fig.7a

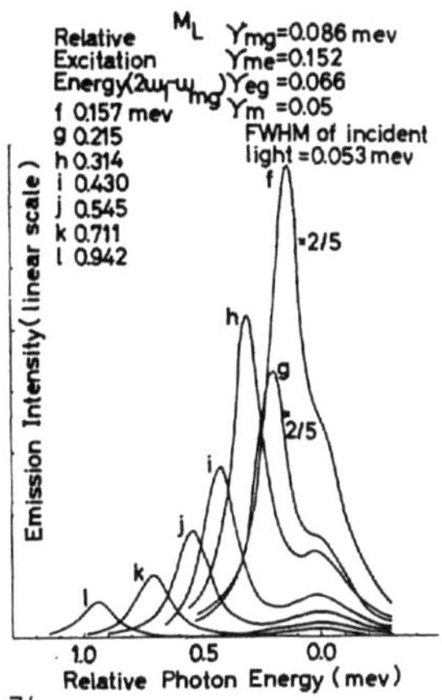

Fig.7b

Fig.7 Secondary emission spectra under the giant two-photon excitation of an excitonic molecule with recoil of a longitudinal exciton a) Experiment by MITA and UETA [26] b) Theoretical [21]

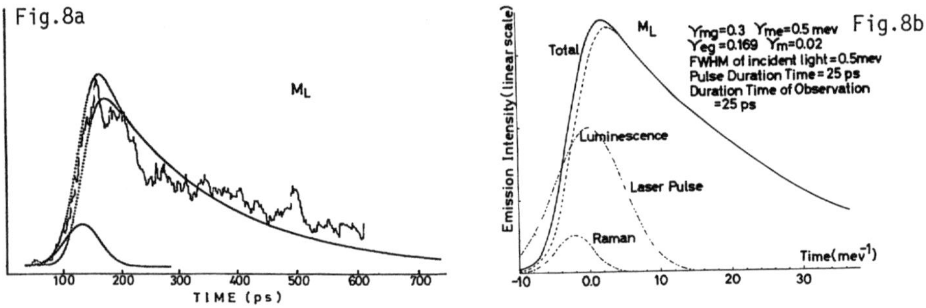

Fig.8 Decay profile of M_L line under just resonant excitation of excitonic molecule. The total emission intensity is decomposed into Raman and luminescence components as shown by dashed lines. a) Experiment by SEGAWA et al. [16] b) Theoretical [21]

SEGAWA et al. [16] observed the transient behavior of the emission intensity from excitonic molecules in CuCl excited by pico-second laser pulse as shown in Fig.8a and found that the Raman scattering component shows the same time dependence as the incident laser pulse. The theoretical time dependence of the peak intensity under just resonant excitation is shown in Fig.8b and is decomposed into the Raman scattering and luminescence according to the diagrams in Fig.4. The agreement between the theory and experiments is remarkably well. Thus our simple damping theory can explain the experimental features rather successfully.

6. A New Formulation of Resonance Raman Scattering in Terms of Keldysh-Schwinger Perturbation Formalism

We have so far developed a general theory of resonant second order optical process making use of the damping theory in the Markovian approximation In this formalism, however, the non-Markovian property of relaxation phenomena cannot be suitably incorporated. In order to incorporate the frequency and wavevector dependence of relaxation phenomena we have reformulated the theory in terms of the Keldysh-Schwinger perturbation scheme and obtained a basic integral equation for the resonant secondary emission spectrum. It will be shown that the new formalism is a complete generalization of our stochastic theory.

The resonant secondary emission spectrum is given by

$$
\begin{aligned}
I(k_1;\omega_1;k_2,\omega_2)=&\int_{-\infty}^{\infty} d(u-u')\int_0^{\infty} d(u-t)\int_0^{\infty} d(u'-t')\, e^{-(\gamma-i\omega_1)(u-t)-(\gamma+i\omega_1)(u'-t')} \\
&\cdot e^{i(\omega_2-\omega_1)(u-u')} \langle T_C \exp_{\rightarrow}[i\int_{t'}^{u'} d\tau' V(\tau')]\exp_{\leftarrow}[-i\int_t^u d\tau V(\tau)] \\
&\cdot a_{k_1}(t')a^+_{k_2}(u')a_{k_2}(u)a^+_{k_1}(t)\rangle \quad , \qquad (6.1)
\end{aligned}
$$

where V is the difference between the Hamiltonians in the ground and excited states, i.e., He-Hg, a_k and a_k^+ the annihilation and creation operators of elementary excitation(exciton or polariton) with wavevector k and ω_1 and ω_2 the incident and scattered photon energies, respectively. γ denotes the wavevector independent part of the longitudinal decay rate of the elementary excitation. The wavevector dependent longitudinal or radiative decay rate can be suitably incorporated in the single-particle Green functions defined in (6.2)-(6.6). All operators in the bracket are in the interaction representation with respect to Hg and T_c is the time-ordering operator introduced by Keldysh [28]. Here the exciton or polariton is described as a single Bose particle with only momentum index for simplicity. This assumption is reasonable when the intraband scattering is the most dominant contribution. In general the band indices must be included especially when a few polariton modes coexist and the inter-branch scattering plays an important role or the scattering between different states of the exciton internal motion is important. For the moment, however, we drop the band or branch indices in order to see the essential points of our theory avoiding complexity of formalism.

Now we introduce four kinds of single-particle Green functions defined as [28]:

$$\phi_k^+(\tau,\tau')=\langle T_c\ a_k(\tau)a_k^+(\tau')U(u,t;u',t')\rangle \quad , \tag{6.2}$$

with

$$U(u,t;u',t')=\exp_{\rightarrow}[i\int_{t'}^{u'}d\tau'V(\tau')]\exp_{\leftarrow}[-i\int_{t}^{u}d\tau V(\tau)] \quad , \tag{6.3}$$

where τ is on the upper branch and τ' on the lower branch,

$$\phi_k^-(\tau,\tau')=\langle T_c\ a_k(\tau)a_k^+(\tau')U(u,t;u',t')\rangle \quad , \tag{6.4}$$

with τ on the lower branch and τ' on the upper branch,

$$\phi_k^c(\tau,\tau')=\langle T_c\ a_k(\tau)a_k^+(\tau')U(u,t;u',t')\rangle \quad , \tag{6.5}$$

with both τ and τ' on the upper branch and

$$\tilde{\phi}_k^c(\tau,\tau')=\langle T_c\ a_k(\tau)a_k^+(\tau')U(u,t;u',t')\rangle \quad , \tag{6.6}$$

with both τ and τ' on the lower branch. These functions are illustrated in Fig.9.

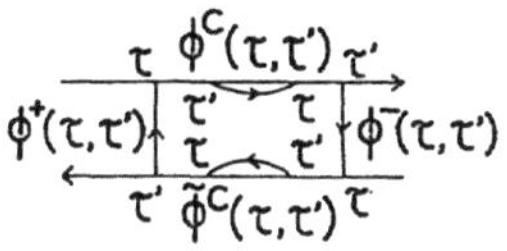

Fig.9 Schematic representation of single-particle Green functions

Then the two-particle Green function in (6.1) can be rewritten as follows making use of the linked cluster theorem:

$$\langle T_c\ a_{k_1}(t')a_{k_2}^+(u')a_{k_2}(u)a_{k_1}^+(t)\exp_{\rightarrow}[i\int_{t'}^{u'}d\tau'V(\tau')]\exp_{\leftarrow}[-i\int_{t}^{u}d\tau V(\tau)]\rangle$$

$$= e^{C(u,t;u',t')}[\delta_{k_1,k_2}\phi^c_{k_1}(u,t)\tilde{\phi}^c_{k_1}(t',u') + \int_t^u d\tau_1\int_t^{\tau_1} d\tau_2\int_{t'}^{u'} d\tau_1'\int_{t'}^{\tau_1'} d\tau_2'$$

$$\cdot\phi^c_{k_2}(u,\tau_1)\phi^c_{k_1}(\tau_2,t)\tilde{\phi}^c_{k_2}(\tau_1',u')\tilde{\phi}^c_{k_1}(t',\tau_2')K_{k_2k_1}(\tau_1,\tau_2;\tau_1',\tau_2')], \tag{6.7}$$

where $C(u,t;u',t')$ denotes the contributions from all the irreducible closed diagrams and $K_{k_2k_1}(\tau_1,\tau_2;\tau_1',\tau_2')$, the complete interaction part of the two-particle Green function [29]. When there is no elementary excitation (exciton or polariton) in the initial state, it can be shown that $C(u,t;u',t')$ is identically zero and the single-particle Green functions ϕ^+ and ϕ^- do not appear in the expression. The first term in the bracket of (6.7) represents the forward Rayleigh scattering, i.e., free propagation of elementary excitation with no perturbation. The second term, shown diagrammatically in Fig.10, describes inelastic Raman scattering and luminescence and also Rayleigh scattering with momentum change ($|k_1|=|k_2|$ but $k_1 \neq k_2$) due to impurity scattering. Thus the function K contains all information about the relaxation processes in the material system and determines the secondary emission spectrum.

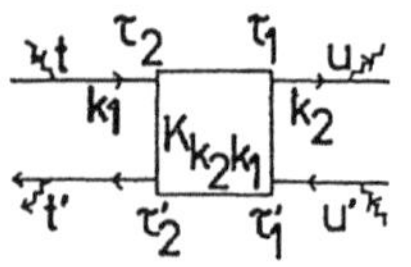

Fig.10 The complete interaction part K of the two-particle Green function

The function K satisfies an integral equation, i.e., the well-known Bethe-Salpeter equation as shown in Fig.11. The function R in Fig.11 represents the irreducible interaction part of the two-particle Green function [29]. For the

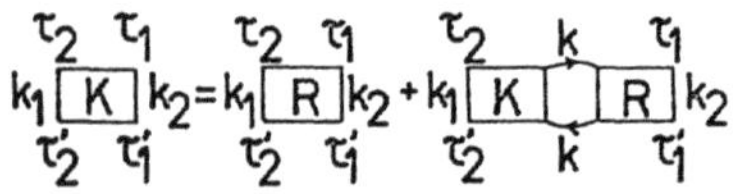

Fig.11 The Bethe-Salpeter equation for the complete interaction part K

sake of convenience in later analysis we carry out Fourier transform with respect to $\tau_1-\tau_1'$ and Laplace transform with respect to $\tau_1-\tau_2$ and $\tau_1'-\tau_2'$ on the functions $K(\tau_1,\tau_2;\tau_1',\tau_2')$ and $R(\tau_1,\tau_2;\tau_1',\tau_2')$ and also introduce Laplace transforms of $\phi^c_k(\tau,\tau')$ with respect to $\tau-\tau'$ and of $\tilde{\phi}^c_k(\tau,\tau')$ with respect to $\tau'-\tau$:

$$K_{k_2k_1}(\tau_1,\tau_2;\tau_1',\tau_2')=\frac{1}{(2\pi i)^2}\,\frac{1}{(2\pi)}\int d\xi\int d\eta\int_{-\infty}^{\infty} d\zeta\; e^{\xi(\tau_1-\tau_2)+\eta(\tau_1'-\tau_2')+i\zeta(\tau_1-\tau_1')}$$

$$\cdot K_{k_2k_1}(\xi,\eta,\zeta)\quad, \tag{6.8}$$

$$R(\tau_1,\tau_2;\tau_1',\tau_2')=\frac{1}{(2\pi i)^2}\,\frac{1}{(2\pi)}\int d\xi\int d\eta\int_{-\infty}^{\infty} d\zeta\; e^{\xi(\tau_1-\tau_2)+\eta(\tau_1'-\tau_2')+i\zeta(\tau_1-\tau_1')}$$

$$\cdot R(\xi,\eta,\zeta)\quad, \tag{6.9}$$

$$\phi^c_k(\tau,\tau')=\frac{1}{2\pi i}\int dz\; e^{z(\tau-\tau')}\,\phi^c_k(z)\quad, \tag{6.10}$$

$$\tilde{\phi}_k^c(\tau,\tau')=\frac{1}{2\pi i}\int dz\; e^{z(\tau'-\tau)}\;\tilde{\phi}_k^c(z) \quad , \tag{6.11}$$

where we note the relation $\phi_k^c(z)^*=\tilde{\phi}_k^c(z^*)$. Then the integral equation for $K_{k_2k_1}(\xi,\eta,\zeta)$ is obtained as

$$K_{k_2k_1}(\xi,\eta,\zeta)=R_{k_2k_1}(\xi,\eta,\zeta)+\frac{1}{2\pi}\int_{-\infty}^{\infty}dz\int d^3k\; R_{k_2k}(\xi+iz,\eta-iz,\zeta-z)\,|\phi_k^c(\xi+iz)|^2$$
$$\cdot K_{kk_1}(\xi,\eta,z). \tag{6.12}$$

On the other hand the secondary emission spectrum given by (6.1) can be rewritten in terms of $K_{k_2k_1}(\xi,\eta,\zeta)$ as

$$I(k_1,\omega_1;k_2,\omega_2)=|\phi_{k_1}^c(\gamma-i\omega_1)|^2[2\pi\delta_{k_1,k_2}\delta(\omega_1-\omega_2)+|\phi_{k_2}^c(\gamma-i\omega_2)|^2\;K_{k_2k_1}(\gamma-i\omega_1,$$
$$\gamma+i\omega_1,\omega_1-\omega_2)]. \tag{6.13}$$

Therefore if the function K is known, the secondary emission spectrum can be calculated.

In the integral equation (6.12) for $K_{kk_1}(\xi,\eta,\zeta)$ the variables ξ and η can be regarded as fixed parameters because they are the same on both sides of the equation and also the momentum k_1 can be fixed. In the case of isotropic crystal we may assume that the matrix element $<k_2|V|k_1>$ depends only on $|k_1|$, $|k_2|$ and $\hat{k_1k_2}$ (angle between k_1 and k_2) and also that $\phi_k^c(z)$ depends only on $|k|$ and z. Correspondingly we can consider that $K_{k_2k_1}(\xi,\eta,\zeta)$ depends on k_1 and k_2 only through $|k_1|$, $|k_2|$ and $\hat{k_1k_2}$. Therefore it is suitable to expand the functions K and R into the following series:

$$K_{kk_1}(\xi,\eta,\zeta)=\sum_{\ell=0}^{\infty}\;K_\ell(\xi,\eta,\zeta;|k|,|k_1|)\;P_\ell(\cos\theta) \quad , \tag{6.14}$$

$$R_{k_2k}(\xi,\eta,\zeta)=\sum_{\ell=0}^{\infty}\;R_\ell(\xi,\eta,\zeta;|k_2|,|k|)\;P_\ell(\cos\Theta) \quad , \tag{6.15}$$

where $\theta=\hat{kk_1}$, $\Theta=\hat{k_2k}$ and $P_\ell(\cos\theta)$ is the ℓ-th order Legendre polynomial. Then we can decompose the integral equation (6.12) into partial waves as follows:

$$K_\ell(\xi,\eta,\zeta;|k_2|,|k_1|)=R_\ell(\xi,\eta,\zeta;|k_2|,|k_1|)+\frac{2}{2\ell+1}\int_{-\infty}^{\infty}dz\int_0^{\infty}d|k|\;\;|k|^2$$
$$\cdot R_\ell(\xi+iz,\eta-iz,\zeta-z;|k_2|,|k|)\,|\phi_k^c(\xi+iz)|^2\;K_\ell(\xi,\eta,z;|k|,|k_1|). \tag{6.16}$$

As mentioned above, the variables ξ, η, $|k_1|$ can be regarded as fixed parameters. Thus the equation (6.12) is reduced to the Fredholm type integral equation of the second kind with two variables z and $|k|$. Once the K_ℓ's are obtained, we can calculate the secondary emission spectrum by means of (6.13) and (6.14).

In the wurtzite-structured semiconductors such as CdS as another example, the piezoelectric electron-phonon interaction is important. This interaction depends on the angle between the uniaxial axis(c-axis) and the phonon wave-vector [30]. In this case we note that the scattering event is invariant under rotation around the c-axis. Therefore we may expand the functions K and R into the following series :

$$K_{kk_1}(\xi,\eta,\zeta)=\sum_{n=-\infty}^{\infty}\sum_{\ell=0}^{\infty}\sum_{\ell_1=0}^{\infty} K_{n,\ell,\ell_1}(\xi,\eta,\zeta;|k|,|k_1|)P_\ell(\cos\theta)P_{\ell_1}(\cos\theta_1)e^{in(\phi-\phi_1)}, \tag{6.17}$$

$$R_{k_2k}(\xi,\eta,\zeta)=\sum_{n=-\infty}^{\infty}\sum_{\ell_2=0}^{\infty}\sum_{\ell=0}^{\infty} R_{n,\ell_2,\ell}(\xi,\eta,\zeta;|k_2|,|k|)P_{\ell_2}(\cos\theta_2)P_\ell(\cos\theta)e^{in(\phi_2-\phi)}, \tag{6.18}$$

where the polar coordinate is used with the c-axis being chosen as z-axis. Then the integral equation (6.12) can be decomposed as

$$K_{n,\ell_2,\ell_1}(\xi,\eta,\zeta;|k_2|,|k_1|)=R_{n,\ell_2,\ell_1}(\xi,\eta,\zeta;|k_2|,|k_1|) + \sum_{\ell=0}^{\infty}\sum_{\ell'=0}^{\infty}\int_{-\infty}^{\infty}dz\int_0^{\infty}d|k|$$
$$\cdot|k|^2\, R_{n,\ell_2,\ell}(\xi+iz,\eta-iz,\zeta-z)\Phi_{\ell\ell'}(\xi+iz)K_{n,\ell',\ell_1}(\xi,\eta,z;|k|,|k_1|) \tag{6.19}$$

with

$$\Phi_{\ell\ell'}(z)=\int_0^{\pi}d\theta\,\sin\theta P_\ell(\cos\theta)|\phi_k^c(z)|^2\,P_{\ell'}(\cos\theta)\;. \tag{6.20}$$

This is a set of linearly coupled integral equations for $\{K_{n,\ell,\ell_1}(\xi,\eta,z;|k|,|k_1|)\}$ with n, ℓ_1, ξ, η and $|k_1|$ being fixed parameters. In this case also the integral equation is reduced to the Fredholm type equation of the second kind with two variables z and $|k|$. Once the solutions of (6.19) are found, the secondary emission spectrum and its angular dependence on k_1 and k_2 can be calculated from (6.13) and (6.17).

We have so far studied relaxation phenomena due to the intra-branch exciton (polariton)- phonon scattering . In the following we will make a brief discussion on the light scattering from systems where relaxation among many branches is present. This situation is often met in real systems. In the case of near resonant excitation the polariton effect becomes very important. Especially in polar semiconductors the photons interact strongly with excitons and their dispersion relations change drastically in the neighborhood of resonance to form two branches of polariton modes [31]. The salient features of this system were demonstrated in the experiments of resonant Brillouin scattering which were reported by ULBRICH and WEISBUCH [32] on GaAs and by WINTERLING and KOTELES [33] on CdS.

In order to discuss such problems we must include the band or branch indices. The integral equation for the complete interaction part $\langle\langle cd|K|ab\rangle\rangle$

with branch indices a, b, c and d is given by almost the same form as (6.12):

$$<<cd|K_{k_2k_1}(\xi,\eta,\zeta)|ab>>=<<cd|R_{k_2k_1}(\xi,\eta,\zeta)|ab>> + \frac{1}{2\pi}\int_{-\infty}^{\infty}dz\int d^3k \sum_{e,f}$$

$$\cdot<<cd|R_{k_2k}(\xi+iz,\eta-iz,\zeta-z)|ef>>\phi^c_{ek}(\xi+iz)\tilde{\phi}^c_{fk}(\xi-iz)<<ef|K_{kk_1}(\xi,\eta,z)|ab>> , \tag{6.21}$$

where $\phi^c_{ek}(z)$ represents the single-particle Green function of the e-branch exciton or polariton with wavevector k. The expression of the secondary emission spectrum is generalized to the following form:

$$I(k_1,\omega_1;k_2,\omega_2)=\underset{a}{\mathrm{Av}}\sum_{f,g}|\phi^c_{ak_1}(\gamma-i\omega_1)|^2[2\pi\delta_{k_1,k_2}\delta_{fa}\delta_{ga}\delta(\omega_1-\omega_2) + \phi^c_{fk_2}(\gamma-i\omega_2)$$

$$\cdot\tilde{\phi}^c_{gk_2}(\gamma+i\omega_2)<<fg|K_{k_2k_1}(\gamma-i\omega_1,\gamma+i\omega_1,\omega_1-\omega_2)|aa>>] , \tag{6.22}$$

where the average with respect to the initial state and sum over the intermediate states are taken into account. In this way we can incorporate the effect of inter-branch scattering or the interference effect in the intermediate states systematically.

We can also rewrite the time-resolved emission spectrum in this formalism. As in §3 we assume the response function of the filter to be given by (3.7). Then the formula (3.6) can be rewritten as

$$I(\omega_1,\omega_2;T)=\frac{1}{(2\pi i)^2}\frac{1}{2\pi\Delta T}\int ds\int ds'\ p(s)p(s')e^{(s+s')T}\ \frac{e^{(s+s')\Delta T}-1}{s+s'}$$

$$\cdot\phi^c_{k_1}(s+\gamma-i\omega_1)\tilde{\phi}^c_{k_1}(s'+\gamma+i\omega_1)[\frac{\delta_{k_2k_1}}{(s+\gamma_0-i(\omega_1-\omega_2))(s'+\gamma_0+i(\omega_1-\omega_2))}$$

$$+\int_{-\infty}^{\infty}d\zeta'\frac{\phi^c_{k_2}(s+\gamma+i(\zeta'-\omega_1))\tilde{\phi}^c_{k_2}(s'+\gamma-i(\zeta'-\omega_1))}{(s+\gamma_0-i(\omega_1-\omega_2-\zeta'))(s'+\gamma_0+i(\omega_1-\omega_2-\zeta'))}K_{k_2k_1}(s+\gamma-i\omega_1,s'+\gamma+i\omega_1,\zeta')], \tag{6.23}$$

where p(s) denotes the Laplace transform of the pulse envelope. It is to be noted that at first sight (6.23) does not look to be symmetric with the interchange of s and s'. However making use of the relations:

$$K^*_{k_2k_1}(\xi,\eta,\zeta)=K_{k_2k_1}(\eta^*,\xi^*,\zeta) \quad \text{and} \quad \phi^c_k(z)^*=\tilde{\phi}^c_k(z^*) , \tag{6.24}$$

we can rewrite (6.23) into a symmetric form. When the function K is known, the time-resolved emission spectrum can be calculated by convoluting this function with p(s).

Our theory is exact and completely general. In practice, however, we are obliged to introduce some approximations. In the following we will make a brief discussion on a possible approximation scheme. As is well known, the irreducible interaction part R of the two-particle Green function is directly related to the self-energy part Σ of the single-particle Green function by the WARD identity [29]. The explicit expression of R can be obtained systematically from the skeleton diagrams of the self-energy part of the single-particle Green function by opening any internal line. For illustration we assume a typical exciton(polariton)-phonon coupling for V(=He-Hg), i.e.,

$$V = \sum_q V(q) a^+_{k+q} a_k (b_q + b^+_{-q})$$

and show the perturbation series of R and Σ up to the third order with respect to V in Fig.12. If we take into account the self-energy correction of the single-particle Green function up to $\Sigma^{(n)}$, we have to incorporate into R the contributions up to $R^{(n)}$ in order to be consistent with the WARD identity. In this way we can construct a general theory which is consistent with the WARD identity to all orders of perturbation series. It is important to remark that our theory is not identical to the usual perturbation theory, e.g., by MARTIN [34], ZEYHER et al. [35] and KLYUCHIKHIN et al. [36] because the integral equation (6.12) is solved self-consistently and multi-phonon processes are incorporated automatically. Therefore our theory can describe the Raman scattering and luminescence from a unified viewpoint.

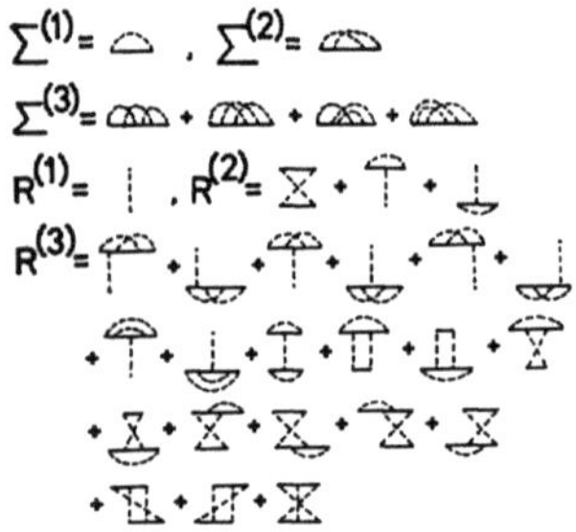

Fig.12 Perturbation series for the self-energy part Σ of the single-particle Green function and the irreducible interaction part R of the two-particle Green function. Solid line represents the exciton propagator and dashed line the phonon propagator.

Now let us look upon our stochastic theory developed in §2 from the viewpoint of this new formalism. In §2 the second order perturbation with respect to V is used and it corresponds to the approximation up to $\Sigma^{(1)}$ and $R^{(1)}$ in Fig.12. Furthermore in §2 the frequency dependence of $\Sigma^{(1)}$ and $R^{(1)}$ is neglected under the Markovian approximation. The three types of diagrams corresponding to the relaxation operator Γ in §2 are shown as follows:

$bb^+\rho(b^+b\rho) \rightarrow$ 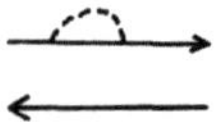 , $\rho b^+ b(\rho b b^+) \rightarrow$  , $b\rho b^+(b^+\rho b) \rightarrow$ 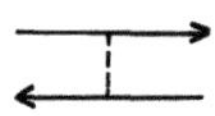 .

Thus it is shown that our theory in §2 is nothing but the lowest order theory of the new formalism with Markovian approximation. In the new formalism we can proceed to higher order approximations systematically with the approximation scheme shown in Fig.12. Thus the new theory is the complete generalization of our stochastic theory in §2.

Finally our theory is completely general and will turn out to be powerful in the study of relaxation phenomena in a variety of materials. This type of theory was first applied to the problem of threshold singularities of the x-ray Raman scattering in metals by NOZIERES and ABRAHAMS [37]. Their theory is concerned only with the asymptotic evaluation of the spectral shape near threshold and cannot give the overall spectral profile to elucidate the competition between the Raman scattering and luminescence. Our theory will most suitably be applied not only to the x-ray Raman scattering in metals but also to multi-phonon processes in polariton or exciton resonance Raman scattering and luminescence [38-40].

Acknowledgments

The author would like to thank very much Professors R. Kubo and E. Hanamura for many valuable discussions. He is very grateful to Dr. Y. Masumoto for enlightening discussions and suggestions on the theory of time-resolved emission spectrum.

References

1 V.V. Hizhnyakov and I.Y. Tehver: Phys. Status solidi 21, 755 (1967); ibid. 39, 67 (1970); I. Y. Tehver and V. V. Hizhnyakov: Sov. Phys.-JETP 42, 305 (1976)
2 D. L. Huber: Phys. Rev. 158, 843 (1967); ibid. 170, 418 (1968); ibid. 178, 93 (1969); ibid. B1, 3409 (1970)
3 M. V. Klein: Phys. Rev. B8, 919 (1973)
4 Y. R. Shen: Phys. Rev. B9, 622 (1974)
5 K. K. Rebane, I. Y. Tehver and V. V. Hizhnyakov: Proc. of the First USSR-USA Seminar of Light Scattering, ed. by V. M. Agranovich and J. L. Birman (Nauka, Moscow 1975) p.467; K. K. Rebane and P. M. Saari: J. Luminescence 12/13, 23 (1976)
6 D. L. Rousseau and P. F. Williams: J. Chem. Phys. 64, 3519 (1976)
7 J. R. Solin and H. Merkelo: Phys. Rev. B12, 624 (1975)
8 S. Mukamel, A. Ben-Reuven and J. Jortner: Phys. Rev. A12, 947 (1975); J. Chem. Phys. 64, 3971 (1976)
9 S. Permogorov: Phys. Status solidi (b) 68, 9 (1975)
10 Y. Toyozawa: J. Phys. Soc. Japan 41, 400 (1976); Y. Toyozawa, A. Kotani and A. Sumi: ibid. 42, 1495 (1977)
11 T. Takagahara, E. Hanamura and R. Kubo: J. Phys. Soc. Japan 43, 802, 811 (1977)
12 T. Takagahara, E. Hanamura and R. Kubo: J. Phys. Soc. Japan 44, 728, 742 (1978)
13 S. Nakajima: Prog. theor. Phys. 20, 948 (1958)
14 R. Zwanzig: Phys. Rev. 124, 983 (1961)
15 R. L. St. Peters, S. D. Silverstein, M. Lapp and C. M. Penney: Phys. Rev. Lett. 30, 191 (1973)
16 Y. Masumoto, S. Shionoya and Y. Tanaka: Solid State Commun. 27, 1117 (1978)
17 Y. Segawa, Y. Aoyagi, O. Nakagawa, K. Azuma and S. Namba: Solid State Commun. 27, 785 (1978)
18 J. H. Eberly and K. Wodkiewicz: J. Opt. Soc. Am. 67, 1252 (1977)
19 M. Born and E. Wolf: Principles of Optics, 5th ed. (Pergamon, Oxford 1975)
20 T. Takagahara, E. Hanamura and R. Kubo: J. Phys. Soc. Japan 43, 1522 (1977)
21 E. Hanamura and T. Takagahara: J. Phys. Soc. Japan 47, 410 (1979)
22 K. K. Rebane: J. Luminescence 18/19, 693 (1979) and references cited therein

23 V. V. Hizhnyakov and I. K. Rebane: Sov. Phys.-JETP 47, 463 (1978)
24 T. D. Lee, K. Huang and C. N. Yang: Phys. Rev. 106, 1135 (1957)
25 W. H. Louisell: Quantum Optics, Enrico Fermi Course XLII ed. by R. J. Glauber (Academic, 1969)
Although his argument is given in the coherent state representation, its transcription in terms of operator form can be easily carried out.
26 T. Itoh, T. Suzuki and M. Ueta: J. Phys. Soc. Japan 42, 1069 (1977); T. Mita and M. Ueta: Solid State Commun. 27, 1463 (1978)
27 E. Hanamura: Solid State Commun. 12, 951 (1973); E. Hanamura and H. Haug: Physics Reports 33c, 209 (1977)
28 L. V. Keldysh: Sov. Phys.-JETP 20, 1018 (1965)
29 P. Nozieres: Theory of Interacting Fermi Systems (W. A. Benjamin, Inc., New York, 1964)
30 G. D. Mahan and J. J. Hopfield: Phys. Rev. Lett. 12, 241 (1964)
31 J. J. Hopfield: Phys. Rev. 112, 1555 (1958)
32 R. G. Ulbrich and C. Weisbuch: Phys. Rev. Lett. 38, 865 (1977)
33 G. Winterling and E. S. Koteles: Solid State Commun. 23, 95 (1977)
34 R. M. Martin: Phys. Rev. B4, 3676 (1971)
35 R. Zeyher, C. S. Ting and J. L. Birman: Phys. Rev. B10, 1725 (1974); R. Zeyher: Solid State Commun. 16, 49 (1975)
36 A. A. Klyuchikhin, S. A. Permogorov and A. N. Reznitskii: Sov. Phys.-JETP 44, 1176 (1976)
37 P. Nozieres and E. Abrahams: Phys. Rev. B10, 3099 (1974)
38 E. S. Koteles and G. Winterling: Phys. Rev. B20, 628 (1979)
39 Y. Oka and M. Cardona: Solid State Commun. 30, 447 (1979)
40 A. Nakamura and C. Weisbuch: Solid State Commun. 32, 301 (1979)

First and Second Order Optical Responses in Exciton-Phonon System

H. Miyazaki and E. Hanamura

Department of Applied Physics, Faculty of Engineering, University of Tokyo
Hongo, Bunkyoku, Tokyo 113, Japan

Abstract

Stochastic theory of the first and second order optical responses is presented for a system of exciton-phonon complex. The dynamical motion of the whole lattice system is projected onto the stochastic motion of the relevant mode which interacts dominantly with the exciton. The dynamical coherent potential approximation is extended to this non-Hamiltonian system and the low energy side of the absorption spectrum due to the exciton is studied. It is shown that the multi-phonon sidebands are merged into the exponential tail with the increase of the temperature or/and transfer of the exciton. This exponential tail originates from the thermal equilibrium distribution of the relevant mode and from the broadening of the phonon sidebands due to the interaction with the reservoir of the irrelevant lattice vibrations or due to the increase of the exciton transfer. The dynamical coherent potential approximation is further extended to the second order optical response in the exciton-phonon system and the emission spectrum of this system is calculated. The broadening of the Raman lines reflects the relaxation of the relevant mode in the final state and the luminescence, on the other hand, are shown to depend sensitively on the exciton transfer as well as the radiative decay rate and the relaxation of relevant phonons. The spectra show how these lines shift to the high energy side and are broadened with the increase of the transfer of the exciton. The relative magnitude of the Raman and luminescence components depends not only on the ratio of radiative decay to the relaxation but also on the incident photon frequency Ω_1. This means that we can observe the coexistence of these components over a wide value of the ratio by choosing Ω_1 suitably and determine the decay and relaxation constant. The study of the second order optical response in the exciton-phonon system is thus shown to provide us with information on the relaxation process and on the characteristic constants of the exciton-phonon system.

1. Introduction

Recent development in the theoretical study of the second order optical process has brought about understanding of the role of the intermediate state interaction of elementary excitations [1-4]. The competitive behavior between the Raman scattering and the luminescence was revealed for the localized electron-phonon system by extending the stochastic theory of the intermediate state interaction [3]. It was shown that the characteristics of the emission spectrum is dominated by the Raman scattering or the luminescence depending upon the ratio of decay constant to the relaxation constant of the intermediate state [4]. On the other hand, the second order optical process has not been studied so extensively up to the present for the exciton-phonon complex where the electronic excitation can propagate in the crystal. The absorption spectrum due to the exciton-phonon complex has shown a wide variety of interesting phenomena. For example, the low energy side of the absorption spectrum I(E) in insulators, semiconductors and molecular crystals obeys Urbach-Martienssen (U-M) rule: the exponential dependence on the incident photon energy E, $I(E)=I_0\exp[-\sigma(E_0-E)/kT]$, over a wide range of the lattice temperature T [5,6]. σ and E_0 are called the steepness parameter and the converging energy respectively and are almost independent of the incident photon energy and the temperature. It is expected that the study of the second order optical response due to the exciton-phonon complex provides us with deep insights into the origin of the U-M rule as well as the characteristics of the exciton-phonon complex.

The study of the optical process due to the exciton-phonon complex is required to satisfy the following conditions: First, the relaxation of the exciton due to the interaction with the phonons is taken into account. Second, the quantum of lattice vibrations is taken into account to enable us to study the Raman scattering in the second order optical response. Third, the interaction between the exciton and phonons is treated with the non-perturbational method. While the exciton interacts weakly with the phonons and the third condition is not necessary in semiconductors, in ionic crystals the coupling between the exciton and phonons is so strong that the exciton is self-trapped by the local lattice deformation [7]. Under such a situation, the perturbational treatment of the interaction is not effective.

We present in this paper the theory of the first and second order optical responses of the exciton-phonon complex satisfying these three conditions. A stochastic model is presented for the exciton-phonon complex in §2 and we study in §3 how the phonon structure in the absorption spectrum is affected by the propagation of the excitation and the relaxation in the excited state of the system. The multi-phonon sidebands are merged into the U-M tail with the increase of the temperature or/and the transfer of the exciton. In §4, we construct the theory of the second order optical response due to the exciton-phonon system by extending the stochastic model under the dynamical coherent potential approximation. It is shown in §5 how the emission spectrum depends on the transfer of the exciton and what kind of information the study of the second order optical response in exciton-phonon complex provides us on the relaxation process and on the characteristic constants of the exciton-phonon system.

2. Model Hamiltonian and Dynamical Coherent Potential Approximation

In our stochastic model of the exciton-phonon complex, we single out the relevant modes among the lattice vibrations. The relevant modes interact dominantly with the exciton and are modeled by Einstein phonons. Other phonons are considered as a reservoir acting stochastically on the relevant modes. Then the model Hamiltonian H consists of three parts, the system Hamiltonian H_S, the reservoir Hamiltonian H_R and their interaction between the system and the reservoir H_{SR}. The system Hamiltonian describes the interacting exciton-Einstein phonon system and is written as

$$H_S = H_{ex} + H_{ph} + H_{ep}, \tag{1}$$

where

$$H_{ex} = \sum_k E_k A_k^\dagger A_k \ , \quad H_{ph} = \sum_n H_{ph}^n = \sum_n \omega b_n^\dagger b_n \quad \text{and}$$

$$H_{ep} = \sum_n H_{ep}^n = \sum_n \omega\sqrt{S}\,(b_n^\dagger + b_n) a_n^\dagger a_n . \tag{2}$$

Here $A_k^\dagger$ is the creation operator of an exciton with momentum k and energy E_k, $a_n^\dagger$ is the creation operator of an exciton at the lattice site R_n and $b_n^\dagger$ is the creation operator of Einstein phonon at R_n with energy ω. $\sqrt{S}$ is the coupling constant between the exciton and Einstein phonons. The reservoir Hamiltonian H_R is given by

$$H_R = \sum_k \omega_k R_k^\dagger R_k , \tag{3}$$

where $R_k^\dagger$ is the creation operator of the reservoir phonon with momentum k and energy ω_k. We assume the anharmonic coupling between Einstein phonons and the reservoir phonons:

$$H_{SR} = \sum_k \sum_{k'} (f_{k,k'} B_{k+k'}^\dagger R_k R_{k'} + \text{h.c.}) \tag{4}$$

with the anharmonic coupling constant $f_{k,k'} = \text{const}/\sqrt{\omega_k \omega_{k'}}$ and $B_k^\dagger = N^{-1/2} \sum_{n=1}^{N} b_n^\dagger \exp(-ikR_n)$. Note that the contribution from the bilinear coupling must be added to the interaction H_{SR} when Einstein phonons are not the normal modes of the crystal.

We introduce the exciton Green function with momentum k and energy E as

$$G(k,E) = \text{Tr}\, A_k \frac{1}{E-L} A_k^\dagger \rho^0 , \tag{5}$$

where Tr means the trace over the states of both the system and the reservoir, L is the Liouville operator $H^\times$ ($H^\times = HX - XH$ for an arbitrary operator X) and ρ^0 is the initial density operator of the exciton-phonon complex. Then the absorption spectrum I(E) due to the exciton is given by the imaginary part of G(0,E) besides an unimportant factor. Here we adopt a stochastic approach to evaluate G(k,E). We project the dynamical motion of the reservoir onto the stochastic motion of the system with the use of the projection technique [8,9]. As a result, the vibrational relaxation of Einstein phonons is described by the relaxation operator Γ represented as

$$\Gamma = \sum_n \Gamma_n , \tag{6}$$

where Γ_n is the relaxation operator acting on Einstein phonon at the n-th lattice site R_n. Γ_n is defined for an arbitrary density operator ρ as

$$\begin{aligned} \Gamma_n \rho = {} & \frac{\gamma_A}{2} (c_n^\dagger \rho c_n - c_n c_n^\dagger \rho + c_n^\dagger \rho c_n - \rho c_n c_n^\dagger) \\ & + \frac{\gamma_B}{2} (c_n \rho c_n^\dagger - c_n^\dagger c_n \rho + c_n \rho c_n^\dagger - \rho c_n^\dagger c_n) \end{aligned} \tag{7}$$

with

$$c_n = b_n + \sqrt{S} a_n^\dagger a_n \quad \text{and} \quad c_n^\dagger = b_n^\dagger + \sqrt{S} a_n^\dagger a_n \tag{8}$$

and

$$\begin{aligned} \gamma_A &= \frac{4\pi}{N} \sum_{k,k'} |f_{k,k'}|^2\, n_k n_{k'} \delta(\omega - \omega_k - \omega_{k'}) \\ \gamma_B &= \frac{4\pi}{N} \sum_{k,k'} |f_{k,k'}|^2\, (n_k + 1)(n_{k'} + 1) \delta(\omega - \omega_k - \omega_{k'}) , \end{aligned} \tag{9}$$

where n_k is the population of the reservoir phonon with energy ω_k. In introducing the relaxation operator, we assume that the width of the energy spectrum of the reservoir is much larger than the magnitude of H_{SR}. and use the second order Born approximation with respect to H_{SR} We also use the Markovian assumption, i.e., the correlation time of the reservoir is much shorter than any time constant of the system. With this projection the exciton Green function is written as

$$G(k,E) = Tr_S \, A_k \frac{1}{E-L_S-i\Gamma} A_k^{\dagger} \rho_{ph}^{0} \, , \tag{10}$$

where Tr_S is the trace over the system, $L_S = H_S^{\times}$ and ρ_{ph}^{0} is the thermal equilibrium density operator of Einstein phonons in the electronic ground state.

The exciton Green function G(k,E) is evaluated under the dynamical coherent potential approximation (DCPA) [10]. We would like to know the characteristics of the optical response over a wide range of the system parameters, from the weak coupling case to the strong coupling case between the exciton and Einstein phonons. For this purpose we resort to the non-perturbational treatment of the coupling. Here we use the DCPA, because this provides a reasonable approximation over a wide range of the system parameters in the exciton-phonon system. The DCPA has been developed in the Hamiltonian system, whilst the present system is non-Hamiltonian in that the stochastic motion of the system is incorporated as a vibrational relaxation of Einstein phonons in the electronic excited state as well as in the ground state. Therefore we extend the DCPA to the present non-Hamiltonian system. Consider an exciton propagating in the crystal with energy E. The exciton is scattered inelastically by Einstein phonons at each lattice site. The dynamical coherent potential V(E) is so determined that the multiple inelastic scattering of the exciton is to be described by V(E) put at each lattice site instead of the scattering terms by Einstein phonons. Note that the dynamical coherent potential depends on the energy of the propagating exciton but not on its momentum and the correlated scatterings among different lattice sites are neglected. Then the dynamical coherent potential is determined as follows: We consider a virtual crystal where all the scattering terms at the lattice sites are replaced by $V(E)a_n^{\dagger}a_n$ except at a particular site, say, the 0-th lattice site R_0, and calculate the diagonal element of the exciton Green function G(E) at the 0-th lattice site. V(E) is determined such that G(E) becomes equal to the diagonal element at the 0-th lattice site of the exciton Green function which describes the exciton propagation in the dynamical coherent potential including the 0-th lattice site:

$$G(E) = \frac{1}{N} \sum_k \frac{1}{E-E_k-V(E)} \, . \tag{11}$$

Then the absorption spectrum I(E) due to the exciton is obtained with the dynamical coherent potential as

$$I(E) = -\frac{1}{\pi} \, \mathrm{Im} \, \frac{1}{E-E_0-V(E)} \, , \tag{12}$$

where Im means taking the imaginary part and E_0 is the value of E_k with k=0.

3. Absorption Spectrum

In the present problem characteristics of the absorption spectrum depends on the temperature T and three system parameters: the exciton band width 2B, the coupling constant S between the exciton and Einstein phonons and the relaxation rate of Einstein phonons $\gamma \equiv \gamma_B - \gamma_A$. In the numerical calculation we take the density of states of the exciton as a hemi-elliptic form with the band width 2B [11]:

$$\rho(E) = \frac{2}{\pi B^2}\sqrt{B^2-E^2} \qquad |E| \leqq B$$
$$= 0 \qquad |E| > B. \qquad (13)$$

The center of the exciton band is taken as the origin of the energy and the Einstein phonon quantum ω is taken as a unit of the energy. The relaxation rate of Einstein phonon in the anharmonic coupling case, (4), depends on the temperature. The values of γ in the present calculation are given in Table I.

Table I Relaxation rate of Einstein phonons γ and steepness parameter $\sigma = kT\times d[\ln I(E)]/dE$ in the case of S=3.0 and B=5.0 for various values of the temperature T.

T	0.0	0.5	0.75	1.0	1.25	1.5
γ	0.02	0.063	0.092	0.122	0.151	0.181
σ	——	1.6	1.4	1.3	1.3	1.1

We are interested in the low energy side of the absorption spectrum over a wide range of the system parameters. In the case of quite large values of S and B, however, the lattice vibrations can be treated classically and the absorption spectrum for that case has been studied by SUMI and TOYOZAWA [6]. So we have performed the numerical calculation for the cases where the quantum treatment of the lattice vibrations is required. The absorption spectra for the cases of S=0.5 and various values of B and T are shown in Figs. 1-a to 1-d. In the small value of B=0.2 in Fig. 1-a, we can see the phonon sidebands with the association of up to three phonons below the sharp zero phonon line. These phonon sidebands are more broadened for higher number of the associating phonons. The relaxation is faster for higher quantum level of Einstein phonon states in the electronic ground state, which contributes mainly to the multi-phonon sidebands. As the temperature is raised, the relaxation rate of Einstein phonons increases as is seen in Table I and the population of Einstein phonons also increases. These result in the broadening and the growing of the phonon sidebands as is seen in Figs. 1-a to 1-d.

Now we consider how the absorption spectrum depends on the transfer of the exciton B. We can see in Figs. 1-a to 1-d that with the increase of B each phonon sideband is broadened and finally there appears a single peak in the absorption spectrum. Besides, the peaks of the phonon sidebands are shifted to the high energy side. It is also seen in Figs. 1-b and 1-c that the peak of the zero phonon line is broadened and shifted to the high energy side with the increase of the temperature. These changes in the absorption spectrum are explained as follows. In Fig. 2 we show the energy diagrams of the exciton-Einstein phonon system for the cases of S and B corresponding to Figs. 1-a to 1-d. The energy of the exciton-Einstein phonon system is plotted on the ordinate as a function of the normalized exciton energy $\varepsilon_k=(E_k-E_0)/2B$. Here the origin of the energy is taken as the lowest energy of the exciton-Einstein phonon system which is given in the parentheses. The solid line in the figure represents the exciton polaron state. The exciton polaron and Einstein phonons are not bound and constitute the phonon-scattering continua which are represented by the shaded regions in the figure. Note that the broadening of the energy levels due to the interaction with the reservoir is

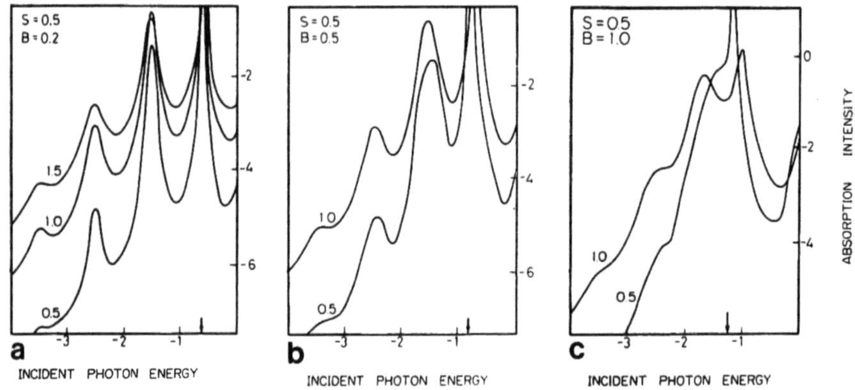

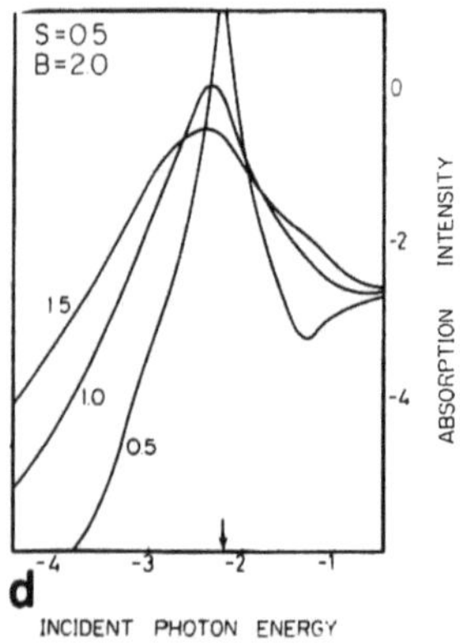

Fig.1 Semi-logarithmic plot of the absorption spectra for S=0.5 and four values of B: (a) 0.2, (b) 0.5, (c) 1.0 and (d) 2.0. The incident photon energy E on the ascissa is in a unit of Einstein phonon quantum ω. The arrow on the abscissa in each figure indicates the fundamental absorption edge. Parameters put on each curve represent the temperature and the relaxation rate γ at each temperature is given in Table I.

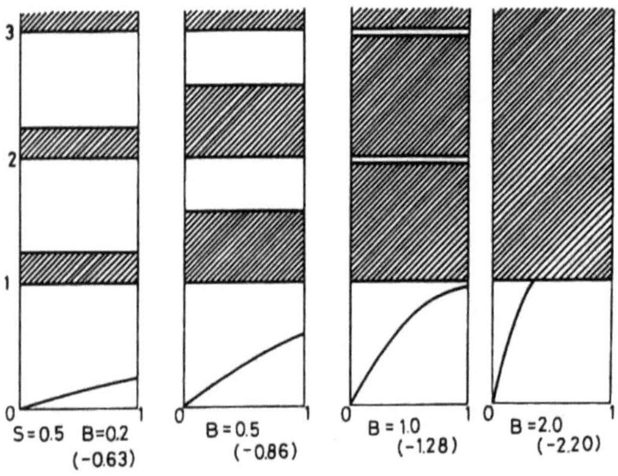

Fig. 2 Energy diagrams of the exciton-Einstein phonon system for S=0.5 and the values of B corresponding to those in Fig. 1. The solid line and the hatched region represent respectively the exciton polaron and the scattering continuum of exciton polaron and Einstein phonons. Energy of the exciton-Einstein phonon system is plotted on the ordinate with the zero taken as the bottom of the exciton polaron band as a function of $\varepsilon_k=(E_k-E_0)/2B$.

not taken into account in the energy diagrams. As the transfer of the exciton increases, the spectrum width of the exciton polaron band increases as well as that of the scattering continua as is seen in Fig. 2. Now the zero phonon line in the absorption spectrum at low temperature is due dominantly to the optical transition from the electronic ground state to the bottom of the exciton polaron band. Therefore the broadening of the zero phonon line at low temperature reflects the relaxation of Einstein phonons. As the temperature is raised, the contribution to the zero phonon line from the optical transition to the scattering continua increases. This results in the shift and broadening of the zero phonon line in the absorption spectrum. The phonon-assisted sidebands at low temperature, on the other hand, are due dominantly to the optical transition to the exciton polaron band with the association of Einstein phonons. In this case the transition is not restricted to the bottom of the exciton polaron band because of the association of Einstein phonons. Therefore the peak of each phonon sideband is shifted to the high energy side with the increase of B even at low temperature. As for the broadening of each phonon sideband, it is governed by the relaxation of Einstein phonons in the case of small excitation transfer, i.e., in the case of small band width of the exciton polaron. With the increase of B the band width of the exciton polaron increases and this contributes to the broadening of the phonon sidebands is addition to the relaxation. The broadening of the phonon sidebands in Figs. 1-a to 1-d is attributed to this broadening of the exciton polaron band as well as that by the relaxation.

Let us see the absorption spectrum for S=0.5 and B=2.0 in Fig. 1-d. A small shoulder is seen in the low energy tail of the absorption spectrum at low temperature (T=0.5). As the temperature is raised, this shoulder is immediately smeared out and there appears an exponential tail in the low energy side of the absorption spectrum. This tendency in the absorption spectra was observed by MARPLE [12] in CdTe. Although the values of B in II-VI semiconductors such as CdTe are much larger, the U-M tail observed in these materials may be explained qualitatively by the present mechanism. This exponential tail also appears in the case of S=3.0 and B=5.0 as shown in Fig. 3.

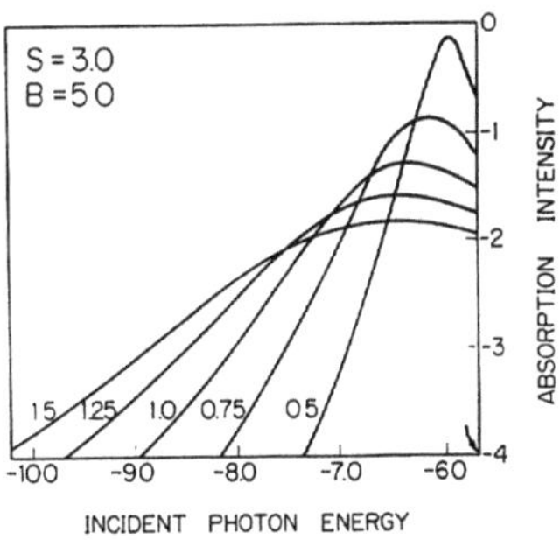

Fig. 3 Absorption spectra for S=3.0 and B=5.0

From the exponential tail region of the absorption spectrum in Fig. 3, we obtain the temperature dependence of the steepness parameter σ. The results are given in Table I. Except at low temperature the steepness parameter σ is almost constant. This shows that the exponential tail below the absorption edge in Fig. 3 almost obeys the U-M rule. As the transfer B is decreased to 0.5 for S=3.0, the phonon sidebands appear as shown in Fig. 4-a. It is noticed in the figure that the peaks of the phonon sidebands are on the exponential line. These phonon sidebands correspond to the multi-phonon-assisted creation of the exciton and the exponential line in Fig. 4-a reflects the population of Einstein phonons which assist the optical creation of the exciton. In Fig. 4-b we show the energy diagram of the exciton-Einstein phonon system for S=3.0 and B=0.5. The lines extending out of the scattering continua denote the vibronic exciton state which is the bound state of the exciton polaron and Einstein phonons. In the present case the band width of the exciton polaron is very small and the broadening of each phonon sideband is

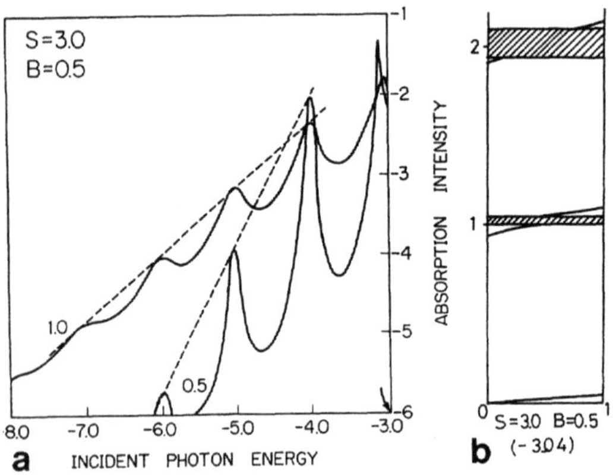

Fig. 4 Absorption spectra for S=3.0 and B=0.5 (a), and corresponding energy diagram of the exciton-Einstein phonon system (b).

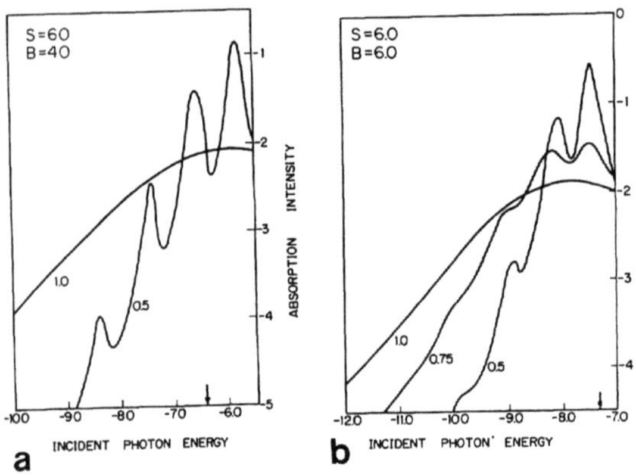

Fig. 5 Absorption spectra in the case of S=6.0 and B=4.0 (a), 6.0 (b).

governed by the relaxation of Einstein phonons. Now the increase of the relaxation rate γ broadens each phonon sideband depending upon the number of the associating phonons but the peaks of the phonon sidebands are still on the exponential line. Then as the relaxation rate γ is increased, the low energy side of the absorption spectrum becomes to obey the U-M rule. The U-M tail appears both when the multi-phonon sidebands are broadened by the increase of B (e.g. Figs. 1-d and 3) and by the relaxation of Einstein phonons (e.g. Fig. 4-a).

The absorption spectra for the case of B=4.0 and 6.0 and fixed value of S=6.0 are shown respectively in Figs. 5-a and 5-b. The corresponding energy diagrams are shown in Figs. 6-a and 6-b. These absorption spectra show an abrupt change against the raise of the temperature. While the phonon sidebands appear at low temperature in Fig. 5-a, they are immediately broadened to form the U-M tail with the raise of the temperature. This change in the absorption spectrum is explained as follows. At low temperature, the phonon sidebands in the absorption spectrum are due to the phonon assisted transition to the exciton polaron band. In the present case the band width of the exciton polaron is small and the broadening of the phonon sidebands is governed by the relaxation rate γ. The situation is similar to that for S=3.0 and B=0.5 and the phonon structure is appreciable in the low energy side of the absorption spectrum. As the temperature is raised to 1.0, the optical

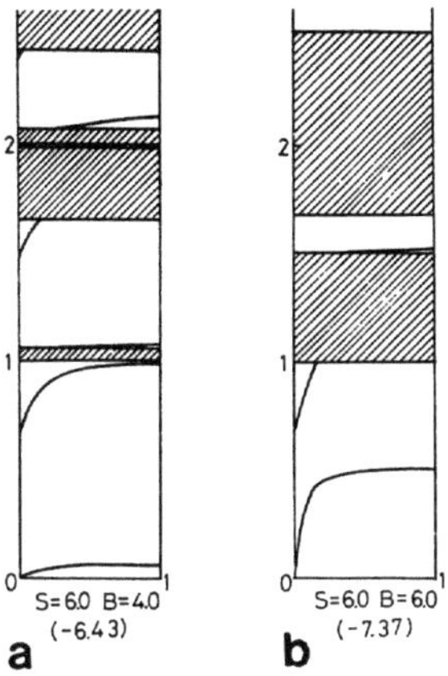

Fig. 6 Energy diagrams of the exciton-Einstein phonon system for S=6.0 and the values of B corresponding to those in Fig. 5

transition to the vibronic exciton bands and the scattering continua takes place. This contributes to the broadening of the phonon sidebands but its contribution is not so large as to smear out the phonon structure, because the band width of the vibronic exciton and the scattering continua is not so wide to smear out the absorption spectrum. Here we consider an additional contribution from the interference between the relaxation and the transfer of the exciton. The vibronic exciton state is broadened by the relaxation of Einstein phonons, i.e., there exists the mixing between the vibronic exciton state and the exciton polaron state via the interaction with the reservoir phonons. At low temperature this mixing is small because of the small value of the relaxation rate γ and the phonon structure in the absorption spectrum is not so affected by the interference. As the temperature is raised, this mixing is enhanced by the increase of the relaxation rate γ and the multiphonon sidebands are smeared out to the U-M tail.

4. Second Order Optical Response in the DCPA

In the previous sections we have presented a stochastic model for the theory of the optical absorption spectrum due to the exciton-phonon complex. There the DCPA was extended to the present non-Hamiltonian system and was used to evaluate the single-particle Green function of the exciton for the absorption spectrum. Here the DCPA is further extended and is applied to evaluate the tetradic Green function of the exciton for the second order optical response of the exciton-phonon system [13].

Consider an exciton-phonon system which absorbs an incident photon Ω_1 and emits a photon Ω_2. The scattering amplitude for this second order optical process at time t is given by

$$\psi^{(2)}(t) = (-i)^2 M_1 M_2^* \int_0^t dt_1 \int_0^{t_1} dt_2\, e^{-i(H+\Omega_2)(t-t_1)} A_0\, e^{-(iH+\gamma_b)(t_1-t_2)} A_0^\dagger \times e^{-i(H+\Omega_1)t_2}\, \psi(0), \tag{14}$$

where M_i is the transition matrix element for the absorption of a photon Ω_i and $\psi(0)$ is the initial state of the exciton-phonon system. The radiative recombination of the exciton is incorporated as the radiative decay rate γ_b in the propagator of the intermediate state in (14). Then the transition probability per unit time $I(\Omega_1,\Omega_2)$ of a second order optical process is obtained by multiplying $\psi^{(2)}(t)$ by its Hermitian conjugate, taking the time derivative and letting the time to infinity:

$$\begin{aligned} I(\Omega_1,\Omega_2) = \lim_{t\to\infty} \frac{\partial}{\partial t} \int_0^t dt_1 \int_0^{t_1} dt_2 \int_0^t dt_1' \int_0^{t_1'} dt_2' \; \mathrm{Tr}\, e^{-i(H+\Omega_2)(t-t_1)} A_0 \\ \times e^{-(iH+\gamma_b)(t_1-t_2)} A_0^\dagger e^{-i(H+\Omega_1)t_2} \rho^0 e^{i(H+\Omega_1)t_2'} A_0 \\ \times e^{(iH-\gamma_b)(t_1'-t_2')} A_0^\dagger e^{i(H+\Omega_2)(t-t_1')}, \end{aligned} \tag{15}$$

where we have omitted an unimportant constant factor. $I(\Omega_1,\Omega_2)$ is expressed by a triple integral in time and we decompose it into six terms according to the chronological order among the time variables. Simultaneously we project the dynamical motion of the system onto the stochastic motion of the system as in the case of the first order optical process in §2. Then the second order optical response is written as

$$I(\Omega_1,\Omega_2) = 2\,\mathrm{Im}[(\mathrm{I}) + (\mathrm{II}) + (\mathrm{III})], \tag{16}$$

where

$$\begin{aligned} (\mathrm{I}) &= \mathrm{Tr}_S\Big(\frac{1}{-\Omega_2+i\gamma_b-L_S-i\Gamma}\Big(\frac{1}{\Omega_1-\Omega_2+i0-L_S-i\Gamma}\Big(A_0\frac{1}{\Omega_1+i\gamma_b-L_S-i\Gamma} A_0^\dagger \rho^0_{ph}\Big)\Big)A_0\Big)A_0^\dagger, \\ (\mathrm{II}) &= \mathrm{Tr}_S\Big(\frac{1}{-\Omega_2+i\gamma_b-L_S-i\Gamma} A_0\Big(\frac{1}{2i\gamma_b-L_S-i\Gamma}\Big(\frac{1}{\Omega_1+i\gamma_b-L_S-i\Gamma} A_0^\dagger \rho^0_{ph}\Big)A_0\Big)\Big)A_0^\dagger\Big), \\ (\mathrm{III}) &= \mathrm{Tr}_S\, A_0\Big(\frac{1}{\Omega_2+i\gamma_b-L_S-i\Gamma}\Big(\frac{1}{2i\gamma_b-L_S-i\Gamma}\Big(\frac{1}{\Omega_1+i\gamma_b-L_S-i\Gamma} A_0^\dagger \rho^0_{ph}\Big)A_0\Big)A_0^\dagger\Big). \end{aligned} \tag{17}$$

These three terms are shown schematically in Fig.7. In the figure the time evolution of the state to the left side starting from ρ^0 in (15) is shown in the upper part of the diagrams and that to the right in the lower part. The wavy line denotes the exciton propagation and the dotted line the exciton vacuum state. Note that the middle propagator of (I) in (17) depends on the incident and emitted photons Ω_1 and Ω_2 in a correlated form $\Omega_1-\Omega_2$. Then the diagram (I) contains the Raman and the Rayleigh scatterings in the second order optical response. Contribution from the diagrams (II) and (III), on the other hand, depends on Ω_1 and Ω_2 in an uncorrelated manner and these two diagrams contribute to the luminescence component in $I(\Omega_1,\Omega_2)$.

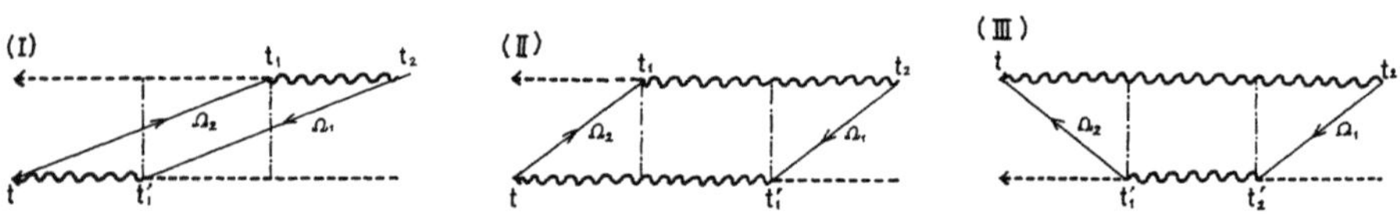

Fig.7 Decomposed propagators of a second order optical process in the exciton-phonon system. Three diagrams are drawn according to the chronological order among the time variables. The time proceeds from right to left. The wavy line and the dotted line denote the exciton propagation in the system and the exciton vacuum state, respectively.

Now we explain our theory of the DCPA, taking the first diagram as an example. In the first diagram (I), the important contribution to the second order optical response comes, e.g., from the process where the exciton in the upper wavy line interacts with Einstein phonon at a lattice R_n and this Einstein phonon interacts with the exciton in the lower wavy line. This process leaves excitation of a finite number of Einstein phonons in the final state such as the Raman scattering. To consider processes like this we define the tetradic propagator $G^{\times}(Z)$ as

$$G^{\times}(Z) = \frac{1}{Z-L_S-i\Gamma}. \tag{18}$$

Then the propagation of the system in the right end and left end of the diagram (I) is described by the tetradic propagator $G^{\times}(Z)$ with $Z = \Omega_1+i\gamma_b$ and $-\Omega_2+i\gamma_b$, respectively. Now we introduce the scattering t-matrix $T^{\times}(Z)$ due to the potential U which is equal to the scattering term of the exciton by Einstein phonons at the lattice sites minus the dynamical coherent potential [13]. $T^{\times}(Z)$ describes the multiple-scattering of the exciton by the potential U. Then the tetradic propagator is expressed with the scattering t-matrix as

$$G^{\times}(Z) = G^{\times}_{eff}(Z) + G^{\times}_{eff}(Z)T^{\times}(Z)G^{\times}_{eff}(Z), \tag{19}$$

where $G^{\times}_{eff}$ is the effective propagator for the virtual system where all the scattering terms of the exciton by Einstein phonons are replaced by the dynamical coherent potential. Now the scattering t-matrix $T^{\times}(Z)$ is expanded in terms of the single site t-matrix $T^{\times}_n(Z)$ as

$$T^{\times}(Z) = \sum_n T^{\times}_n(Z) + \sum_{n\neq m}\sum T^{\times}_n(Z)G^{\times}_{eff}(Z)T^{\times}_m(Z) + \cdots. \tag{20}$$

Here the single site t-matrix $T^{\times}_n(Z)$ describes the multiple-scattering of the exciton by the potential U at the lattice site R_n, and the exclusion in the summation over the lattice sites in (20) prohibits the exciton from the successive multiple-scattering at the same lattice site. Note that $T^{\times}_n(Z)$ satisfies the equation:

$$\mathrm{Tr}_n\, a_n\, T^{\times}_n(Z)a^{+}_n\rho^0_n = 0, \tag{21}$$

where Tr_n is the trace over the states of Einstein phonon at R_n and ρ^0_n is the

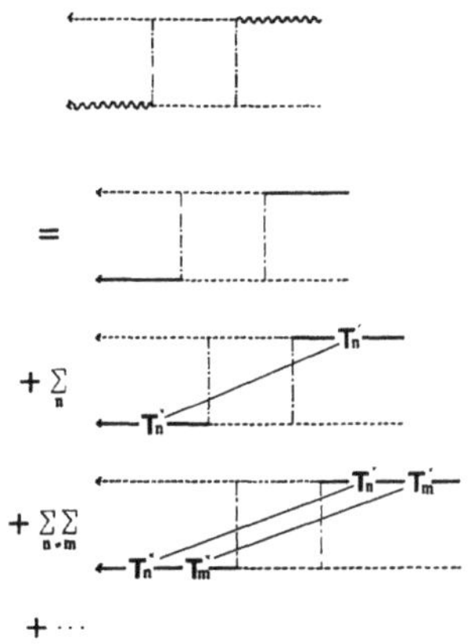

Fig.8 Schematic representation of the diagram (I) within the framework of the DCPA. The thick line denotes the propagation of the exciton according to the effective propagator and $T^{\times}_n$ is the single site t-matrix associated with the lattice site R_n.

thermal equilibrium density operator of this Einstein phonon in the electronic ground state. This equation is equivalent to the condition to determine the dynamical coherent potential, (11). By combining (19) and (20) we obtain the multiple-scattering expansion for the tetradic propagator. Then we substitute this multiple-scattering expansion into the propagator at the right end and left end of the diagram (I) in (17) and neglect the correlated scatterings of the exciton among different lattice sites obeying the spirit of single site DCPA. Then the contribution from the diagram (I) to the second order optical response is expressed in terms of the single site t-matrix $T_n^\times(Z)$ as shown schematically in Fig.8. In the figure the upper or lower thick line denotes the propagation of the system according to the effective propagator $G_{eff}^\times$ and lines connecting the upper and the lower $T_n^\times(Z)$ is the multi-phonon line of Einstein phonon at the lattice site R_n.

Now we consider the sum of the ladder type diagrams where $T_n^\times$ is put at the upper and lower left end and denote it by K_n. K_n satisfies the equation which is shown schematically in Fig.9, and is solved in the framework of the DCPA as follows. In the diagram involving K_m in Fig.9, all Einstein phonons at the other lattice sites are assumed to be in the thermal equilibrium during the multiple-scattering of the exciton at the lattice site R_n denoted by the line connecting the $T_n^\times$'s in the upper and the lower states. We denote K_n in the DCPA as $\tilde{K}_n$. The contribution from the diagram (I) to the second order optical response $I(\Omega_1,\Omega_2)$ is expressed with $\tilde{K}_n$ as shown in Fig.10. Note that the first term in Fig.10 contributes to the Rayleigh component in $I(\Omega_1,\Omega_2)$.

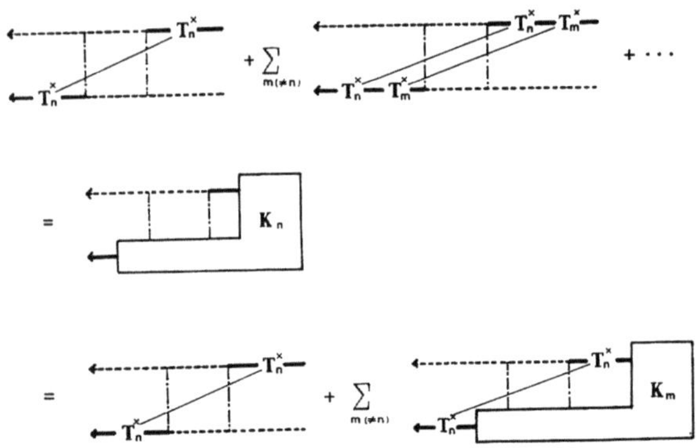

Fig.9 Schematic representation of the equation for the sum of the ladder type diagrams K_n.

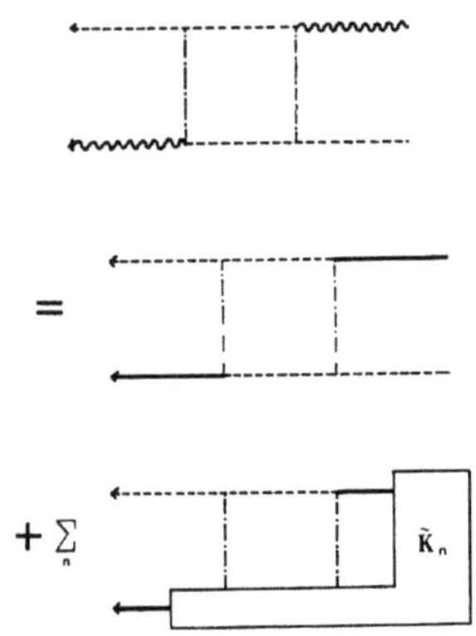

Fig.10 Equation for the diagram (I) in the DCPA.

On the other hand, the second term containing $\tilde{K}_n$ contributes to the Raman component leaving behind the excitation of finite Einstein phonons in the final state.

The evaluation of the diagrams (II) and (III) follows the same line as the diagram (I) although the situation is rather complicated because of the fact that the exciton is propagating both in the upper and the lower states in the intermediate state of the system in these diagrams. We consider eight types of K_n in the second diagram and seven in the third and can solve them by repeating the similar procedure.

5. Emission Spectrum

The characteristics of the emission spectrum in the present exciton-phonon complex depends on the four system parameters: the exciton band width 2B, the coupling constant S between the exciton and Einstein phonons, the relaxation rate γ of Einstein phonons and the radiative decay rate γ_b of the exciton and on the two variables: the lattice temperature T and the incident photon energy Ω_1. We have performed the numerical calculation and investigate the dependence of the emission spectrum on B, γ and Ω_1 for fixed values of S=0.5, T=1.0 and γ_b=0.1. We study the Raman and Brillouin scatterings and the luminescence and skip the Rayleigh component in the numerical calculation. The density of states of the exciton is taken as a hemi-elliptic form, (1.3). In the figures the origin of the energy is taken at the fundamental absorption edge.

In Fig.11-a the emission spectrum is shown for the case of B=0.2, Ω_1=2.5 and $\gamma=3.16\times10^{-2}$ at T=1.0. We can see the coexistence of the Raman lines at half the integer points and the luminescence lines at the integer points. The broadening of the Raman lines reflects the broadening of the energy levels in the final state, i.e., it is governed by the relaxation of Einstein phonons in the electronic ground state. Now the ratio of the total inte-

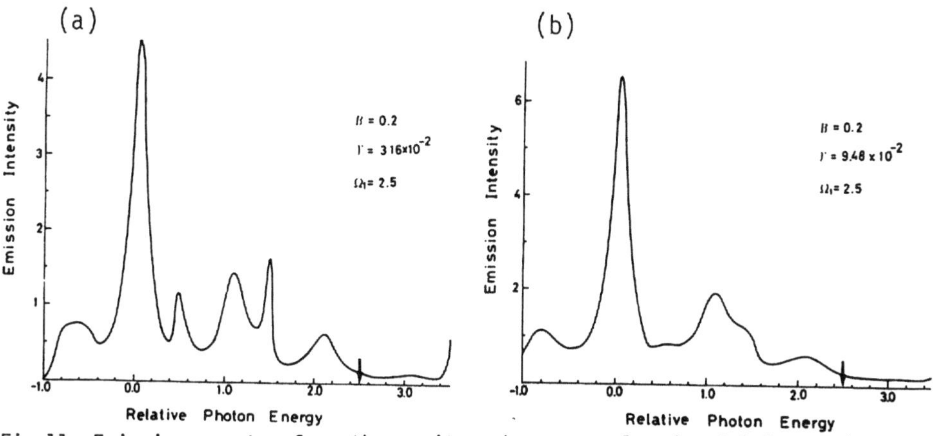

Fig.11 Emission spectra from the exciton-phonon complex for S=0.5, γ_b=0.1, T=1.0, B=0.2, Ω_1=2.5 and γ= (a) 3.16×10^{-2}, (b) 9.48×10^{-2}. The emission intensity is plotted in an arbitrary unit. The origin of the emitted photon energy is taken at the fundamental absorption edge which lies 0.6313 below the center of the exciton band for S=0.5 and B=0.2. The arrow in the figure indicates the position of the excitation energy Ω_1.

grated intensity of the Raman component to that of the luminescence is determined by the ratio γ_b/γ. In the case of $\gamma_b>\gamma$, the optically created exciton is radiatively annihilated before the relaxation in the excited state takes place and the Raman lines are clearly observable in the emission spectrum even for the excitation Ω_1 of the incident photon energy above the fundamental absorption edge. In the opposite case the luminescence lines are dominant. In Fig.11-b the emission spectrum is shown for the case of γ three times larger than in Fig.11-a. The Raman lines are indeed reduced in intensity and they appear as shoulders above the luminescence lines. In these two cases the value of B is less than the value of S and the emission spectra are expected to be similar to those in the localized electron-phonon system [4]. Therefore these results guarantee that our theory of the second order optical response due to the exciton-phonon system reproduces well the characteristic feature of the emission spectrum even for the nearly localized exciton-phonon system.

Now we consider how the emission spectrum depends on the transfer of the exciton B. The emission spectra for B=0.5, Ω_1 =2.5 and γ= (a) 3.16×10^{-2} and (b) 9.48×10^{-2} are shown respectively in Figs. 12-a and 12-b. While the Raman lines are unchanged against the increase of B, the luminescence lines are broadened and shifted to the high energy side. Besides, the luminescence line below the fundamental absorption edge increases in intensity with the increase of B. These changes of the luminescence lines are explained as follows. Let us see the energy diagrams in Fig. 2 for B=0.2 and B=0.5. The luminescence line in the emission spectrum is a sum of the emission from the lowest exciton polaron state and from the scattering continua. The increase of B broadens the spectrum width of the lowest exciton polaron band as well as that of the scattering continua as is seen in Fig. 2. This causes the peak shift and the broadening of the luminescence lines in the emission spectrum as in the case of the phonon sidebands in the absorption spectrum in §3. Now the luminescence lines are in general composed of the hot luminescence and the ordinary luminescence. The former is the emission from the excited state during the course of the vibrational relaxation while the latter is that after the completion of the relaxation in the excited state. The hot luminescence contributes mainly to the lines above the fundamental absorption edge. The luminescence lines on and below the fundamental absorption edge, on the other hand, consists mainly of the ordinary luminescence.

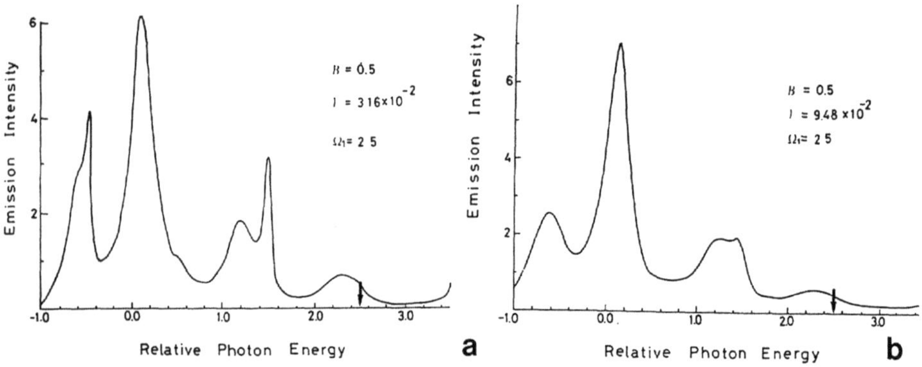

Fig.12 Emission spectra from the exciton-phonon complex. The transfer of the exciton B is increased to 0.5. The center of the exciton band for S=0.5 and B=0.5 lies 0.8571 above the fundamental absorption edge.

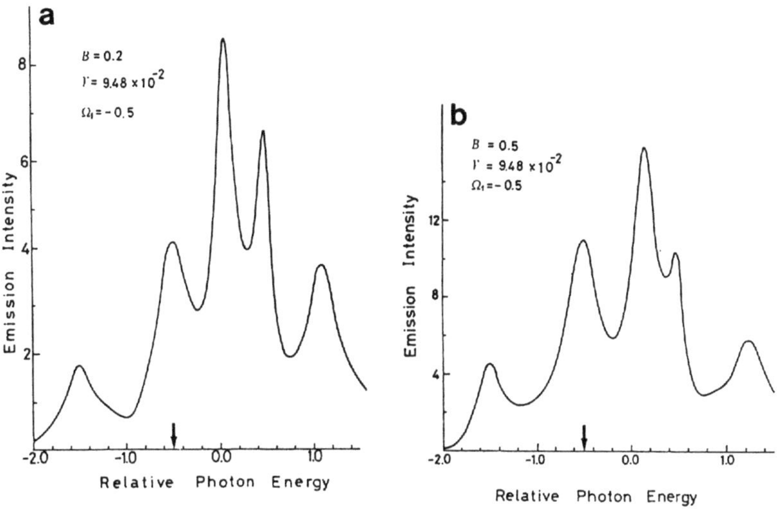

Fig. 13 Emission spectra from the exciton-phonon complex under the excitation below the fundamental absorption edge, $\Omega_1=-0.5$, for B=0.2 (a) and 0.5 (b).

The increase of B reduces the contribution from the hot luminescence while it increases that from the ordinary luminescence and as a result the luminescence line below the fundamental absorption edge increases in intensity. This effect is attributed to the interference between the relaxation and the transfer of the exciton noticed in §3. The exciton in the lowest scattering continuum relaxes to the exciton polaron state by the interaction with the reservoir. The probability that this relaxation process takes place depends on the relaxation rate γ and the energy separation between the lowest scattering continuum and the exciton polaron band. In the case of B=0.2, the energy separation is large and the probability is small. As B is increased to 0.5, the energy separation between them is reduced and they are smeared out by the relaxation effect, which causes the rapid dissipation of the extra energy to the reservoir in the excited state. This reduces the contribution from the hot luminescence and increases that from the ordinary luminescence.

Finally we show the emission spectra under the excitation of the photon energy $\Omega_1=-0.5$ in Figs.13-a and 13-b in the case of $\gamma=9.48\times10^{-2}$ and B=0.2 and 0.5, respectively. It is seen that the Raman lines and the luminescence lines are dominant below and above the fundamental absorption edge, respectively. From the comparison of Figs.13-a and 13-b with Figs.12-a and 12-b, we can see that the Raman lines are enhanced under the excitation below the fundamental absorption edge. In this respect, the freedom of the incident photon energy Ω_1 is advantageous in determining the relaxation constant from the competitive behavior of the Raman and the luminescence components in a large range of the ratio γ_b/γ.

6. Summary and Discussions

We have presented the theory of the first and second order optical responses in the exciton-phonon system where an exciton is interacting dominantly with Einstein phonons embedded in the reservoir phonons. The dynamical motion of the reservoir is projected onto the stochastic motion of the system, i.e., the vibrational relaxation of Einstein phonons. Then the dynamical coherent potential approximation was extended to the present non-Hamiltonian system. We have studied how the relaxation due to the interaction with the reservoir is reflected on the low energy side of the absorption spectrum due to the exciton. The relaxation of Einstein phonons donimates the broadening of the phonon-assisted sidebands for the strong coupling between the exciton and Einstein phonons (S>B). In the opposite case the broadening is mainly governed by the band width of the exciton polaron. The phonon sidebands in the low energy side of the absorption spectrum are merged into the U-M tail as the temperature is raised or/and the transfer of the exciton increases. The peaks of the phonon sidebands also are shown to be on the exponential line even when they are clearly resolved. This suggests that in the present model of the U-M rule the population of Einstein phonons plays an essential role.

The dynamical coherent potential approximation was further extended and applied to the second order optical response due to the exciton-phonon system. The numerical calculation was performed and the dependence of the emission spectrum on the transfer of the exciton was studied. It is shown that the broadening of the Raman lines is determined by the relaxation in the final state and is independent of the transfer of the exciton. The luminescence lines, on the other hand, depend sensitively on the transfer of the exciton. With the increase of the transfer the luminescence lines are broadened and shifted to the high energy side and the interference between the relaxation and the transfer causes the reduction of the hot luminescence component with the simultaneous increase of the ordinary luminescence component. The relative magnitude of the Raman and the luminescence components depends not only on the ratio of the radiative decay of the exciton to the relaxation but also on the incident photon energy. This means that we can observe the coexistence of these two components over a wide range of the ratio by choosing the incident photon energy suitably and determine the decay and the relaxation rate.

Now we discuss on the validity of the relaxation operator Γ in (6). In the derivation of Γ, we have used the Markovian assumption with respect to the reservoir. In order that the Markovian assumption is valid the correlation time of the reservoir phonons should be much shorter than the representative time constant of the system. The former is of the order of the inverse of the spectrum width of the reservoir phonons. The latter in the present problem is of the order of the inverse of the transfer rate of the exciton polaron. Therefore we restrict ourselves to the case of S>B or B<1. In the opposite situation, the characteristic feature of the absorption spectrum due to the exciton is almost reproduced by the spectrum width of the exciton polaron without the relaxation as is noticed in Figs.1-b to 1-d and Fig.2. Hence our theory of the first order optical response in the exciton-phonon system can explain the characteristics of the absorption spectrum from weak coupling S<B to the strong coupling case S>B if we take the value of γ to be zero or finite depending upon the relative value of S and B. In the second order optical response in the exciton-phonon system, on the other hand, the relaxation in the excited state plays the key role. Therefore our theory of the second order optical response is applicable to the exciton-phonon system with S>B or B<1.

Finally we remark about what is known from the study of the second order optical response of the exciton-phonon complex. First, the relaxation rate of Einstein phonons γ is known from the broadening of the Raman lines in the emission spectrum. Second, the ratio of the integrated intensity of the Raman component to the luminescence component is determined by the ratio of the radiative decay rate of the exciton γ_b to the relaxation rate γ. Therefore the values of γ_b and γ are determined from the emission spectrum. Third, the peak shift and the broadening of the luminescence lines give the information on the transfer of the exciton. The numerical calculations were confined to the case of small S and B. When we perform calculations over a wide range of the system parameters, we will be able to get deeper understanding about the mechanism of self-trapping of the exciton and the origin of the U-M rule. These works are in progress.

References

1 V.V. Hizhnyakov and I.M. Tehver: phys. status solidi 21, 755 (1967); ibid. 39, 67 (1970)
2 Y. Toyozawa: J. Phys. Soc. Jpn. 41, 400 (1976); A. Kotani and Y. Toyozawa: ibid. 41, 1699 (1976)
3 R. Kubo, T. Takagahara and E. Hanamura: Lecture Notes in Physics 57 (Springer-Verlag, Heidelberg, 1976) p.304; J. Phys. Soc. Jpn. 43, 802, 811 and 1522 (1977)
4 T. Takagahara, E. Hanamura and R. Kubo: J. Phys. Soc. Jpn. 44, 728 and 742 (1978)
5 F. Urbach: Phys. Rev. 92, 1324 (1953); W. Martienssen: J. Phys. Chem. Solids 2, 257 (1957); for review article, see V. Kurik: phys. status solidi A8, 9 (1971)
6 H. Sumi and Y. Toyozawa: J. Phys. Soc. Jpn. 31, 342 (1971)
7 Y. Toyozawa: Prog. Theor. Phys. 26, 29 (1961); A. Sumi and Y. Toyozawa: J. Phys. Soc. Jpn. 35, 137 (1973)
8 S. Nakajima: Prog. Theor. Phys. 20, 948 (1958)
9 Z. Zwanzig: Phys. Rev. 124, 983 (1961)
10 H. Sumi: J. Phys. Soc. Jpn. 32, 616 (1972); ibid. 36, 770 (1974)
11 B. Velicky, S. Kirkpatrick and H. Ehrenreich: Phys. Rev. 175, 747 (1968)
12 D. T. F. Marple: Phys. Rev. 150, 728 (1966)
13 B. Velicky: Phys. Rev. 184, 614 (1969); K. Levin, B. Velicky and H. Ehrenreich: Phys. Rev. B2, 1771 (1970)

Vibronic Spectra of Molecular Crystals: Dynamic Theory and Comparison with Experiment

E.I. Rashba

L.D. Landau Institute for Theoretical Physics, Academy of Sciences of the USSR
Chernogolovka, 142432, USSR

ABSTRACT

The general principles of the dynamic theory of vibronic spectra of molecular crystals are reviewed. This theory describes quantitatively both the energy spectrum and the absorption spectrum in the vibronic region, provided that the exciton band is narrow, as compared with the phonon frequency. This criterion is fulfilled, e.g., in aromatic molecular crystals (benzene and so on). The dynamic theory is closely connected with the theory of local exciton absorption spectra. Therefore, it allowes us to describe from a unified point of view the extensive experimental data on the spectra of perfect and doped crystals. As an example, the interpretation of the vibronic spectrum of naphthalene crystal is discussed briefly in the framework of the dynamic theory.

1. Introduction

In this paper we consider the vibronic absorption spectra of molecular crystals, i.e., the absorption spectra, corresponding to simultaneous creation of an electronic exciton (below-exciton) and one or several intramolecular phonons. The structure of this spectrum –the number of bands, their positions, distribution of intensities, and so on,– is determined by exciton-phonon coupling. In what follows it will be seen that side by side with the linear exciton-phonon coupling the quadratic coupling is also of importance; the effect of the latter is actually no less than that of the former. Thus, the problem is very complicated and a sufficiently effective approximation has to be found for its successful treatment.

First of all, let us single out the basic parameters; a competition between them determines the fundamental properties of the energy spectrum of the interacting exciton-phonon system. These parameters are: the half-width W^0 of the exciton band in a rigid lattice, the characteristic phonon frequency ν, and the Franck-Condon energy E_{FC} of the lattice relaxation around the exciton fixed at some lattice site. The parameter γ, defined by the equality $E_{FC}=\gamma^2\nu$, has the sense of the nondimensional coupling constant. It is convenient to consider two limiting cases:

$$W^0 \ll \nu \quad \text{(a)}, \qquad W^0 \gg \nu \quad \text{(b)}. \tag{1}$$

Following [1], we shall say that an exciton is heavy or light, depending on which criterion, (a) or (b), is fulfilled.

A heavy exciton slowly moves through the lattice; the lattice deformation follows it, which diminishes additionally the width of the exciton band.

This effect was originally pointed out by PEIERLS and FRENKEL. It is large only at $\gamma \gtrsim 1$, but the theory may be developed at any value of γ.

The situation is much more complicated for light excitons. Such an exciton at small γ remains free, but at large γ it is self-trapped according to Landau. For the general case the rigorous theory cannot be developed. But there are some interpolation schemes which allow us to draw some conclusions on the behavior of spectrum in a wide range of values of the parameters γ and W^0/ν; they have been developed mostly by Japanese theoreticians (see [2-4]).

In what follows we shall assume the condition (1a) to be fulfilled. This is our basic assumption. We impose no restrictions on the value of the linear exciton-phonon coupling parameter γ, and consider explicitly the quadratic exciton-phonon coupling side by side with the linear coupling. We treat relatively easy problem of the energy spectrum renormalization only briefly and concentrate our attention on a more complicated and much more interesting problem of the spectrum of an excited exciton-phonon system, i.e., the problem of an exciton interacting with one (or several) real phonons.

In organic molecular crystals there are two types of vibrations: intramolecular and intermolecular (internal and external). They usually have different scales of frequencies; the frequencies of the former are roughly one order of magnitude larger than those of the latter. As a result these vibrations are nearly totally independent, and for intramolecular vibrations the criterion (1a) is fulfilled rather well. Therefore, everywhere below the intramolecular phonons will be understood as phonons. The dispersion of these phonons is neglected. The interaction of an exciton with intramolecular phonons is designated as vibronic interaction. The interaction of an exciton with intermolecular phonons is neglected.

In conclusion, the main assumption of the theory is as follows:

$$W^0/\nu \ll 1, \qquad (1')$$

and the ratio W^0/ν will be taken to be a small parameter of the theory.

2. Dynamic Theory of Vibronic Spectra

The Hamiltonian of the exciton-phonon system is

$$H = H^0_{ex} + H_{ph} + H_{ex\text{-}ph}, \qquad (2)$$

where

$$H^0_{ex} = \sum_{n \neq m} M^0_{nm} a^+_n a_m, \quad H_{ph} = \nu \sum_n b^+_n b_n, \qquad (3)$$

$$H_{ex\text{-}ph} = \gamma\nu \sum_n a^+_n a_n (b^+_n + b_n) + \frac{\delta}{2} \sum_n a^+_n a_n (b^+_n + b_n)^2. \qquad (4)$$

Here H^0_{ex} and H_{ph} are the Hamiltonians of free excitons and phonons. The Hamiltonian $H_{ex\text{-}ph}$ involves the intramolecular electron-phonon interaction only; a relatively weak intermolecular exciton-phonon interaction is neglected. The first term in $H_{ex\text{-}ph}$ describes the linear interaction, and the second term, the interaction which is quadratic in phonon amplitudes. The last term is responsible for the change in vibrational frequencies under the electronic excitation of the molecule:

$$\nu^* = \nu\sqrt{1+2\delta/\nu}\ , \qquad \Delta\nu \equiv \nu^*-\nu \simeq \delta; \tag{5}$$

the last equality holds at $|\delta|<<\nu$. Usually $\delta<0$; thus, $\Delta_\nu<0$.

To make use of the advantages arising due to existence of the small parameter W^0/ν, it is necessary first to transform the Hamiltonian, eliminating the first term in H_{ex-ph}, which does not conserve the number of phonons and enters at $\gamma \gtrsim 1$ with a large coefficient. This may be achieved by a canonical transformation $\exp(\tau U)$, which diagonalizes the Hamiltonians of separate molecules in phonon operators. The operator U equals

$$U = \sum_n U_n, \quad U_n = \{(b_n-b_n^+)+\xi(b_n^2-b_n^{+2})\}a_n^+a_n, \tag{6}$$

where the paramters τ and ξ are determined by the formula

$$e^{-8\xi\tau} = 1+\frac{2\delta}{\nu}, \quad \frac{1}{\xi} = \frac{2\gamma}{(1+\frac{2\delta}{\nu})(1-e^{2\xi\tau})} \ . \tag{7}$$

As a result of this transformation the last two terms in (2) are simplified significantly. With an accuracy to the additive constant, which is about the Franck-Condon energy shift, we get

$$H_{ph}+H_{ex-ph} \Rightarrow \nu\sum_n b_n^+b_n + \Delta_\nu\sum_n a_n^+a_nb_n^+b_n . \tag{8}$$

Transforming the first term in (2), we make use of the fact that usually $|\delta|/\nu << 1$, and thus $\tau \simeq -\gamma$, $\xi\tau \simeq -\delta/4\nu << 1$, and the quadratic term in U_n may be omitted. Then

$$H_{ex}^0 \Rightarrow \sum_{n\neq m} M_{nm}a_n^+a_m\, e^{\gamma(b_n^+-b_m^+)}\, e^{-\gamma(b_n-b_m)} \ , \tag{9}$$

where the renormalized transfer integrals are

$$M_{nm} = M_{nm}^0 \exp(-\gamma^2); \tag{10}$$

the exciton half-width is transformed in the same way

$$W = W^0 \exp(-\gamma^2).$$

The principal difficulties are connected now with the operator (9) - it includes both the excitation transfer between lattice sites inherent in exciton and the linear exciton-phonon interaction, which does not conserve the phonon numbers. Since the terms, which do not conserve the number of phonons, enter now into the Hamiltonian with the coefficients M_{nm} [cf. (9)], which are small as compared with the frequency ν entering into energy of free phonons [cf. (8)], the perturbation theory may be applied to eliminate these terms.

The transformed operator H_{ex}^0 may be divided into two parts: $H_{ex}^0=(H_{ex}^0)_d+(H_{ex}^0)_{nd}$. The first of them $(H_{ex}^0)_d$ is diagonal in phonons and thus conserves the number of phonons. Taken in the matrix representation, it includes the mean value of H_{ex}^0 over the phonon vacuum, the matrix elements of H_{ex}^0 connecting single-phonon states, and so on. The second part $(H^0_{ex})_{nd}$ includes the matrix elements of H^0_{ex} connecting the states with a noncoinciding number of phonons. Of course, we have in mind here the number of real phonons; the virtual phonons forming a "clothed" exciton are excluded by transformation (6). The diagonal and nondiagonal parts of H^0_{ex} have the same order of magnitude, but their effect on the energy spectrum is quite different. To show this explicitly let us designate

$$H_0 = H_{ph} + H_{ex-ph} + (H^0_{ex})_d , \qquad H_1 = (H^0_{ex})_{nd} . \tag{11}$$

Since, according to (8), H_0 includes the terms of the order of ν, and H_1 of the order of W only, they have quite different scales: $H_0 \gg H_1$. This allows one to perform the canonical transformation by the operator exp(T) with $T \ll 1$,

$$H \Rightarrow e^T He^{-T} \simeq H_0 + \{H_1+(TH_0-H_0T)\} + \cdots ; \tag{12}$$

the operator T has to be chosen so as to cancel H_1, and turn, thus, to zero the part of the operator enclosed in braces which is nondiagonal in the number of phonons. Therefore, designating the eigenstates by indices α and β, we get

$$T_{\alpha\beta} \simeq (H_1)_{\alpha\beta}/(E_\alpha - E_\beta), \quad E_\alpha \neq E_\beta . \tag{13}$$

E_α are the eigenvalues of H_0; the separation between them is about ν; thus $T \sim W/\nu$. Therefore, the operator H_1 introduces the corrections to the energy spectrum in the order of $(W/\nu)^2$ only; we shall neglect them.

Thus, the problem is reduced to separation of the part of the operator (9) acting in the subspace with a fixed number of phonons. Zero-, one-, two-, etc., phonon Hamiltonians arise as independent parts of the same operator. We restrict ourselves here by writing out the one-phonon Hamiltonian H_ν of the vibronic interaction

$$H_\nu = \gamma^2 \sum_{n\neq m} M_{nm} a^+_n a_m (b^+_n b_m + b^+_m b_n - b^+_n b_n - b^+_m b_m) + \Delta_\nu \sum_n a^+_n a_n b^+_n b_n . \tag{14}$$

With the Hamiltonians of free excitons and free phonons

$$H_{ex} = \sum_{n\neq m} M_{nm} a^+_n a_m , \qquad H_{ph} = \nu \sum_n b^+_n b_n \tag{15}$$

it constitutes the total Hamiltonian of the system

$$H = H_{ex} + H_{ph} + H_\nu . \tag{16}$$

The eigenvalues of H_{ex} constitute the energy spectrum E(K) of a free exciton.

It is an important peculiarity of the Hamiltonian H_ν that in distinction to the initial interaction Hamiltonian H_{ex-ph} it conserves the number of phonons. This considerably simplifies the problem - it is reduced from the field theoretical problem to the usual quantum mechanical one, i.e., to the problem of the dynamics of two interacting particles. The theory of vibronic interaction based on the Hamiltonian conserving the number of particles is known as the dynamic theory of vibronic spectra; it was originally proposed and developed in [5a,b].

It is necessary to stress that no unknown parameters enter into the Hamiltonian H_ν. The excitation transfer integrals M_{nm} are completely determined by the energy spectrum of an electronic exciton. The constants γ^2 and Δ_ν are determined by the intramolecular electron-phonon interaction, and, at least in principle, they may be found from the spectrum of an isolated molecule.

Separate terms of H_ν have a clear physical sense; the corresponding diagrams are shown in Fig.1. The first term corresponds to the joint transfer of exciton and phonon (Fig.1a); the second, to the exchange transfer (Fig.1b); and the two next terms describe the phonon corrections to the exciton transfer rate (Fig.1c,d). The last term describes the exciton-phonon contact interaction arising due to a change of the vibrational frequency (Fig.1e).

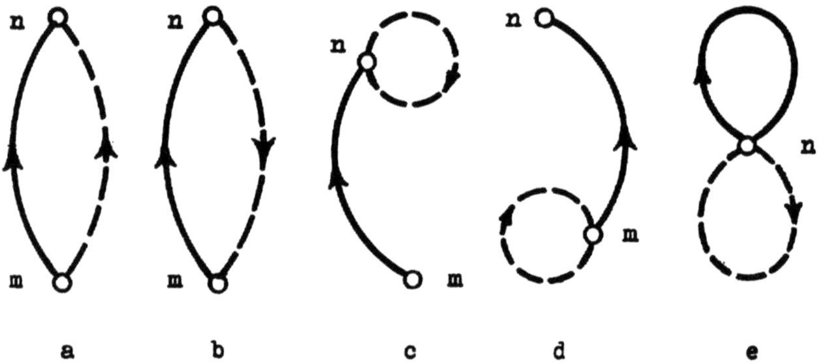

Fig.1 Diagrams corresponding to separate terms of the Hamiltonian H_ν.

It is seen from (14) why in the problem under consideration such a unique situation arises that the effect of the quadratic interaction may have a comparable order of magnitude with that of the linear interaction. The quadratic interaction enters into H_ν in the first order of the perturbation theory, and Δ_ν competes with W rather than with ν. The effect of the quadratic interaction is large when $|\Delta_\nu| \gtrsim W$; under these conditions $|\Delta_\nu|$ remains small as compared to ν ($|\Delta_\nu| \ll \nu$). On the contrary, the linear interaction enters into H_ν through the factor γ^2, i.e., in the second order only, and is multiplied by the small transfer integrals M_{nm}.

3. Energy Spectrum and Spectrum of Optical Absorption

There is one restriction to the form of the Hamiltonian (4) which is connected with the symmetry requirements only and is, thus, quite general. We suppose the excited state of the molecule to be nondegenerated. Then the term with γ^2 enters into the H_{ex-ph} for the totally symmetric (TS) phonons only. The absence of this term for non-totally symmetric (NTS) phonons is equivalent to the absence of Jahn-Teller effect for nondegenerated levels. Accordingly, the term with γ^2 enters into H_ν for TS phonons only. On the contrary, there are no symmetry restrictions to the last term in H_ν proportional to Δ_ν; thus, it enters into H_ν for both TS and NTS phonons. Below we shall refer to the γ-mechanism and Δ_ν-mechanism of the vibronic interaction.

The energy spectrum of a pair of interacting particles, exciton + phonon, includes two types of states. The unbound, i.e., dissociated, states of the pair always exist; we shall designate this spectrum as two-particle. For dispersionless phonons the width of the two-particle spectrum coincides with the exciton spectrum width $2W$.

However, when the vibronic coupling is sufficiently strong, exciton-phonon bound states arise. We shall designate such bound states as vibrons. They constitute a one-particle spectrum. The number of branches in this spectrum depends on the mechanism of vibronic interaction and on the value of the coupling constant. The vibron energy depends on the momentum of the vibron K.

Since H_ν has quite different forms for NTS and TS phonons, we consider them separately.

NTS phonons; Δ_ν-mechanism

In this case $\gamma = 0$. Thus, H_ν is reduced to the last term in (14) and $b_n^+ b_n$ commutes with H [cf. Eqs.(14)-(16)]. Therefore, the eigenstates correspond to the localization of a phonon on one of the lattice sites (below n=0) and to motion of an exciton in the field of this phonon. The effective exciton Hamiltonian is

$$H_{eff} = \sum_{n \neq m} M_{nm} a_n^+ a_m + \Delta_\nu a_0^+ a_0 . \tag{17}$$

It totally coincides with the Hamiltonian of the exciton moving in the field of isotopic impurity, which has the isotopic shift Δ_ν and is placed at the site $n = 0$. Thus, the problem of determination of vibronic levels is reduced to another problem, to the motion of exciton in the field of the impurity (or defect)[1]. This problem was studied intensively, both theoretically and experimentally. Therefore, the obtained results may be used.

The position E_ν of the level corresponding to the exciton-phonon bound state is determined by the well-known equation of I. LIFSHITZ and KOSTER-SLATER

$$\Delta_\nu G_0(E_\nu) = 1, \quad G_0(E) = \int \frac{\rho(E')}{E - E'} \, dE' . \tag{18}$$

Here $\rho(E)$ is the normalized density of states in the exciton energy spectrum. This equation has an isolated root only when Δ_ν is sufficiently large: $|\Delta_\nu| > \Delta_{cr}$. Usually. $\Delta_{cr} \simeq W/2$. The quantity E_ν is the energy of one-particle states; in the lowest approximation the one-particle branch is dispersionless. At $|\Delta_\nu| >> W$, i.e., for deep levels, $E_\nu \simeq \Delta_\nu$. On the contrary, for shallow levels, when $|\Delta_\nu| \to \Delta_{cr}$, $|E_\nu| > |\Delta_\nu|$ always, i.e., the level is "repelled" from the spectrum edge.

The wave function of exciton $\psi(n)$, when the phonon is fixed at the lattice site $m = 0$, may be found too. We shall write here only

---

1) In the allied problem of a bi-phonon such as approach was advanced in [6].

$$\psi_0^2 = |\psi(n{=}0)|^2 = \frac{1}{\Delta_\nu^2 \, dG_0(E_\nu)/dE_\nu} \tag{19}$$

and

$$|\psi(k{=}0)|^2 = \frac{1}{(E_\nu - E(k{=}0))^2} \frac{1}{dG_0(E_\nu)/dE_\nu} \, . \tag{20}$$

The quantity ψ_0 is the contribution to the wave function of the vibron of the joint configuration, i.e., such configuration that exciton and phonon are at the same lattice site. At $|\Delta_\nu| \gg W$ this configuration dominates. Such vibron may be called a molecular vibron; it was studied by DAVYDOV [7], CRAIG and WALMSLY [8], and others. On the contrary, at $|\Delta_\nu| \to \Delta_{cr}$ the value $\psi_0^2 \to 0$, and the contribution of separated configurations, in which exciton and phonon occupy different lattice sites ($n \neq m = 0$), is progressively growing. This "swelling" of wave function is displayed in the strong dependence of $|\psi(k{=}0)|^2$ on Δ_ν at $|\Delta_\nu| \to \Delta_{cr}$; the second factor in (20) tends to zero in this limit. The behaviour of the first factor depends critically on whether the point $k = 0$ is positioned at that band edge (at $\Delta_\nu < 0$ at the bottom of the band) which the vibronic level is approaching. In general, the behaviour of $|\psi(k{=}0)|^2$ corresponds to a gigantic change in the oscillator strength [9,10], known in the spectroscopy of local excitons.

TS phonons; Δ_ν- and γ-mechanisms

Here in H_ν all the terms are present and this changes the picture significantry. A TS phonon, even distersionless, becomes movable due to its interaction with an exciton: it jumps from one site to another together with the exciton [first term in (14)], or interchanges the sites with it [second term in (14)]. Although the problem of two interacting particles (exciton and TS phonon) even in this case may be reduced to a one-particle one, this problem cannot be solved exactly. Though, in general, the picture resembles that described above, as applied to the NTS phonon, there are significant distinctions too. For example, at $\gamma^2 \sim 1$ several one-particle branches may arise, their dispersion being rather high - of the order of W ; these branches, as a rule, exist in a limited part of the K-space only.

In the framework of the dynamic theory the spectrum of vibronic absorption either may be found in an explicit form, for NTS phonons, or can be calculated numerically by solving relatively easy equations, for TS phonons. We restrict ourselves to the temperature $T = 0$.

In both cases the absorption spectrum consists of one (or several) narrow band, corresponding to transition to a one-particle state, and of a wide band corresponding to transitions to two-particle states. The integral intensity of the spectrum is determined by the oscillator strength of the intramolecular vibronic transition. The relative intensity of the one-particle absorption equals ψ_0^2.

For NTS phonons the conductivity $\sigma(\omega)$, expressed in convenient units, is

$$\sigma(\omega) = \frac{\rho(\omega)}{(1-\Delta_\nu G_0(\omega))^2 + (\pi\Delta_\nu \rho(\omega))^2} \; ; \tag{21}$$

G_0 is determined by (18). If the expression in the first parentheses turns to zero, i.e., a one-particle state arises, the $\sigma(\omega)$ is peaked δ-functionally. Except for this peak, $\sigma(\omega) \neq 0$ only where $\rho(\omega) \neq 0$, this is the region of two-particle absorption. Expression (21), considered as an equation for $\rho(\omega)$, may be solved explicitly [11]

$$\rho(\omega) = \frac{\sigma(\omega)}{(1 + \Delta_\nu \int \frac{\sigma(\omega')}{\omega - \omega'} d\omega')^2 + (\pi \Delta_\nu \sigma(\omega))^2} . \tag{22}$$

Equation (22) allows one to solve the inverse problem to deduce the density of states $\rho(E)$ from the experimental data on the absorption spectra.2)

For TS phonons the calculation is more complicated. So, without going into detail, we reproduce in Fig.2 the calculated absorption spectrum for the crystal of the naphthalene type, found in [13], to illustrate the general features of the spectrum; the calculations have been done for a number of values of the parameters γ^2 and Δ_ν 3). It is seen that the shape of the spectrum is changed widely from the asymmetric curve covering the exciton band to narrow peaks. When $\gamma^2 \to 1$, the absorption spectrum is progressively narrowed. To elucidate the causes of this phenomenon, it should be noted that the point $\gamma^2 = 1$ is remarkable by the fact that for it the wave function

$$\Psi_{K=0} \propto \sum_n a_n^+ b_n^+ |vac\rangle \tag{23}$$

is the exact solution of the problem. This function includes joint configurations only, so the transition to the state $\Psi_{K=0}$ absorbs the total oscillator strength of vibronic transition, since the wave functions of all other states with $K = 0$ are orthogonal with respect to $\Psi_{K=0}$ and thus include only separated configurations of exciton and phonon. If the energy of the state (23) is outside of the two-particle continuum, this state is one-particle; but if its energy gets inside the two-particle continuum, it is quasi-one-particle (non-stable).

The dynamic theory was also applied to many-phonon states [5a,15,13], but due to complexity of the problem only limited results have so far been obtained.

In addition to the process of vibron creation, discussed above, the production of vibron may also take place at $T \neq 0$ by adding an optically produced exciton to a thermal intramolecular phonon [16]. At low temperatures $T \lesssim \nu$ the intensity of such absorption is proportional to the factor $\exp(-\nu/T)$, and at $\Delta_\nu < 0$ it has to be shifted to the low frequency side from the exciton band. However, the frequency dependence of this absorption has nothing to do with the Urbach one, especially in the area of the one-phonon peak. The intensity of this absorption is proportional to $|\psi(k=0)|^2$ [cf.(20)], i.e., it has to be gigantic, provided that the point $k = 0$ is at the bottom of the exciton band.

---

2) An analysis of vibrational exciton spectrum in $NaNo_3$ [12] may serve as an example of successful application of this method.

3) The density of states $\rho(E)$ for naphthalene crystal used in these calculations is also shown in the figure: it has been chosen according to [14,11]. The transfer integral for nearest neighbours is $M_{12} \approx 20\ cm^{-1}$.

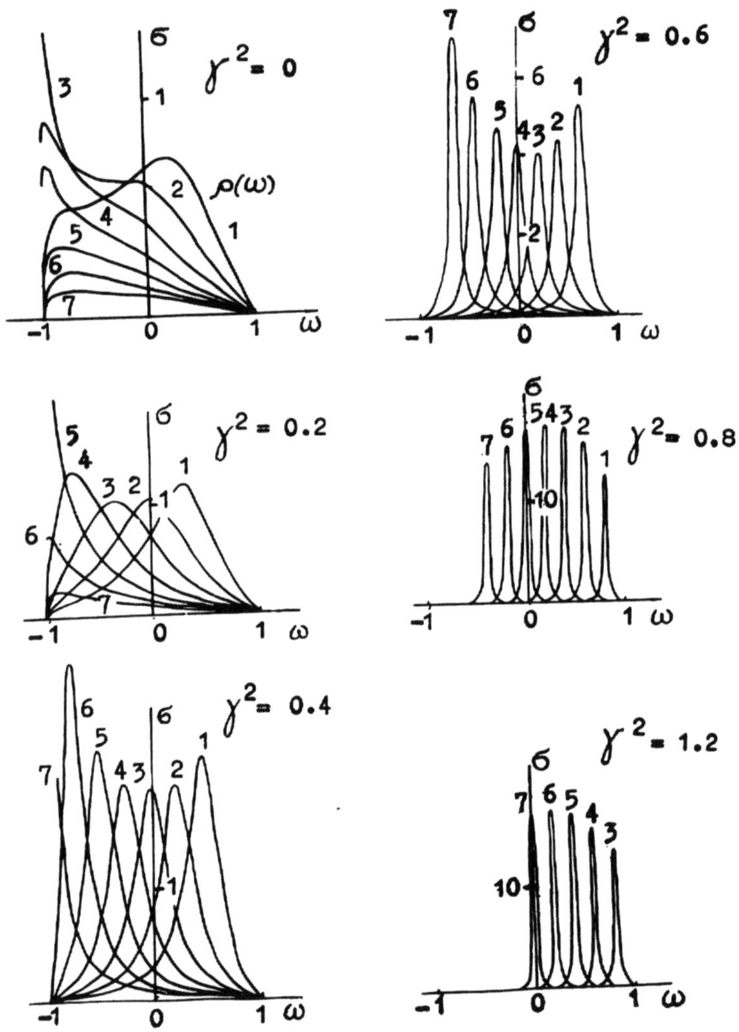

Fig.2 Calculated spectrum of two-particle vibronic absorption of naphthalene crystal (monoclinic b-component) at $\Delta_\nu < 0$. Marking of curves corresponds to the following values of the parameter $|\Delta_\nu|/W$: 1 - 0; 2 - 0.2; 3 - 0.4; 4 - 0.6; 5 - 0.8; 6 - 1.2.

4. Relationship of Vibronic Spectra and Spectra of Local Excitons

The fact that the Hamiltonian of vibronic interaction may be reduced to the form (14) expressing explicitly the phonon number conservation shows the existence of an intimate connection between the theory of vibronic spectra and the theory of local excitons, i.e., of the excitons trapped by impurities or lattice defects. Really, we have already seen in Sect.3 that the theory of vibrons, including NTS phonons, is indentical to the theory of local excitons bound to isotopic impurities. For TS phonons the connection to the theory of local excitons is not so immediate. It is important, however, that both theories are formulated in the same terms. They include a set of transfer integrals M_{nm} as the basic data. These data should be supplemented by the values of the parameters γ^2 and Δ_ν for vibronic spectra, and the values of isotopic shifts and of some other paramters for local excitons. The calculational methods are also the same for both theories; it is the theory of degenerated perturbations. Thus, the unified description and treatment of vibronic spectra and local exciton spectra significantly broaden the possibilities of deducing the exciton energy spectrum from the experimental data [17]. Such a unified description is especially effective as applied to the spectra of doped crystals [5a, 18, 19, 13].

Owing to its visuality, the dynamic theory allows one to imagine the main features of quantum states from the intuitive physical consideration and to obtain estimations even when the straightforward solution of the equations is quite cumbersome. Therefore, the formulation of the dynamic theory, based on the Hamiltonian H_ν in the form (14), is especially convenient for treatment of the experimental data.

Some results of the dynamic theory were later rederived by a slightly different technique, but no new physical results were obtained; these papers are reviewed in [20] (see §50). In this way only some one-dimensional models were investigated.

5. Vibronic Spectra of Naphthalene Crystal

A qualitative analysis of vibronic spectra of aromatic molecular crystals based on the dynamic theory [5a, b; 18] was given by SHEKA [21]. She showed that for three crystals, benzene, naphthalene, and anthracene, the theory is in a good qualitative agreement with the experimental data. This analysis allowed significant advancement of the interpretation of the origin of some bands.

We restirict ourselves to the experimental data on naphthalene only, but include the results of the last quantitative treatment of spectra [13].

A primitive cell of naphthalene crystal contains two molecules. So the exciton doublet consists of two bands (A_0 and B_0); DAVYDOV splitting is about 150 cm^{-1}. The width of the exciton band is $2W \simeq 180$ cm^{-1}. The low frequency band A_0 of the exciton doublet is two orders of magnitude weaker than the B_0-band; the latter is polarized along the monoclinic b-axis.

The first intensive vibronic band corresponds to the NTS phonon $\nu = 509$ cm^{-1}; $\Delta_\nu = -86$ cm^{-1}. Thus $|\Delta_\nu| \simeq W > \Delta_{cr}$; so, a one-particle state arises. The strong absorption band, known as the M-band (Fig.3), corresponds to it. The M-band is followed by a wide band absorption, its total intensity is about that of the M-band. According to [11], the wide band absorption is roughly equally distributed between the two-particle absorption (dissociated states: D_M-band) and the phonon sideband of the M-band; this sideband corresponds to external phonons with $\nu \sim 60$ cm^{-1}. In Fig.3a both these contributions are shown separately. This interpretation may be confirmed by the spectra of strained crystals, in which the exciton band is about two times narrower. As a result, ψ_0 grows essentially [cf.(19)]; therefore, the one-particle absorption increases, and the two-particle one decreases. Figure 3b is in agreement with this conclusion. The M-band is strengthened, and the high-frequency wing is weakened; practically it is completely reduced to a phonon sideband.

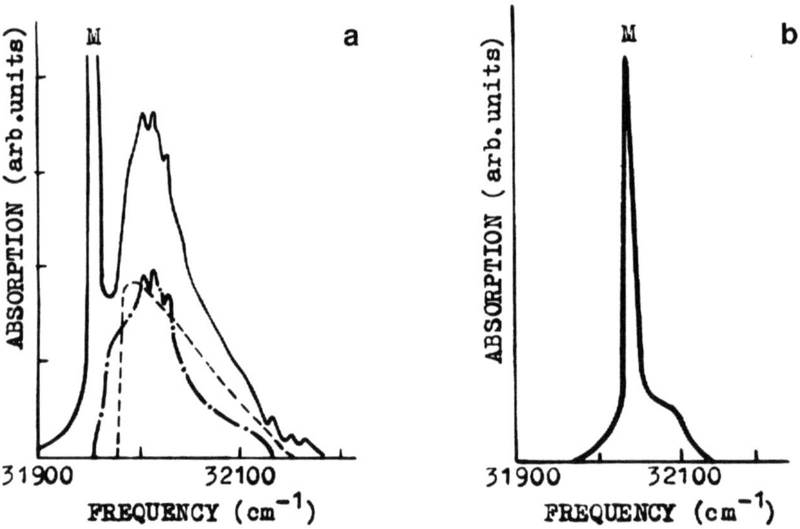

Fig.3 Vibronic absorption with NTS phonon $\nu = 509$ cm^{-1} in naphthalene. a. Free crystal. Solid curve - experiment [22], dashed curve - D_M - band, dash-dotted curve - phonon sideband [23]. b. Strained crystal [24b].

From TS phonons the phonon with $\nu = 758$ cm^{-1} contributes mostly to the vibronic spectrum. In Fig.4 are shown the corresponding absorption spectrum [22] and its interpretation, according to [13]. The narrow band in the a-component lies outside of the two-particle spectrum, being thus one-particle (A_1-band). In the strong b-component of the spectrum almost all the absorption is concentrated in a two-particle region; it has to be interpreted as D_b-band. We stress that the multiplet of one-particle vibronic bands is incomplete; it includes the a-component only. The whole spectrum, the position of the A_1-band, and the shape of the D_b-band are described quite well for a set of parameter values $\Delta_\nu = -57$ cm^{-1}, $\gamma^2 = 0.20$; they are close to their values in free molecule. In strained crystals the one-particle band arises also in the b-component due to a decrease of W ; thus, a complete vibronic exciton multiplet $A_1 - B_1$ is seen.

The two-phonon vibronic spectra were investigated experimentally too; their detailed theory was not yet developed. Although the spectrum becomes

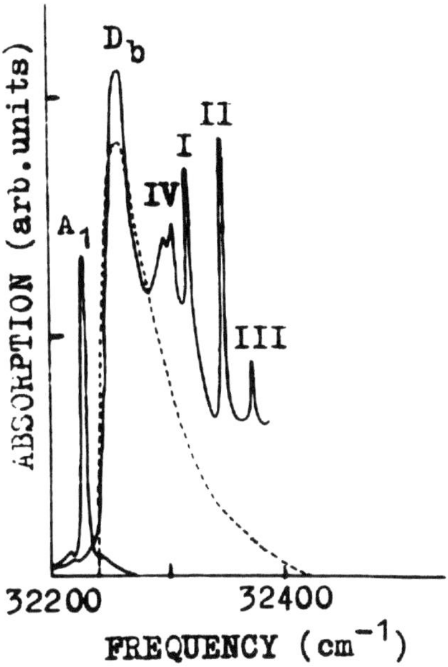

Fig.4 Vibronic absorption with TS phonon ν=758 cm^{-1} in naphthalene; spectrum is shown in two polarizations (a and b). Solid curve - experiment, dashed curve - theory [13]. Bands I-IV are of outside origin

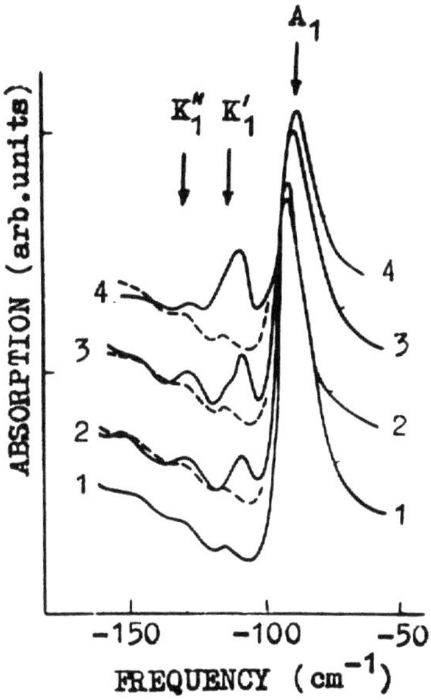

Fig.5 Vibronic absorption of octadeuteronaphthalene doped with naphthalene [25]. Numbers at the curves show the concentration of impurity: 1 - 1%; 2 - 2%, 3 - 5%; 4 - 10%. Successive curves are shifted upward without change in scale; the dashed curves duplicate the absorption of the host crystal. A_1 - intrinsic vibronic absorption; K_1', K_1'' - vibronic impurity bands

very complicated, some common tendency is seen from general considerations. It consists in the strengthening of the one-particle band, which follows from the appearance of a strong attractive field for exciton due to adding the quantities Δ_ν for both phonons.

The vibronic spectra of isotopic and even more complicated impurity centers were investigated too. When the isotopic impurity is in the lattice site $n=0$, the main part of the impurity contribution to the Hamiltonian is

$$H_{imp} = \Delta_{ex}\, a_0^+ a_0 + \Delta_{ph}\, b_0^+ b_0 . \tag{24}$$

Here Δ_{ex} and Δ_{ph} are the intramolecular isotopic shifts of the electron level and the vibrational frequency. The spectrum of impurity vibronic states depends significantly on the ratio of Δ_ν, Δ_{ex} and Δ_{ph} and on their signs.

As an example, Fig.5 shows the impurity vibronic spectrum ($\nu = 758$ cm^{-1}) of octadeuteronaphthalene doped with naphthalene. Here $\Delta_{ex} = -115$ cm^{-1}, $\Delta_{ph} = 69$ cm^{-1}. The band K_1' is the most pronounced one in the impurity spectrum: a dominating contribution to the wave function of this state is made by such a configuration when both exciton and phonon are on the guest site. The second band K_1'' is seen most distinctly when the impurity concentration is about 5%. The configurations in which the exciton is on the guest site and the phonon on one of the neighbouring host sites dominate in the corresponding wave function.

An analysis of the whole variety of experimental data on vibronic spectra of naphthalene shows that these spectra are utterly described by the dynamic theory.

6. Conclusion

The review of the dynamic theory of vibronic spectra given above shows that it allows us to classify the states and calculate both the energy and the optical absorption spectrum of the system, provided that the main applicability condition of the dynamic theory $W << \nu$ is fulfilled. This theory is applicable for intrinsic as well as for extrinsic absorption spectra. The theory turned out to be effective for a quantitative analysis of spectra of several crystals up to describing the fine details of these spectra.

1 E.I. Rashba, Opt. Spektrosk. 2, 75, 88 (1957); Izv. AN SSSR, Ser. Fiz. 21, 37 (1957)
2 K. Cho, Y. Toyozawa, J. Phys. Soc. Japan 30, 1555 (1971)
3 H. Sumi, loc. cit. 36, 770 (1974); 38, 825 (1975)
4 A. Sumi, Tech. Rep. of ISSR Ser.A N 971, June 1979
5 E.I. Rashba, (a) Zh. Eksp. Teor. Fiz. 50, 1064 (1966) [Sov. Phys. - JETP, 23, 708 (1966)]; (b) loc. cit. 54, 542 (1968) [27, 292 (1968)]
6 J. Van-Kranendonk, Physica 25, 1080 (1959)
7 A.S. Davydov, Light Absorption in Molecular Crystals (Acad. Sci. Ukr. SSR Publ., Kiev, 1951) [Molecular Excitons McGraw-Hill Publ. Corp., NY, 1962]
8 D.P. Craig, S.H. Walmsley, Molec. Phys. 4, 114 (1961)
9 E.I. Rashba, Opt. Spectrosk. 2, 568 (1957)
10 V.L. Broude, E.I. Rashba, E.F. Sheka, Dokl. AN SSSR, 139, 1085 (1961) [Sov. Phys. - Dokl. 6, 718 (1962)]
11 N.V. Rabin'kina, E.I. Rashba, E.F. Sheka, FTT 12, 3569 (1970) [Sov. Phys. - Solid State 12, 2898 (1971)]
12 M.V. Belousov, D.E. Pogarev, A.A. Shultin, FTT 18, 521 (1976)
13 T.A. Krivenko, E.F. Sheka, E.I. Rashba, Mol. Cryst. Liq. Cryst. 47, 119 (1978)
14 S.D. Colson, D.M. Hanson, R. Kopelman, G.W. Robinson, J. Chem. Phys. 48, 2215 (1968)
15 M.R. Philpott, J. Chem. Phys. 47, 2534, 4437 (1967)
16 E.I. Rashba, FTT 12, 1801 (1970) [Sov. Phys. - Solid State 12, 1246 (1970)]
17 V.L. Broude, E.I. Rashba, Pure and Appl. Chem. 37, 21 (1974)
18 V.L. Broude, E I. Rashba, E.F. Sheka, Phys. stat. sol. 19, 395 (1967)
19 M.R. Philpott, J. Chem. Phys. 53, 136 (1970)
20 A.S. Davydov, Solid State Theory (Nauka, Moscow, 1976)
21 E.F. Sheka, Usp. Fiz. Nauk 104, 593 (1971) [Sov. Phys. - Uspekhi 14, 484 (1971)]
22 A.F. Prikhot'ko, M.S. Soskin, Opt. Spectrosk. 13, 522 (1962)
23 E.F. Sheka, Mol. Cryst. Liq. Cryst. 29, 323 (1975)
24 (a) V.L. Broude, A.K. Tomashchik, Ukr. Fiz. Zh. 9, 39 (1964); (b) A.F. Prikhot'ko, M.S. Soskin and A.K. Tomashchik, Opt. Spectrosk. 16, 615 (1964)
25 T.A. Krivenko, A.V. Leiderman, E.I. Rashba, E.F. Sheka, Pis'ma Zh. Eksp. Teor. Fiz., 25, 538 (1977) [Sov. Phys. - JETP Lett. 25, 538 (1977)]

Ultrafast Vibrational Relaxation Processes of Polyatomic Molecules, Investigated by Picosecond Light Pulses

A. Laubereau

Physikalisches Institut der Universität Bayreuth
D-8580 Bayreuth, Fed. Rep. of Germany

1. Introduction

For several decades frequency domain spectroscopy has been the main source of information on vibrational dynamics. The analysis of the spectroscopic data, however, faces several important difficulties. The finite instrumental resolution and the problem of overlapping neighboring bands are to be mentioned here; contributions of overtones and hot bands of other vibrational modes, although not accurately known, have to be taken into account in many practical cases. These factors represent a serious source of experimental uncertainty. Once a band contour is determined the question arises about its interpretation. In general, many physical processes may contribute to an observed band shape [1]. Rotational motion and various kinds of vibrational dephasing and of energy redistribution processes affect a vibrational band. The different contributions to the band contour cannot be separated unambiguously in conventional spectroscopy.

Some of these difficulties are overcome using time resolved techniques. The advance in laser technology enables us to generate intense picosecond light pulses in the spectral region extending from the near uv to the infrared [2,3]. These pulses allow direct studies of ultrafast molecular processes with an exceptional time resolution of a fraction of a picosecond. The present chapter presents several recent experiments to illustrate the state of the art and the potential of our new techniques.

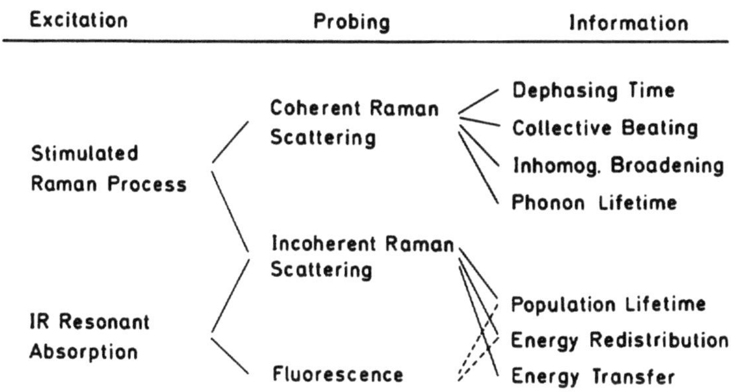

Fig.1 Physical processes for the excitation and probing of molecular vibrations in liquids. The experimental information is listed on the right.

Figure 1 outlines the experimental methods and the various results obtained [4]. The excitation of the molecules is achieved by an intense laser pulse via stimulated Raman scattering or by resonant absorption of an infrared pulse. After the passage of the first pulse the excitation process rapidly terminates and free relaxation of the excited mode occurs. A second weak pulse, which is properly delayed with respect to the first one, monitors the instantaneous state of the excited system. Varying the delay time, the time evolution of the system is observed. Three probing processes were used to obtain the dynamical information listed on the right side of Figure 1.

The present paper concentrates on experiments using the first two probing processes shown in Figure 1. The fluorescence probing technique, which is also indicated in the figure, was found very useful for investigations of large molecules in highly diluted systems [5] and has been recently applied for a novel two-pulse spectroscopy [6]; these investigations will not be discussed below.

2. Excitation Process

Some remarks on transient stimulated Raman scattering are given first. It is convenient to treat the vibrational system as a two-level model; transitions involving higher excited states are negligible on account of the anharmonic frequency shift. In this picture the coherent interaction of a vibrational mode with a non-resonant light pulse via stimulated Raman scattering is quite analoguous to the well-known case of magnetic dipole transitions of a spin system. In the following it is advantageous to consider the more general case of an ensemble of two-level systems with a narrow distribution of vibrational transition frequencies. Molecules in a small frequency interval of this distribution have transition frequency ω_j and number density $f_j \times N$, where $\Sigma f_j = 1$. N denotes the total number density. In this way the case of isotopic line splitting and inhomogeneous line broadening is included in the present calculation.

Treating the light field classically, the Hamiltonian for the molecular subensemle j interacting with the light field $\mathbb{E}$ has the form:

$$H = H^o - \frac{1}{2} q_j \frac{\partial\alpha}{\partial q} E^2 \qquad (1)$$

q_j is the normal mode operator. $\partial\alpha/\partial q$ represents the change of polarizability with vibrational amplitude.

We note that the molecular orientation does not explicitly enter (1) when $\partial\alpha/\partial q$ is a scalar. This approximation holds for the experiments to be discussed here. A careful analysis including the anisotropy of the Raman scattering tensor shows that the molecular rotation has a negligible effect on the excitation and probing processes, even for a moderately large anisotropy, corresponding to a depolarization factor $\rho_s \lesssim 0.2$ [4]. This point is quite favorable as compared to conventional Raman spectroscopy where deconvolution procedures are required [7].

The vibrational excitation of the molecules is described by the expectation values of the normal mode operators, $\langle q_j \rangle$, and the excited state population, n_j. For the two level approximation the equations of motion have the form [4,8,9]

$$\left(\frac{\partial^2}{\partial t^2} + \frac{2}{T_2}\frac{\partial}{\partial t} + \omega_j^2\right)\langle q\rangle = \frac{1}{2m}\frac{\partial\alpha}{\partial q} E^2 \,[1-2\, n_j] \tag{2}$$

$$\left(\frac{\partial}{\partial t} + \frac{1}{T_1}\right)(n_j-\bar{n}) = \frac{1}{2\hbar\,\omega_o}\frac{\partial\alpha}{\partial q} E^2 \frac{\partial}{\partial t}\langle q_j\rangle \tag{3}$$

$\bar{n}$ is the thermal equilibrium population of the upper levels. The time constants T_1 and T_2, the reduced mass m and the coupling parameter $\partial\alpha/\partial q$ are assumed to be equal for all subensembles j.

Of special interest are the two relaxation times T_1 and T_2 introduced in (2) and (3). T_2 denotes the dephasing time of the vibrational amplitude of a molecular group with transition frequency ωj. T_1 represents the population lifetime of the first excited vibrational state.

Eqs. (2) and (3) assume exponential decay of $\langle q_j\rangle$ or n_j after the excitation process has terminated. This time dependence is not expected to hold for times shorter than τ_c [10] where $\tau_c \sim 10^{-13}$ s is a correlation time connected to intermolecular interactions, e.g. rapid translational motion [11,12]. The exponential time constants are well supported by experimental results for times >> τ_c [4]. On the other hand, finite experimental time resolution prevents information on times $\lesssim \tau_c$. Eqs. (2) and (3) are supplemented by the classical wave equation for the light field:

$$\Delta E - \frac{1}{c^2}\frac{\partial^2}{\partial t^2}(\mu E) = \frac{4\pi}{c^2}\frac{\partial^2}{\partial t^2} P^{NL} \tag{4}$$

μ denotes the refractive index of the medium; the nonlinear polarization P^{NL} represents the stimulated scattering off the vibrational mode of interest. The coherent amplitude $\langle q_j\rangle$ leads to the induced dipole moment $\partial\alpha/\partial q \langle q_j\rangle E$ of the individual molecule. Summing over the molecular subensembles with number density $f_j N$ yields

$$P^{NL} = N \frac{\partial\alpha}{\partial q} E \sum_j f_j \langle q_j\rangle \tag{5}$$

Eqs. (2) to (5) represent a complete set describing the generation of an intense Stokes pulse and of a coherent material excitation by stimulated scattering.

We have made numerical studies of (2) to (5) using material parameters relevant for the experiments below. Some of the findings are briefly summarized here:

Starting from quantum noise, a rapidly rising and decaying Stokes field develops in the high gain stimulated scattering process. Our calculation shows that the Stokes pulse is highly coherent; i.e. a field of constant phase builds up. This point is important for the generation of a material excitation with well defined wave vector. Examples of the time dependence of the generated Stokes pulse are shown in Fig. 2. The intensity of the Stokes light was evaluated and plotted as a function of time in units of t_p, the half-power pulse width of the incident pulse.

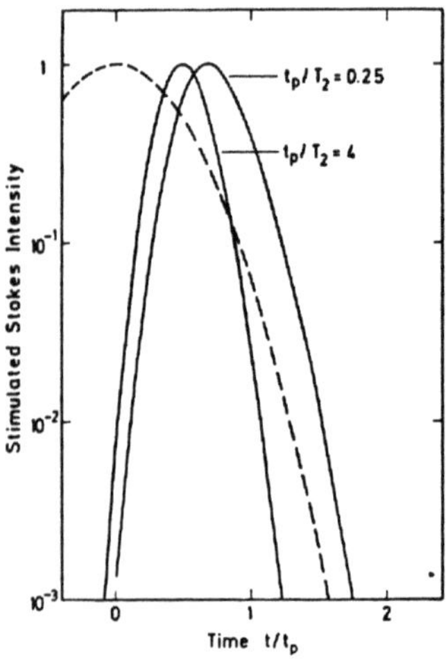

Fig.2 Generated Stokes pulse vs time for a moderately transient. $t_p/T_2 = 4.0$, and a highly transient, $t_p/T_2 = 0.25$, situation. The incident laser pulse of duration t_p is assumed to be Gaussian in the calculation (broken curve).

Fig.3 Calculated coherent excitation of a vibrational system with a Gaussian distribution of transition frequencies (bandwidth $\delta\omega$). The two curves respectively represent the total system and a subensemble with negligible spread of transition frequencies.

Fig.4 Excess population n of the first excited vibrational state vs time. n decays exponentially with the population relaxation time T_1. Parameters are the ratios t_p/T_2 and T_1/T_2, where t_p and T_2 are the duration of the pump pulse and the dephasing time, respectively.

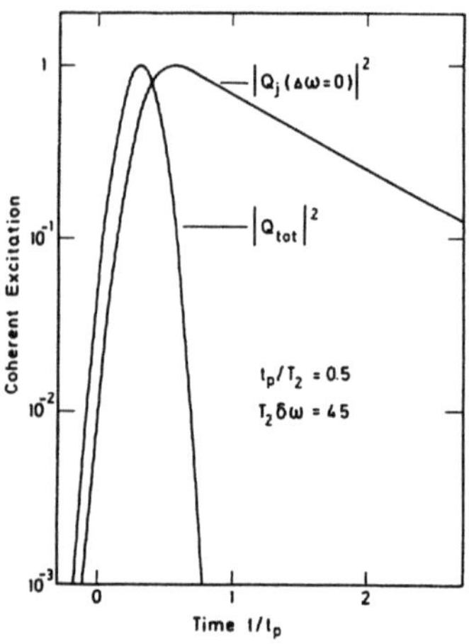

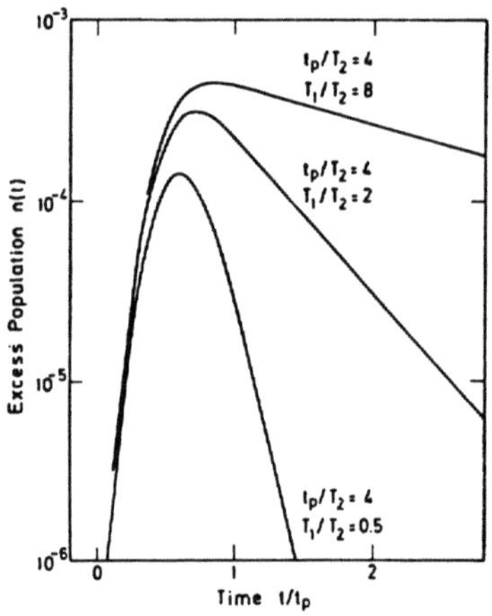

The time evolution of the vibrational system depends on the distribution of transition frequencies. For a homogeneously broadened system ($\Delta\omega_j = \omega_0 - \omega_j = 0$) simple exponential decay with time constant $T_2/2$ is predicted for the excitation $|\langle q\rangle|^2$. A vibrational system with a distribution of transition frequencies is considered in Fig. 3. The total excitation $|Q_{tot}|^2 = |\Sigma f_j \langle q_j\rangle|^2$ steeply rises during the highly nonlinear generation process and disappears very rapidly with non-exponential time dependence, due to the broad frequency distribution. The individual component, on the other hand, decays more slowly with time constant $T_2/2$ similar to the time dependence of the homogeneous situation.

The second aspect of the vibrational excitation is illustrated in Fig. 4. We recall that for each Stokes photon emitted one molecule is promoted from the ground to the first vibrational state. The excess population of the upper vibrational state is plotted versus normalized time. The numerical examples are evaluated for homogeneous line broadening. Similar results were obtained

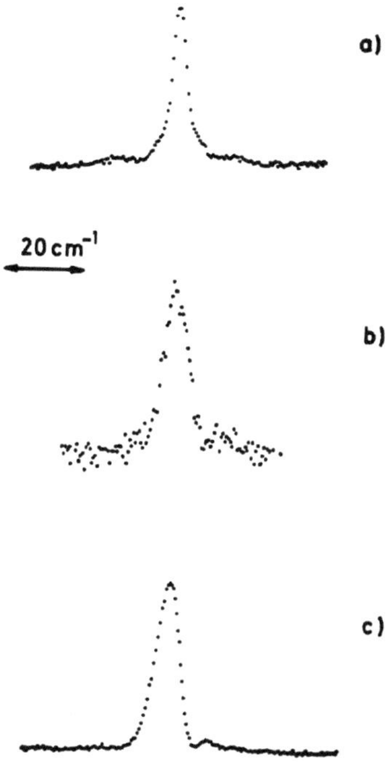

Fig.5 Optical spectra of three different pulses: (a) incident laser pulse; (b) stimulated Stokes pulse generated by the $\tilde{\nu}$ = 2835 cm^{-1} mode in CH_3OH; (c) stimulated Stokes pulse generated by the $\tilde{\nu}$ = 2935 cm^{-1} mode in $(CH_2OH)_2$.

for the inhomogeneous situation. Most important are the exponential parts of the curves in Fig. 4 for larger values of t. The individual energy relaxation time T_1 determines the decay of n. The exponential asymptote is quickly reached after the maximum for $T_1 \gtrsim t_p/2$.

Our theoretical results on the monochromaticity of the Stokes pulse are substantiated by the experimental findings of Fig. 5 [13]. Three spectra taken with a grating spectrometer are depicted. The spectral width of the input laser pulse (Fig. 5 a) is ≃ 5 cm^{-1}, the Fourier transform of the time dependence with $t_p \simeq 3.5$ ps. Figs. 5 b and c show stimulated Stokes spectra in methanol and ethylene glycol, respectively. Both spectra have the same bandwidth of ≃ 8 cm^{-1}. This value is determined by the duration of the Stokes pulse which is somewhat shorter than the laser (see Fig. 2). Theory predicts the spectra in Fig. 5 b and c to be approximately the same in spite of the very different spontaneous Raman bandwidths of ∿ 20 cm^{-1} and ∿ 60 cm^{-1}, respectively.

3. Coherent Raman Probe Scattering

The coherent excitation discussed in Section 2 is monitored by a delayed probe pulse. The amplitudes $\langle q_j \rangle$ give rise to a macroscopic polarization $P = N\partial\alpha/\partial q\, E_2 \Sigma f_j \langle q_j \rangle$, where E_2 denotes the electromagnetic field of the probe pulse [4,14]. The polarization emits a scattering emission, which is shifted by the vibrational frequency to larger (anti-Stokes) or smaller (Stokes) frequencies. This process is called coherent Raman scattering. The molecules vibrate with a defined spatial phase relation described by a wave vector $k_0 = k_L - k_S$, prepared in the excitation process by the laser and stimulated

Stokes pulse. The subensembles j behave like oscillating three-dimensional phase gratings. Scattering-off these gratings occurs when the k-matching condition for the wave vectors of the incident probe pulse (k_{L2}), the scattered light and the vibrational components is fulfilled [4]. For Stokes scattering we have the condition:

$$|\Delta k_j|\Delta\ell = |k_{L2}-k_{Sj}-k_o|\Delta\ell \leq 0.5 \tag{6}$$

k_{Sj} denotes the wave vector of the Stokes component generated by subgroup j; $\Delta\ell$ is the effective interaction length.

A similar condition holds for anti-Stokes scattering of the probe pulse. Due to dispersion effects in condensed matter, Δk_j in (6) is different for different subgroups j. Those molecules with approximate k-matching, $\Delta k_j \Delta\ell \leq 0.5$, determine the coherent scattering signal. Components with large mismatch $\Delta k_j \Delta\ell >> 0.5$ give negligible contributions, only. The frequency interval of "allowed" transition frequencies depends on the interaction length $\Delta\ell$. We wish to distinguish two limiting cases (13,14):

(i) Selective k-matching. For large values of $\Delta\ell$($\gtrsim$ 1 cm) a highly selective k-vector geometry can be adjusted. In this way probe scattering of a molecular subensemble with negligible spread of transition frequencies may be observed. This subgroup is expected to decay with a characteristic dephasing time T_2.

(ii) Non-selective k-matching. For a small interaction length ($\Delta\ell$ << 1 cm), the k-matching is not sensitive to the different transition frequencies. In this case, coherent superposition of various subgroups occurs and fast dephasing by the distribution of frequencies can be observed; i.e., the total excitation of Fig. 3 may be studied.

Experimentally we detect scattering signals of several 10^{-12} s which is considerably shorter than the time resolution of conventional photo-detectors. Therefore, time integrated scattering signals $S^{coh}(t_D)$ are measured at the appropriate Stokes (or anti-Stokes) position versus time delay between excitation and probing pulse. For the simple case of homogeneous broadening $\omega_j = \omega_o$ and for a probe pulse of amplitude $E_{L2}(t-x/v)$ we find the coherent scattering signal at the end of the sample ($x = \ell$):

$$S^{coh}(t_D) \propto \frac{\Delta\ell^2}{1+(2\Delta k\Delta\ell)^2} \int_{-\infty}^{\infty} dt |E_{L2}(\ell, t-t_D) \ \langle q\rangle|^2 \tag{7}$$

Eq. (7) represents a convolution of the coherent excitation with the probe intensity. S^{coh} reproduces the time dependence of the coherent excitation for sufficiently small values of probe pulse duration.

Our experimental system for coherent probe scattering with collinear wave vector geometry is depicted in Fig. 6 [13]. For the generation of the ultrashort light pulses we use a Nd:glass laser system consisting of a mode-locked laser oscillator, an electrooptic switch to select a single pulse, and a laser amplifier. We emphasize the need to work with pulses of well defined duration, pulse shape and peak intensity. Bandwidth limited pulses [15] are required to achieve high k-vector resolution. After frequency doubling in a KDP crystal we work with pulses of frequency ν_L = 18,910 cm^{-1}, width 5 cm^{-1}, duration $t_p \simeq$ 3.5 ps and approximately Gaussian shape. The probing pulse is generated

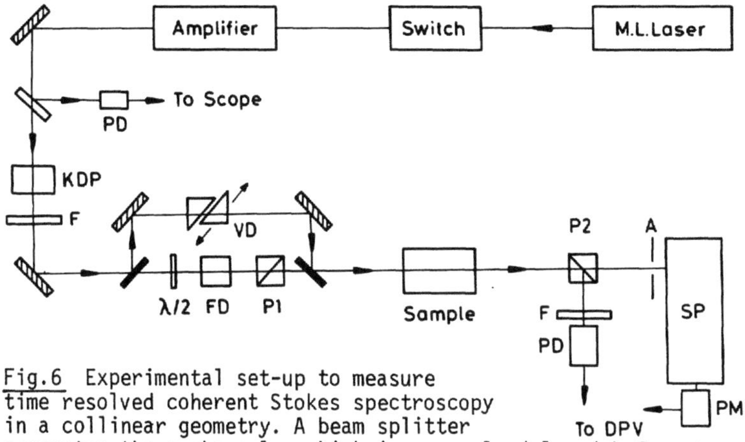

Fig.6 Experimental set-up to measure time resolved coherent Stokes spectroscopy in a collinear geometry. A beam splitter generates the probe pulse which is properly delayed before travelling collinearly with the exciting pulse. The scattered signals of the two pulses are separated by the two polarizers P1 and P2.

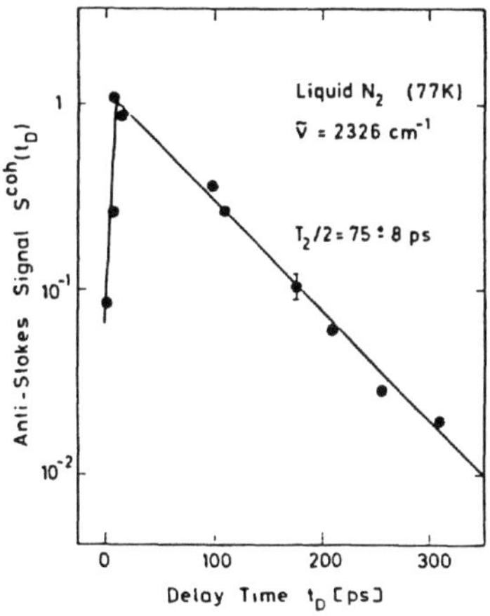

Fig.7 Coherent probe scattering signal of liquid N_2 versus delay time t_D. The dephasing time of the fundamental mode at $\tilde{\nu}$ = 2326 cm^{-1} is found to be T_2 = 150 ps.

by a first beam splitter and delayed in a variable delay system. A second beam splitter directs the probe pulse collinearly with the exciting pulse through the medium. The intensity of the probe pulse is less than 1 % of the excitation pulse. The pump pulse passed a $\lambda/2$ plate and polarizer P1 defining the plane of polarization of the excitation process. The stimulated Stokes emission of the pump pulse is effectively blocked by a factor of $\geq 10^5$ with the help of the second polarizer P2. The probe pulse and the coherent Stokes signal pass P2 without significant attenuation. An aperture, A, determines the angle of acceptance γ of the detection system consisting of a spectrometer and a photomultiplier.

An experimental result for the simple case of homogeneous line broadening is shown in Fig. 7. The fundamental vibrational mode of liquid N_2 at 2326 cm^{-1} is investigated [16]. From the decay of the signal curve we deduce a dephasing time of T_2 = 150 ps.

It is interesting to relate the exponential relaxation with time T_2 to the line shape observed in conventional Raman spectroscopy. For the homogeneous system the exponential decay corresponds to a Lorentzian line with full width (FWHM in units of cm^{-1})

$$\delta\tilde{\nu}_{hom} = (\pi c T_2)^{-1} \tag{8}$$

For liquid N_2 we deduce from (8) and the measured T_2 a dynamic line broadening of 0.070 cm^{-1}. This number is in excellent agreement with the linewidth of the isotropic scattering component observed in spontaneous Raman spectroscopy [17]; i.e. the dephasing time T_2 fully accounts for the spectroscopic line broadening.

It is interesting to note that an exceptionally long population lifetime T_1 of the fundamental mode of approximately one minute was reported by several authors [18]. As a result, population decay does not effect the dephasing process of liquid N_2. Similar results were reported for mixtures of N_2 with liquid Ar. A comparatively small variation of T_2 with concentration and temperature was found [19] in good agreement with spectroscopic findings. In Table I dephasing times T_2 for a number of vibrational modes are listed. The experimental values are compared with numbers evaluated from spectroscopic data using (8).

There are vibrational modes where evidence exists of a distribution of transition frequencies. For instance, liquids with strong hydrogen bonding show extended OH-bands which are believed to be inhomogeneously broadened. Spontaneous spectroscopic techniques do not allow a separation of rapid dephasing processes and longer-lived frequency shifts of liquid molecules. Our novel picosecond Raman technique allows for the first time to study this problem.

Table I. Measured dephasing times of several modes and molecules. Comparison with spontaneous Raman linewidth.

	$\tilde{\nu}_o$ [cm^{-1}]	$T_2/2$ [10^{-12} sec] measured	$T_2/2$ [10^{-12} sec] from linewidth
N_2 (77K)	2326	75 ± 8	79 ± 8
CCl_4	459	3.5 ± 0.4	3.8 ± 0.5
$SiCl_4$	425	3.0 ± 0.5	2.8 ± 0.5
$SnCl_4$	368	2.8 ± 0.3	2.5 ± 0.5
CH_3CCl_3	2939	1.1 ± 0.2	1.2 ± 0.1
$SnBr_4$	221	3.0 ± 0.3	> 1.6
CH_3OH	2837	2.3 ± 0.5	> 0.2
$(CH_2OH)_2$	2935	3.0 ± 0.5	> 0.1

To demonstrate the potential of the technique we have studied vibrational bands with well-known discrete substructure due to isotopic multiplicity. Many liquids are composed of several isotopic species. $SnBr_4$ is mentioned here as an example. Two isotopes, ^{79}Br and ^{81}Br, give rise to five lines of the totally symmetric vibration around 221 cm^{-1}. The isotopic structure is not resolved, in the conventional Raman spectrum. With our coherent probing technique with selective wave-vector geometry we have observed exponential dephasing of a single isotope component with time constant T_2 = 6 ps [14]. This time constant corresponds to a line broadening of the single isotope species of 1.8 cm^{-1}, significantly smaller than the total Raman bandwidth of 3.2 cm^{-1} of $SnBr_4$.

In addition, coherent probe scattering with non-selective k-matching was investigated. In this measurement coherent superposition of the $SnBr_4$ isotope components was observed yielding a frequency spacing of $\Delta\omega/2\pi c \approx 0.7$ cm^{-1} in accordance with theoretical estimates. Additional support of the picosecond data is obtained by a comparison with the spontaneous Raman band shape. The results clearly demonstrate that the selective k-matching technique allows to distinguish dephasing via rapid processes from the loss of phase correlation by a distribution of vibrational frequencies. Estimates show that the experimental k-resolution is equivalent to a frequency resolution of better than 0.5 cm^{-1}.

We have investigated several liquids with our picosecond k-matching technique. Two types of measurements are performed to study the coherent scattering of inhomogeneously broadened Raman lines. First, we investigate the spectra of different scattered light pulses, and second, we measure the time dependence of the relaxation process.

The following spectral data give direct information on the material excitation by transient stimulated Raman scattering. The CH_2 stretching mode of ethylene glycol at 2935 cm^{-1} was chosen to demonstrate the strong effect of the transient stimulated Raman excitation of a very broad (60 cm^{-1}) vibrational band. A large Stokes divergence, typical for nonselective k-matching geometry, was used in a short cell of 1 cm length. Under these conditions the probe pulse interacts with many excited vibrational components with good scattering efficiency. In Fig. 8 four different spectra are presented [4].
(a) The spectrum of the incident laser pulse at $\tilde{\nu}_L$ = 18910 cm^{-1} shows a linewidth of $\delta\tilde{\nu}_L \approx 5$ cm^{-1}, which is the Fourier transform of the pulse duration of 3.5 ps.
(b) The spectrum of the stimulated Stokes emission has a bandwidth of $\delta\tilde{\nu} \approx$ 8 cm^{-1}. Theory predicts that the molecules of the vibrational distribution are driven with the frequency $\omega_L - \omega_S$ during the early part of the excitation process. As a result, the stimulated Stokes spectrum of the excitation process does not reflect the molecular distribution.
(c) The spectrum of the Stokes scattered probe light, $S^{coh}(t_D=0)$, is taken at the same time as the previous spectrum (b) using a perpendicular polarization direction. As expected from the isotropic Raman tensor of the normal mode investigated here, the spectrum of $S^{coh}(t_D=0)$ has the same bandwidth as the previous one in (b).
(d) The Stokes scattered pulse $S^{coh}(t_D=5$ ps) sees a completely different situation at the delay time t_D = 5 psec. For times $t \gtrsim t_D$ the molecules relax freely with their own resonance frequencies. A bandwidth of ~18 cm^{-1} is deduced from the spectrum of Fig. 8 (d). This bandwidth will be discussed below after the time-resolved investigations.

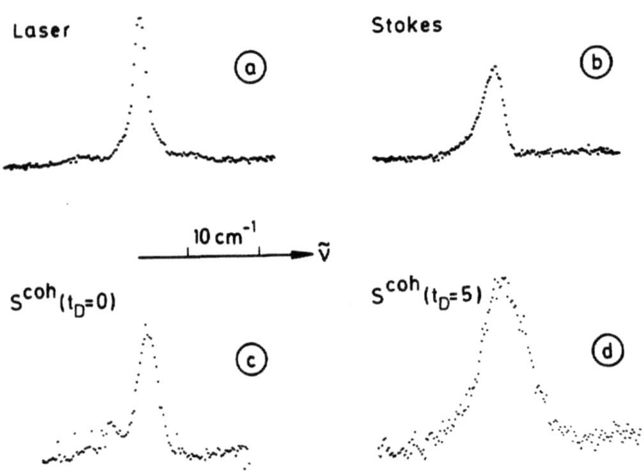

Fig.8 Optical spectra of four different pulses: (a) incident laser pulse; (b) generated Stokes pulse; (c) Stokes scattered probe pulse $S^{coh})t_D=0)$ without delay time; (d) Stokes scattered probe pulse $S^{coh}(t_D=5\ ps)$ taken 5 ps after the maximum of the exciting laser pulse. Note the broad spectrum of the freely relaxing molecular system. Measurements were made with the CH_3-stretching mode at $\tilde{\nu}$ = 2935 cm^{-1} of ethylene glycol, cell length 1 cm.

We have measured the spectra of the Stokes probe scattering for the same delay times of t_D = 0 and t_D = 5 psec, but with a selective k-vector geometry. The spectrum for t_D = 0 is found to be $\delta\tilde{\nu}$ = 8 cm^{-1}, which, in analogy to Fig. 8 (c), is the Fourier transform of the ultrashort Stokes pulse. The spectrum measured with a delay time of t_D = 5 psec showed a spectral width of 8 cm^{-1} in contrast to Fig. 8 (d). This spectrum is considered to result from the interaction of the probe pulse with a relaxing molecular group of a small spread of transition frequencies.

Fig. 9 presents an example for the time evolution observed for an inhomogeneously broadened mode. The CH_3-stretching mode of methanol at 2835 cm^{-1} is studied [13]. The spontaneous Raman band of this vibration gives a linewidth (FWHM) of $\delta\tilde{\nu}_{spon} \approx 20$ cm^{-1}. In Fig. 9 a highly selective k-matching situation is used with a sample length of ℓ = 10 cm and with a small divergence of the Stokes beam of γ = 3 mrad. The data indicate an exponential decay with a time constant of T_2 = 4.6 ± 1 ps. The time dependence is different for less selective k-matching (Fig. 9 b) using a shorter cell of 1 cm and a larger Stokes divergence of 10 mrad. Under these experimental conditions the coherent scattering signal disappears rapidly.

The comparison of Fig. 9 a and b provides direct evidence that the Raman band of width 20 cm^{-1} represents a distribution of vibrational frequencies. Our value of T_2 suggests a homogeneous linewidth which is smaller than the Raman band by a factor of approximately ten. Similar results were also obtained for the CH_2-mode of $(CH_2OH)_2$ at 2935 cm^{-1} where we measure T_2 = 6±1 ps.

While inhomogeneous broadening is not surprising for the broad Raman bands of systems with strong hydrogen bonding, the question arises about the situation in weakly associated liquids. As an example we briefly summarize results on deuterated benzene C_6D_6 [20]. The totally symmetric ring vibration at 945 cm^{-1} gives rise to an intense and very narrow Raman line of width 1.5 cm^{-1},

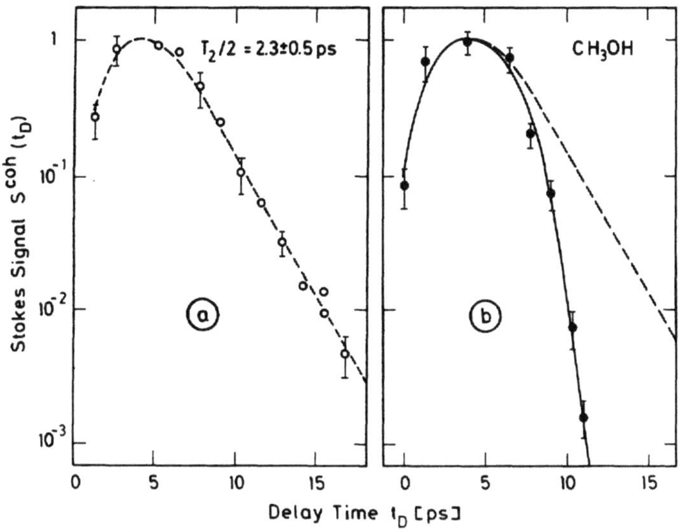

Fig.9 (a) Coherent Stokes scattered probe signals vs delay time measured in a selective k-vector geometry. The vibrational mode of CH_3OH at 2835 cm^{-1} is investigated. We observe a dephasing time $T_2/2$ = 2.3 psec corresponding to a homogeneous linewidth of 2.3 cm^{-1}. (b) Coherent Stokes signals for a less selective k-vector geometry. The signal decays rapidly on account of destructive interference of excited molecules with a distribution of transition frequencies.

which has been studied repeatedly by conventional Raman equipment [21]. We investigated this mode with our picosecond apparatus of Fig. 6 and measured the coherent probe scattering under a selective k-vector geometry. Monitoring the signal decay over a factor of 200 we observe approximately exponential relaxation with a time constant of T_2 = 10 ± 1 ps. For non-selective k-matching conditions, on the other hand, we find a notably accelerated signal decay indicating additional dephasing by a distribution of vibrational frequencies. For an assumed Gaussian frequency distribution, a theoretical curve was fitted to the picosecond data in order to determine the linewidth of the inhomogeneous broadening, yielding a value of $\delta\tilde{\nu}_{inh}$ = 0.6 ± 0.2 cm^{-1}. The picosecond results are supported by a comparison with the spontaneous Raman band. The value of T_2 = 10 ps corresponds to a homogeneous broadening of Lorentzian shape and width $\delta\tilde{\nu}_{hom} \approx$ 1.0 cm^{-1}. Convolution of the homogeneous (1.0 cm^{-1}) and inhomogeneous (0.6 cm^{-1}) contributions leads to a Voigt profile which agrees well with the measured Raman band contour (polarized scattering component) of this mode.

4. Incoherent Raman Probe Scattering

Spontaneous anti-Stokes probe scattering allows the study of the instantaneous population of a vibrational level. With this technique it was possible, for the first time, to observe population lifetimes, energy transfer and energy redistribution [22-26]. Time constants between 1 ps and 100 ps were measured for different dynamical processes in a number of polyatomic molecules. The experimental method consists of two steps. First, a powerful short light pulse traverses the sample and excites the vibrational mode of interest. Stimulated Raman scattering or resonant infrared absorption can produce an excess population $n_i(x,t)$ of the excited vibrational state. A second weak pulse of different frequency probes the vibrational excitation via spontaneous anti-Stokes Raman scattering. The scattering intensity observed under a large scattering angle is proportional to the instantaneous population of the upper vibrational state and to the incident laser intensity. Experimentally, we observe a time-integrated signal denoted by S_i^{inc} [4]:

$$S_i^{inc}(t_D) = \frac{c\mu_A}{8\pi} N \left(\frac{d\sigma}{d\Omega}\right)_i \Delta\Omega \int_{-\Delta x/2}^{+\Delta x/2} dx \int_{-\infty}^{\infty} dt\, n_i(x,t) |E_{L2}(t-t_D)|^2 \qquad (9)$$

$\vec{x}$ is the propagation direction of the excitation and probe pulse. $\Delta\vec{x}$ represents the observed local interval in the sample. N is the number density of the molecules in the liquid. $(d\sigma/d\Omega)_i$ denotes the spontaneous Raman cross section of the observed mode i and $\Delta\Omega$ to the solid angle of acceptance of the probing pulse. (9) represents a convolution of the probe pulse of delay time t_D with the vibrational population n_i.

The experimental system for measuring the population lifetime T_1 is shown in Fig.10 [26]. A powerful single picosecond pulse at 9455 cm^{-1} is generated with a mode-locked Nd:glass laser system. In a first kind of experiments, the laser pulse traverses the sample excitating the normal mode with largest Raman gain via stimulated Raman scattering. In a second type of experiments, the laser pulse passes through two $LiNbO_3$ crystals (shown in the Figure) producing an infrared pulse of tunable frequency [27]. At 3000 cm^{-1}, the pulse duration is ~3 ps and the pulse energy corresponds to 10^{15} quanta. The infrared pulse with adjusted frequency is focused into the sample cell, and absorbed by the infrared-active molecular vibration, generating the desired

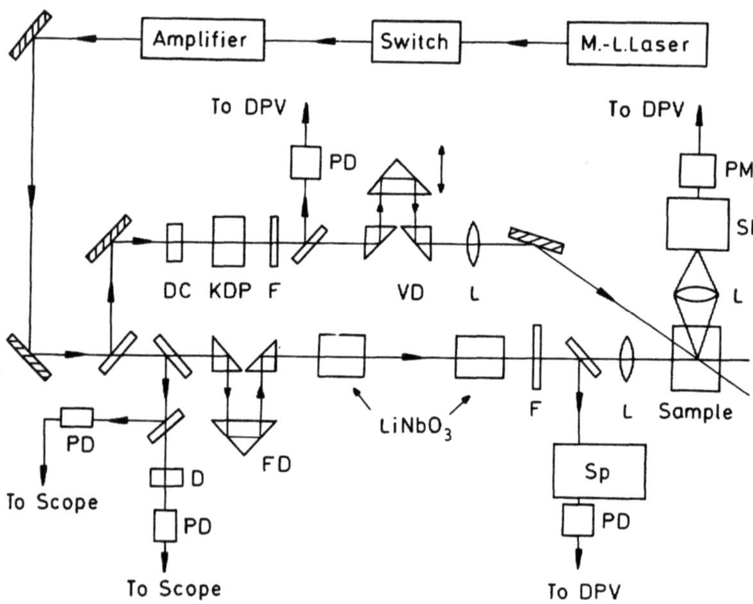

Fig.10 Schematic of the experimental system to measure population lifetimes. A picosecond pulse of a Nd:glass laser system serves directly as excitation pulse or generates ultrashort infrared pulses in a single pass through two $LiNbO_3$ crystals. Tuning of the infrared frequencies is achieved by adjusting the crystal orientation. Part of the laser pulse is converted to the second harmonic frequency, properly delayed and interrogates the vibrationally excited molecules.

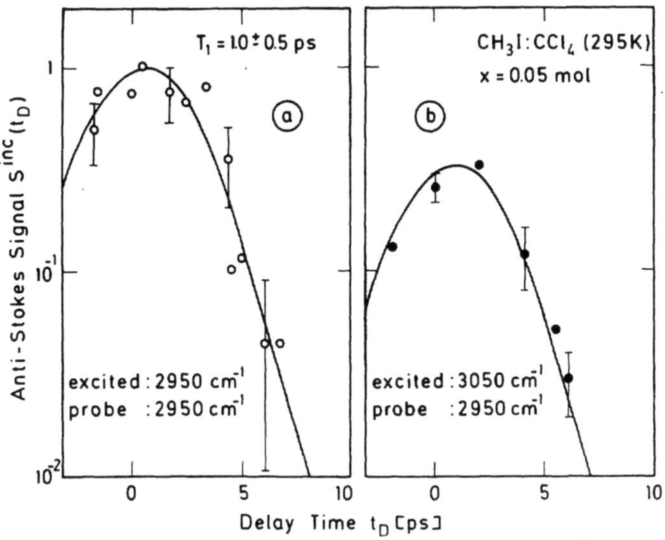

Fig.11 Incoherent anti-Stokes probe scattering $S^{inc}(t_D)$ of the symmetric CH_3-stretching mode at 2950 cm^{-1} of CH_3I dissolved in CCl_4, mole fraction 0.05. (a) The mode at 2950 cm^{-1} is directly populated by the resonant infrared excitation pulse. (b) The asymmetric CH_3-stretching mode at 3050 cm^{-1} is excited by the tunable pump pulse. The excess population of the vibration at 2950 cm^{-1} is observed indicating rapid energy redistribution between the neighboring modes.

excited-state population. A beam splitter in the input beam produces a weak pulse, which is converted to the second-harmonic frequency and serves as a probe pulse with variable delay t_D. The spontaneous anti-Stokes Raman signal $S^{inc}(t_D)$ is observed at a scattering angle of 90° with a monochromator or a set of dielectric filters. With approximately 10^{15} photons in the probe pulse, one estimates for an occupation number $n \sim 10^{-3}$ an anti-Stokes signal of 10^3 photons in a solid angle $\Delta\Omega \simeq 0.2$ sr.

We have investigated the population lifetime and energy relaxation processes in a number of pure liquids and in liquid mixtures. Fig. 11 shows data of a diluted solution of CH_3I [26]. The symmetric ν_1 or asymmetric ν_4 CH_3-stretching mode is excited by resonant absorption (Figs. 11 a and b, respectively); probe scattering of the ν_1-mode is observed in the two measurements. The results indicate rapid decay of the population of the ν_1-mode with T_1 = 1 ps and energy redistribution between the two neighboring modes with a time constant of $\sim$ 1.5 ps. For an explanation of the fast vibrational depopulation rotational coupling, Fermi resonance, and Coriolis coupling are expected to be important [28]. Theoretically, rapid equilibration between the ν_1 and ν_4 modes and a fast rate ($k - 5\times10^{12}$ s) for the transition from the ν_1 mode to the adjacent overtone $2\nu_5$ of the CH_3 bending vibration are predicted.

Experimental evidence for energy transfer between CH-stretching and bending modes is presented in Fig. 12. The symmetric CH_3 vibration of CH_3CCl_3 at ν_H = 2939 cm^{-1} was excited by stimulated Raman scattering and the spontaneous anti-Stokes probe scattering measured at two frequencies [24].

The upper curve (open circles) in Fig. 12 corresponds to a frequency shift of 2939 cm^{-1}; it gives information on the degree of occupation of the primary

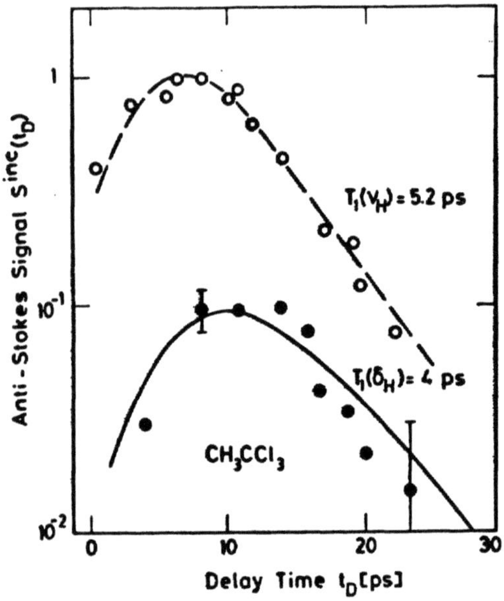

Fig.12 Incoherently scattered probe signals measured in CH_3CCl_3. The upper curve (o) represents the temporal occupation of the laser-excited ν_H vibration at 2939 cm^{-1}. The lower curve (•) represents the occupation of the δ_H vibration at 1450 cm^{-1}. The curves through the experimental points are calculated.

excited vibrational mode. The value of T_1 = 5.2 psec is the population lifetime in the neat liquid. Of interest are the scattering data at a frequency shift of ~1450 cm^{-1} (full circles). These data points indicate an excess population of the bending mode(s) generated by an internal transfer process. Comparing model calculations with the measured ratio of the maxima of the initial and the lower mode suggest a high quantum yield; i.e. the transfer $\nu_H \rightarrow 2\delta_H$ is found to be the dominant decay channel of the higher ν_H mode.

5. Discussion of the Relaxation Processes

The dephasing data are discussed first. Exponential decay is observed over several orders of magnitude represented by the time constant T_2. To understand this behavior in terms of stochastic theory [29], relaxation in the fast modulation limit has to be considered. In other words, homogeneous line broadening with motional narrowing is suggested by the experimental data. We note that a time scale of ≲ 1 ps is required for the individual interaction event in the fast modulation limit. The small amplitude translational motion of the molecules in their liquid cages (time scale 10^{-13} s) and/or the fast molecular rotation are important factors for an exponential dephasing time T_2 [11,12, 31]. For a recent review, we refer to [30]. Some general aspects of dephasing are discussed in [37].

The following processes have been suggested to contribute to the measured T_2 values of polyatomic molecules:

(i) "Direct" dephasing by quasi-elastic interaction of nearest neighbors via the intermolecular potential [11,12,31-34]. Rough numerical estimates neglecting the attractive part of the interaction and off-axis collisions agree with experimental results within a factor of three [11]. Improved agreement is reported when the coupling with rotational motion and vibrational anharmonicities are included [33]. Good agreement with experimental data was reported for computer simulations of liquid N_2 [34].

(ii) Indirect dephasing by anharmonic coupling to low frequency vibrational modes [35,36].

(iii) Resonant exchange of vibrational quanta via the attractive part of the intermolecular potential of transition dipole-dipole interaction [38,39].

(iv) Energy transfer to neighboring vibrations [40,41].

Processes (iv) are distinguished from the mechanisms (i) to (iii) measuring the time evolution of the excited state population (T_1-techniques). For the symmetric CH-stretching mode of CH_3CCl_3 at 2939 cm^{-1} (see Table 1), theoretical estimates suggest process (i) as dominant T_2 mechanism while process (ii) appears to be unimportant [36]. The small concentration dependence of the linewidth rules out a significant contribution of (iii). Approximately 20 % of the total line broadening is due to population decay (iv). For many vibrational modes the dominant mechanism of T_2 cannot be stated definitely at the present time.

Different mechanisms have been considered for the energy decay and redistribution process mentioned above. We wish to distinguish collision-induced intramolecular transitions and intermolecular transitions which can be short-range collision induced or long-range resonance transfer processes.

(i) The intramolecular collision-induced vibrational transitions have been interpreted in terms of extended SSH models including mode matching and rotational coupling; i.e. details of the normal mode displacements and of the rotational motion are considered [40,41]. Calculations show that a large amount of the energy mismatch of a vibrational transition is transferred to molecular rotation.

(ii) Fermi resonance and Coriolis coupling. Due to anharmonic coupling normal modes with same symmetry can mix. As a result the transition probability between stretching modes and bending modes of molecular groups such as CH_3- is considerably augmented. Coriolis coupling additionally allows coupling of vibrational states of different symmetry [42].

(iii) Resonance transfer. Concentration studies of systems, e.g. CH_3CCl_3 and C_2H_5OH show that intermolecular transitions are involved. Long range transfer occurs via dipole-dipole coupling [43].

We turn now to the question of inhomogeneous broadening. It was discussed above that for the rapid dephasing processes the fast translational and rotational motion of the molecules in their liquid cages appears to be important. In addition, the picosecond data present evidence for a distribution of vibrational frequencies in several strongly and weakly associated liquids. The frequency distribution is found to exist on the picosecond time scale, possibly not changing notably for several 10^{-11} s or longer. The different frequencies of individual molecules may be interpreted as variations of the solvent shift produced by the specific molecular environment. Frequency shifts of the band centers are well understood [44] and have been related to inhomogeneous line broadening in the slow modulation limit [45]. The important point here is the slow time scale of the interaction required to explain the experimental findings. A preliminary physical picture suggests that long range intermolecular interaction and/or sufficiently long-lived cluster of molecules may be responsible for a slowly varying frequency distribution in strongly and weakly associated liquids.

6. Concluding Remarks

This chapter discusses several experiments which study ultrafast vibrational processes and time-resolved Raman spectroscopy in liquids. It is shown that new experimental methods have been devised which benefit from the specific molecular excitation and time resolution provided with high quality picosecond light pulses. New and detailed information on different aspects of molecular dynamics is obtained.

References

1. R.T. Bailey, in "Molecular Spectroscopy", Vol. 2 (The Chemical Society, London, 1974)
2. A.J. DeMaria, D.A. Stetser and J. Heyman, Appl. Phys. Lett. 8, 174 (1966); W. Schmidt and F.P. Schäfer, Phys. Lett. A26, 558 (1968); D.J. Bradley, A.J.F. Durant, F.O'Neill and B. Sutherland, Phys. Lett. 30A, 535 (1969); E.P. Ippen, C.V. Shank, and A. Dienes, Appl. Phys. Lett. 21, 348 (1972).
3. A. Laubereau, L. Greiter, and W. Kaiser, Appl. Phys. Lett. 25, 87 (1974); A.H. Kung, ibid 25, 653 (1974); for recent summaries see "Picosecond Phenomena", eds. C.V. Shank, E.P. Ippen and S.L. Shapiro, Springer (Berlin 1978).
4. A. Laubereau and W. Kaiser, Rev. Mod. Phys. 50, 607 (1978).
5. A. Laubereau, A. Seilmeier, and W. Kaiser, Chem. Phys. Lett. 36, 232 (1975).
6. A. Seilmeier, W. Kaiser, A. Laubereau and S.F. Fischer, Chem. Phys. Lett. 58, 225 (1978); A. Seilmeier, W. Kaiser and A. Laubereau, Optics Commun. 26, 441 (1978).
7. R.G. Gordon, J. Chem. Phys. 40, 1973 (1964); 42, 3658 (1965); 43, 1302 (1965); S. Bratos and E. Marechal, Phys. Rev. A4, 1078 (1971); L.A. Nafie and W.L. Peticolas, J. Chem. Phys. 57, 3145 (1972).
8. J.A. Giordmaine, W. Kaiser, Phys. Rev. 144, 676 (1966).
9. M. Maier, W. Kaiser, and J.A. Giordmaine, Phys. Rev. 177, 580 (1969).
10. C.P. Slichter, "Principles of Magnetic Resonance" (Harper and Row, New York, 1963).
11. S.F. Fischer, A. Laubereau, Chem. Phys. Lett. 35, 6 (1975).
12. W.G. Rothschild, J. Chem. Phys. 65, 2958 (1976).
13. A. Laubereau, G. Wochner, and W. Kaiser, Chem. Phys. 28, 363 (1978).
14. A. Laubereau, G. Wochner, and W. Kaiser, Phys. Rev. A13, 2212 (1976).
15. W. Zinth, A. Laubereau, W. Kaiser, Opt. Commun. 22, 161 (1977).
16. A. Laubereau, Chem. Phys. Lett. 27, 600 (1974).
17. W.R.L. Clements, B.P. Stoicheff, Appl. Phys. Lett. 12, 246 (1968); M. Scotto, J. Chem. Phys. 49, 5362 (1968).
18. W.F. Calaway and G.E. Ewing, Chem. Phys. Lett. 30, 485 (1975); J. Chem. Phys. 63, 2842 (1975); S.R. Brueck, R.M. Osgood, Jr., Chem. Phys. Lett. 39, 568 (1976).
19. H.M.M. Hesp, J. Langelaar, B. Bebelaar, J.D.W. van Voorst, Phys. Rev. Lett. 39, 1376 (1977); S.A. Akhmanov, N.I. Koroteev, R. Yu. Orlov, I.L. Shumay, JETP Lett. 23, 276 (1976).
20. A. Laubereau, G. Wochner and W. Kaiser, to be published
21. M.J. Colles and J.E. Griffiths, J. Chem. Phys. 56, 3384 (1972).
22. A. Laubereau, D. von der Linde, and W. Kaiser, Phys. Rev. Lett. 28, 1162 (1972).
23. R.R. Alfano and S.L. Shapiro, Phys. Rev. Lett. 29, 1655 (1972).
24. A. Laubereau, L. Kirschner and W. Kaiser, Opt. Comm. 19, 182 (1973).

25. A. Laubereau, G. Kehl and W. Kaiser, Optics Commun. 11, 74 (1974); P.R. Monson, L. Patumtevapibal, K.J. Kaufmann, and P.W. Robinson, Chem. Phys. Lett. 28, 312 (1974).
26. K. Spanner, A. Laubereau, and W. Kaiser, Chem. Phys. Lett. 44, 88 (1976).
27. A. Seilmeier, K. Spanner, A. Laubereau, and W. Kaiser, Opt. Commun. 24, 237 (1978).
28. A. Laubereau, S.F. Fischer, K. Spanner and W. Kaiser, Chem. Phys. 31, 335 (1978).
29. R. Kubo, in "Fluctuation, Relaxation and Resonance in Magnetic Systems", ed. D.ter Haar (Oliver and Boyd, Edinburgh, 1962).
30. D.W. Oxtoby, Adv. Chem. Phys., 40, 1 (1979) and to be published.
31. P.A. Madden, R.M. Lynden-Bell, Chem. Phys. Lett. 38, 163 (1976).
32. D.W. Oxtoby, S.A. Rice, Chem. Phys. Lett. 42, 1 (1976).
33. R.K. Wertheimer, Mol. Phys. 35, 257 (1978).
34. D.W. Oxtoby, D. Levesque and J.J. Weis, J. Chem. Phys. 68, 5528 (1978).
35. S. Bratos, J. Chem. Phys. 63, 3499 (1975); C.B. Harris, R.M. Shelby and P.A. Cornelius, Phys. Rev. Lett. 38, 1415 (1977).
36. S.F. Fischer, A. Laubereau, Chem. Phys. Lett. 55, 189 (1978).
37. S. Mukamel, Chem. Phys. 37, 33 (1979).
38. T. Tokuhiro and W.G. Rothschild, J. Chem. Phys. 62, 2150 (1975).
39. D. Döge, R. Arndt, A. Khuen, Chem. Phys. 21, 53 (1977); M. Gilbert, M. Drifford, J. Chem. Phys. 66, 3205 (1977).
40. A. Miklavc, and S.F. Fischer, Chem. Phys. Lett. 44, 209 (1976).
41. A. Miklavc and S.F. Fischer, Chem. Phys. 69, 281 (1978).
42. R. Zygan-Maus and S.F. Fischer, Chem. Phys. 41, 319 (1979).
43. T. Förster, Z. Naturforsch. A4, 321 (1949).
44. A.D. Buckingham, Proc. Ray. Soc. London A248, 169 (1958).
45. S. Bratos. J. Rios, Y. Guissany, J. Chem. Phys. 52, 439 (1970).

Part III

Nonlinear Optical Phenomena

Relaxation Study in the Extremely Short Time Region by Stationary and Transient Nonlinear Optical Spectroscopy

T. Yajima, H. Souma, Y. Ishida, and Y. Taira

Institute for Solid State Physics, The University of Tokyo
Roppongi, Minato-ku, Tokyo 106, Japan

Abstract

An outline of new nonlinear spectroscopic methods which are developed for measuring ultrashort relaxation times of excited states in a range of 10^{-13}-10^{-14} s. They are all based on the third-order nonlinear optical processes in resonant media with two kinds of input light beams. Both the frequency-domain (stationary) and time-domain (transient) methods are developed with particular emphasis on the measurement of dephasing time T_2. Resonant Rayleigh-type mixing and resonant Kerr effect are typical of the former, while transient spatial parametric effect is typical of the latter. A description of subpicosecond dye laser work necessary as a source for the time-domain measurement is also given. All the theories are based on the semiclassical description of the system with density matrix formalism incorporating phenomenological relaxation times. The $S_0 \rightarrow S_1$ electronic transitions of dye molecules in both liquids and solids were selected as main objects of the present study.

1. Introduction

There are diverse fields of photo-physical, chemical and biological processes in which relaxations of excited states of materials are extremely fast, their time constant being in the subpicosecond and femtosecond range. Although the development of picosecond laser technique advanced revolutionarily the study of ultrafast phenomena [1,2], it is still not in a stage sufficient for making possible to observe directly the extremely fast processes mentioned above.

In view of this situation, we have explored in recent years various new spectroscopic methods utilizing nonlinear optical effects which provide information on extremely short relaxation times in a range of 10^{-13}- 10^{-14} s associated with, in particular, electronic excited states of condensed matter. The emphasis is placed mainly on measuring dephasing times or spectral diffusion times which could not easily be known from conventional transient absorption or emission spectroscopy. This paper describes an outline of a series of our works concerned with this problem including both the theory and experiment on liquids and solids.

It is well known that relaxation phenomena can also be studied in the frequency-domain by observing spectral broadening effects. Because the broadening is inversely proportional to the relaxation time, the frequency-domain methods are essentially powerful for measuring shorter relaxation times. Conventional absorption and emission spectra, however, include generally additional broadening effects not originated from relaxation such as the inhomogeneous broadening and others which should be eliminated or separated for the study of relaxation processes. This can be accomplished only by utilizing some nonlinear optical effects within a material under study. Various types of saturation spectroscopy and Doppler-free two-photon spectroscopy [3] are developed in gaseous materials for this purpose, but they are usually not suited to condensed matter. The hole burning and fluorescence narrowing effects have been applied to solids and liquids to measure picosecond relaxation times [4-6]. These methods, however, do not give unambiguous results when applied to materials with very fast spectral diffusion which encounters often in condensed matter. As a result of seeking more reliable new method essentially suitable for condensed matter, we found the resonant Rayleigh-type mixing as will be described in 3.1. As a related frequency-domain method, the resonant Kerr effect will be described and compared with the mixing method in 3.2.

The time-domain method by which relaxation processes are observed directly as transient phenomena is still important to complement and check the results obtained by the frequency-domain method, because the latter is of indirect nature and requires careful interpretation of the results. In order to know information on relaxation times of 10^{-13}- 10^{-14} s by this means, the efforts along two ways are necessary. One is to realize shorter pulses beyond the limit presently available by subpicosecond lasers. The other is to explore a new means capable of providing relaxation information even for $\tau_r \ll t_p$ where τ_r is the relaxation time, and t_p is the pulse width. Our works along these lines will be described in 4.

2. Basic Model on Energy Levels and Relaxation

Since we are considering general methods which could be applied to a wide variety of materials, some idealized common model of materials is required to clarify essential features of the methods.

In most cases studied here, we consider the situation that a material under study is irradiated by two incident light beams whose frequencies are resonant or near resonant with one transition of the material. Under this circumstance, the material can well be described, in the first approximation, as a two-level atomic system, the corresponding transition being generally inhomogeneously broadened. Spatial variation of atomic or molecular environmental fields is one of main origins of the inhomogeneous broadening in both liquids and solids. Multi-level structure with dense sublevels in molecules and crystals can also be treated effectively as a kind of inhomogeneous broadening in certain approximations.

For this kind of two-level system, the relaxation properties are conventionally describable in terms of phenomenologically introduced three kinds of constants, longitudinal (T_1), transverse (T_2) and cross (T_3) relaxation times. Here, T_1^{-1} represents the rate of population relaxation between the two levels constituting an optical transition, T_3^{-1} denotes the rate of spectral diffusion or energy transfer between neighbouring levels or transitions constituting the inhomogeneous broadening, and T_2^{-1} is the total rate of dephasing arising from all kinds of relaxation mechanisms including the origins of both T_1 and T_3. Therefore, T_2 is generally shorter than both T_1 and T_3. When three- or four-level model should be used to describe a material system, an increased number of relaxation parameters may generally be required to describe its dynamical behavior. But the basic concept of using above three kinds of relaxation times can still be applied even in these cases.

The problem is then divided into two parts, i.e., the experimental determination of the values of T_1, T_2 and T_3, and the investigation of individual relaxation mechanisms underlying T_1, T_2 and T_3. The limitations and precautions of using this type of relaxation model will be discussed in 5.

3. Nonlinear Frequency-Domain Spectroscopy

3.1. Resonant Rayleigh-Type Optical Mixing

3.1.1. Theory

When two coherent light beams with different frequencies ω_1 and ω_2 are incident on a material, new light components with frequencies $\omega_3 = 2\omega_1 - \omega_2$ and $\omega_4 = 2\omega_2 - \omega_1$ are generated through the third-order nonlinear effect. The dispersive properties of the nonlinear susceptibility, $\chi^{(3)}(\omega_3 = 2\omega_1 - \omega_2)$, describing one of these processes are useful to provide spectroscopic information of the material when some resonant conditions are present.

The well-known CARS (coherent anti-Stokes Raman spectroscopy) is the case of $|\omega_1 - \omega_2| \simeq \omega_r$ where ω_r is the vibrational or other low transition frequency of materials, and it provides essentially similar information as conventional Raman scattering does. The resonant Rayleigh-type mixing is concerned with a different type of resonant conditions, $\omega_1, \omega_2 \simeq \omega_0$ and $|\omega_1 - \omega_2| \simeq \tau_r^{-1}$, where ω_0 is an optical transition frequency and τ_r symbolizes a relaxation time of various types. The corresponding third order nonlinear susceptibility can conveniently be calculated by a density matrix formalism incorporating the phenomenological relaxation times. For a simple two-level atomic model with broadly inhomogeneous transition neglecting cross relaxation, the result is given by [7,9]

$$\chi^{(3)} = \frac{iK}{(1 + i\omega_d T_1)(1 + i\omega_d T_2)} \tag{1}$$

where $\omega_d = \omega_2 - \omega_1$, and K is a real coefficient nearly independent of ω_d. The dispersive property of $\chi^{(3)}$ as a function of ω_d is thus determined only by T_1 and T_2 independent of the inhomogeneous broadening. The values of T_1 and T_2 are then obtained from a careful analysis of the dispersion curve measured in a range from $\omega_d \ll T_1^{-1}$ to $\omega_d \gg T_2^{-1}$. This range, however, cannot always be covered because of the presence of a spectral resolution limit and a background contribution from the nonresonant susceptibility term. Even in these cases we can determine useful upper and lower limits of T_1 and T_2 which are not easily known by other means.

For the determination of T_1 and T_2 in real materials further considerations should be taken into account. First, various factors which obscure the true dispersive behavior of $\chi^{(3)}$ must be eliminated or separated from the measured dispersive properties. These are the higher-order saturation effect, the interference effect between resonant and nonresonant susceptibility terms and the wave propagation effect governed by a coherence length.

Secondly, the theory of resonant term of $\chi^{(3)}$ must be refined so as to fit the more realistic model of materials. The cross relaxation effect can be incorporated by using a generalized density matrix formalism which describes properly an atomic system in a continuously distributed energy level scheme constituting the inhomogeneous broadening. The resultant $\chi^{(3)}$ consists of two terms. The first term is the dominant one with the same form as (1) except that T_1 is replaced by an effective longitudinal relaxation time $T_1' = (T_1^{-1} + T_3^{-1})^{-1}$. The second term arises from the inverse spectral diffusion effect and can make a significant contribution at low values of ω_d in certain circumstances. The effect of additional energy levels other than the resonant two-levels has also been examined by considering a simple three-level system. The result showed similar effects as the cross relaxation does, i.e., the replacement of T_1 by an effective population relaxation time in (1) and the appearance of an additional susceptibility term.

Summarizing the above examination, it has been revealed that the dispersive property of $\chi^{(3)}(\omega_d)$ is relatively model-insensitive in the higher ω_d region contributing to the determination of T_2, and that it is somewhat model-dependent in the lower ω_d region contributing to the determination of longitudinal relaxation time. Therefore, the resonant Rayleigh-type mixing is particularly suitable for the determination of T_2. Detailed discussions on these problems have been given in the literature [9].

3.1.2. Experiment

In the experiment, pulsed light sources are not necessary in principle, but usually used to enhance the nonlinear optical effect. Even with the pulsed sources, the nonlinear process is essentially quasi-stationary as far as the pulse width is much longer than the relaxation times under study.

Earlier experiments were done by using two independently tunable nanosecond dye lasers pumped by a nitrogen laser as light sources [8,10]. Special care was taken to eliminate a strong stray component of the input beams because the output frequency is very close to the input one, the separation $\omega_3 - \omega_1$ $(= \omega_1 - \omega_2)$ being in a range of 0.5 - 500 cm^{-1} corresponding to the relaxation-time range of 10^{-11}- 10^{-14} s. As a material transition to be studied, the $S_0 \rightarrow S_1$ electronic transition of several dye molecules in liquid solutions was selected because of the flexibility in selecting the sample conditions as well as its basic interest in molecular physics and quantum electronics. Among these, measurements on 1,1'-diethyl-4,4'-quinocyanine iodide have been made in detail. The measured frequency characteristics of $\chi^{(3)}$ as a function of ω_d were found to be well fitted to the theoretical formula of (1). The asymmetric nature and the occurrence of a dip in the dispersive curve of $\chi^{(3)}(\omega_d)$ were also observed and attributed to the interference effect between resonant and nonresonant susceptibility terms. Thus, T_2 in a range of 0.02 - 2 ps and $T_1' = (T_1^{-1} + T_3^{-1})^{-1}$ in a range of 0.1 - 5 ps have been determined. These values were found to be largely dependent on the spectral position inside the inhomogeneous broad absorption band and the sample preparation and deterioration as well as on the kind of dye materials and solvents.

In order to study the relaxation mechanisms underlying phenomenologically determined relaxation times, it is worth-while to see the effect of drastic change of environmental conditions of a specific molecule. This has been done by comparing the result of a liquid dye solution with that of a high-polymer solid sheet containing the same dye molecule [11,12]. This experiment has been made possible first by using a picosecond laser system instead of nanosecond one, because the measurement in solids by the nanosecond source is seriously limited by the optical damage of the sample due to strong absorption. Two independently tunable picosecond dye lasers synchronously pumped by the second harmonic of a mode-locked Nd:YAG laser were used. In this experimental system special care must be taken on both the temporal and spatial coincidences of the two input pulses. The results revealed that for rhodamine B and erythrosine B dyes, T_2 is much shorter (<< 0.1 ps) in liquid by more than one order of magnitude than in solid. The temperature dependence of T_2 for rhodamine B in solid has been measured in a range 10 - 300 K, and showed no appreciable change. The result for another dye, DQCI (1,3'-diethyl-2,2'-quinolythia carbocyanine iodide), showed different type of character that the values of T_2 for liquid and solid samples are of the same order of magnitude and not so short (> 0.1 ps). It is suggested from these results that in the former type of dye (rhodamine B and erythrosine B) the dephasing processes of intra- and inter-molecular origins are dominated in solid and liquid, respectively, and that in the latter type of dye (DQCI), the inter-molecular perturbations are screened out even in liquid for some reasons, and intra-molecular processes are contributing to the dephasing in both solid and liquid.

The resonant Rayleigh-type mixing method can also be applied to other kind of materials. The interband transition of semiconductors such as Ge and GaAs has been examined by a reflection-type mixing experiment [13]. A preliminary result indicated the T_2 value to be in the order of 10^{-13} s.

3.1.3. Other Related Methods

(1) The theory of resonant Rayleigh-type mixing has been extended to the case with two-photon resonance condition, i.e., $2\omega_1$, $2\omega_2 \simeq \omega_0$ [14]. This scheme can suitably be applied to a pair of levels which, in one-photon transition, is forbidden by the selection rule or shows a strong absorption to prevent the experimental observations. The result of analysis based on an inhomogeneously broadened three-level model revealed that the Rayleigh-type dispersive behavior similar to that in (1) also appears in the higher-order nonlinear susceptibility. It was found that the dispersions governed by the longitudinal and transverse relaxation times appear in $\chi^{(n)}(n \geq 5)$ and $\chi^{(n)}(n \geq 7)$, respectively. Thus, the two-photon resonant Rayleigh-type mixing has also been confirmed to serve as a means of relaxation study. However, its experimental study is left for future study.

(2) SIEGMAN [15] proposed a similar mixing method for the determination of relaxation times in which one additional probe light at ω_p is used to produce an output light at $\omega_3 = \omega_p + \omega_2 - \omega_1$. From an analysis based on the formalism similar to ours, it was found that $\chi^{(3)}$ for this process does not contain the factor $(1 + i\omega_d T_2)^{-1}$, and the frequency response is mostly determined by the longitudinal relaxation time. Therefore, this method serves as a means complementary to our method which is primarily suitable to the study of transverse relaxation time.

3.2. Resonant Kerr Effect

There exists other type of third-order nonlinear optical process with two incident beams at frequencies ω_1 and ω_2. In term of nonlinear susceptibility, it is expressed by $\chi^{(3)}(\omega_2 = \omega_1 - \omega_1 \mp \omega_2)$, the effect of which can be observed by measuring the change of transmission characteristics of the ω_2 beam in the presence of the ω_1 beam. This change can be detected sensitively if the ω_2 beam passes through a pair of crossed polarizers between which the sample is placed. When the resonant conditions similar to those in resonant Rayleigh-type mixing are present, this nonlinear process also serves as a method for studying relaxation processes. This is called resonant Kerr effect or polarization spectroscopy.

The principle was first conceived for high resolution atomic spectroscopy [16], and later applied to condensed phases by SONG et al. [17]. Their analysis showed two types of nonlinear susceptibility terms. The first one is essentially the same as that for the resonant Rayleigh-type mixing. The second one represents essentially the same effect as the hole burning and has been used to estimate the value of T_2. We have examined carefully the effect of spectral cross relaxation on the resonant Kerr effect [18], because it generally affects sensitively the hole burning behavior. An analysis based on the same formalism as in the Rayleigh-type mixing revealed that there occurs an additional third-type of susceptibility term originated from cross relaxation which have a broad frequency width primarily determined by the inhomogeneous width. As a result, the combined behavior of the second- and third-type of terms does not reflect the homogeneous broadening when the cross relaxation is fast as in dye materials. Even when $T_3/T_2 = 10$, the apparent width is broadened by an order of magnitude compared with the homogeneous width. The steady-state polarization spectroscopy, therefore, is not generally preferable for measuring the T_2 values. We have found, however, if one introduces a partially transient condition, $T_1 >> t_p >> T_2, T_3$, the effect of cross relaxation can be eliminated in the measurements, and the true T_2 value could be obtained by this method. These characteristics have been verified experimentally by observing the time-resolved nonlinear dispersion spectra of resonant Kerr effect in rhodamine B and DODCI dye solutions [18].

4. Nonlinear Time-Domain Spectroscopy

4.1. Development of Subpicosecond Light Sources

For the time-domain work on the extremely fast phenomena, one requires light pulses as short as possible. The shortest pulses are presently obtainable from passively mode-locked cw dye lasers which have been extensively studied by several research groups [19-21]. The shortest pulse width hitherto reported in the literature is 0.16 ps [21]. We have also attempted to get over the present limit along this lines, and for this purpose, constructed a rhodamine 6G mode-locked dye laser pumped by a cw argon ion laser. The laser consists of four cavity mirrors and a single jet stream of a mixed solution containing both amplifying and mode-locking dyes. With careful adjustment of various laser parameters to optimum conditions, we have observed the shortest pulse which was measured to be below 0.15 ps [22]. The pulses below 0.2 ps are, however, not stable and very critical to reproduce, and therefore require some special means of stabilization to be used as a practical light source.

Another new approach we are attempting to obtain shortest pulses is the external compression of picosecond dye laser output pulses. Our laser system consists of a dye laser oscillator synchronously pumped by the second harmonic of a mode-locked Nd:YAG laser and a two-stage pulse compressor containing dye absorber and amplifier chain. With this system 2 MW, 0.8 ps pulses have already been achieved with a compression ratio of 20 [23]. This scheme is attractive in its high power character, the wide tunability and the potentiality of further shortening. For practical use, however, it also requires a stabilizing means particularly in the compression stage.

4.2. Transient Spatial Parametric Effect

The relaxation of excited states can directly be observed by various transient optical phenomena, such as transient absorption or emission after the pulsed excitation and a variety of coherent transient processes. It is generally required to have light pulses much shorter than the relaxation times to observe these phenomena. Presently available laser sources are not sufficient to satisfy this condition for relaxation times in the range of 10^{-13} - 10^{-14} s. This situation is particularly severe for short dephasing times T_2. In view of this circumstances, we explored a method which provides some information on T_2 even when $T_2 \lesssim t_p$ (pulse width). This is the transient spatial parametric effect as described below [24].

When two noncollinear light pulses with the wave vectors $\vec{k}_1$, $\vec{k}_2$ and the same frequency ω are incident on a material, new light components with $\vec{k}_3 = 2\vec{k}_1 - \vec{k}_2$ and $\vec{k}_4 = 2\vec{k}_2 - \vec{k}_1$ are produced by the third order nonlinear mixing process in $\vec{k}$-space. When ω is one-photon resonant with a material transition, the time behavior of this process is governed by and therefore provides information on the relaxation associated with this transition. A transient analysis based on a density matrix formalism revealed that the correlation trace (output light energy with $\vec{k}_3$ or $\vec{k}_4$ vs. the time separation between two input pulses) shows a decay characteristic determined only by T_2 when $t_p \ll T_2$, and directly gives the value of T_2. The decay rates are $2T_2^{-1}$ and $4T_2^{-1}$ for homogeneously and inhomogeneously broadened transitions, respectively. When $t_p \gtrsim T_2$, the correlation trace reflects dominantly the incident pulse characteristics rather than the relaxation, but even in this case, we can estimate reliably an upper limit of T_2. This case can also be applied to the measurement of ultrashort pulse width with the time resolution limited by T_2 [24,25].

This spatial parametric effect includes the well-known photon echo phenomena as a limiting case, but serves as a more general and flexible method in the sense that we can always observe a signal giving some information on T_2 in a variety of situations including the cases of both the homogeneously and inhomogeneously broadened transitions, for incident pulses both separated and overlapped in time, and for both regions $t_p > T_2$ and $t_p < T_2$.

The present method is also closely related to the laser-induced transient grating method [26] in which one time-delayed probe pulse is used in addition to two coincident noncollinear pulses. The latter process is, however, shown to be governed primarily by the longitudinal relaxation time rather than T_2. This relation between the two cases of two and three input pulses are the same as in the frequency-domain spectroscopy mentioned in 3.1.3.

The experimental demonstration has been made first by irradiating the 10^{-11} s pulses of the second harmonic of a mode-locked Nd:YAG laser into a rhodamine 6G dye solution [24]. Recently the subpicosecond pulses (0.2 - 0.3 ps) from a cw mode-locked dye laser as described in 4.1 were used in our work [27] to investigate the subpicosecond dephasing processes in several dye solutions. The measured correlation traces showed symmetric character for all the examined dye samples (DODCI, brilliant green, pinacyanol chloride in solutions and DQCI in a solid sheet) indicating the pulse-width-limited situation. The upper limit of T_2 in these samples was then carefully estimated to be 0.5 ps by the help of the calculated correlation traces representing the convolution of pulse and relaxation characteristics. The results are consistent with those from the frequency-domain methods. It is also found from the calculation that there occurs a peak shift of the correlation curve which is a more sensitive measure of the relaxation effect than the change of the shape, and therefore can serve for the determination of a relaxation time which is comparable or shorter than the pulse width. An experimental attempt along this line is also under way.

5. Discussion and Conclusion

We have shown that both the frequency- and time-domain spectroscopic methods based on the third-order nonlinear processes with two kinds of incident light beams are useful for the measurement of ultrashort relaxation times of excited states in a range of 10^{-13}- 10^{-14} s. This kind of approach is still in an early stage and much is left for further study to get complete understanding of the concerned relaxation phenomena.

In order to have more reliable data of relaxation times, the consistency between the results obtained from various independent methods must carefully and extensively be examined. Especially the comparison between the results of the frequency- and time-domain methods is important. The accomplishment of this problem is largely dependent upon further development of subpicosecond laser technique.

The other problem is to examine in what extent the description of relaxation properties in terms of T_1, T_2 and T_3 is valid. In certain cases of material transition other types of combination of relaxation parameters may become adequate. More basically, we must re-examine the validity of describing a relaxation process by a rate constant, which is based on an assumption of weak and rapid random perturbations. This consideration will be of increased importance when we proceed further into shorter time region. The refinement of the theory and the accumulation of the experimental data of the nonlinear spectroscopic methods as described here will gradually answer these questions.

References

1. S.L. Shapiro ed.: Ultrashort Light Pulses (Springer-Verlag, Berlin, Heidelberg, New York, 1977)
2. C.V. Shank, E.P. Ippen and S.L. Shapiro ed.: Picosecond Phenomena (Springer-Verlag, Berlin, Heidelberg, New York, 1978)
3. K. Shimoda ed.: High-Resolution Laser Spectroscopy (Springer-Verlag, Berlin, Heidelberg, New York, 1976)
4. G. Mourou: IEEE J. Quantum Electron. **11**, 1 (1975)
5. T. Kushida and E. Takushi: Phys. Rev. **B12**, 824 (1975)
6. F. Keilman: IEEE J. Quantum Electron. **12**, 592 (1976)
7. T. Yajima: Opt. Commun. **14**, 378 (1975)
8. T. Yajima, H. Souma and Y. Ishida: Opt. Commun. **18**, 150 (1976)
9. T. Yajima and H. Souma: Phys. Rev. **A17**, 309 (1978)
10. T. Yajima, H. Souma and Y. Ishida: Phys. Rev. **A17**, 324 (1978)
11. H. Souma, Y. Taira and T. Yajima: in Picosecond Phenomena ed. C.V. Shank, E.P. Ippen and S.L. Shapiro (Springer-Verlag, Berlin, Heidelberg, New York, 1978) p.224
12. H. Souma, T. Yajima and Y. Taira: ISSP Technical Report No.1020 (1980); submitted to J. Phys. Soc. Japan
13. H. Souma, T. Yajima and Y. Ishida: presented at the Meeting of the Physical Society of Japan, April 1977 (unpublished)
14. T. Yajima: J. Phys. Soc. Japan **44**, 948 (1978)

15. A.E. Siegman: Appl. Phys. Lett. 30, 21 (1977)
16. C. Wieman and T.W. Hänsch: Phys. Rev. Lett. 36, 1170 (1976)
17. J.J. Song, J.H. Lee and M.D. Levenson: Phys. Rev. A17, 1439 (1978)
18. Y. Taira and T. Yajima: to be published
19. I.S. Ruddock and D.J. Bradley: Appl. Phys. Lett. 29, 296 (1976)
20. E.P. Ippen and C.V. Shank: in Picosecond Phenomena ed. C.V. Shank, E.P. Ippen and S.L. Shapiro (Springer-Verlag, Berlin, Heidelberg, New York, 1978) p.103
21. J.-C. Diels, E.W. Stryland and D. Gold: ibid. p.117; J.-C. Diels: Laser Focus 15, Nov. p.9 (1979)
22. Y. Ishida and T. Yajima: unpublished work
23. Y. Taira and T. Yajima: Opt. Commun. 29, 115 (1979)
24. T. Yajima and Y. Taira: J. Phys. Soc. Japan 47, 1620 (1979)
25. Y. Ishida and T. Yajima: to be published
26. D.W. Phillion, D.J. Kuizenga and A.E. Siegman: Appl. Phys. Lett. 27, 85 (1975)
27. T. Yajima, Y. Ishida and Y. Taira: to be published

Thermodynamics and Transient Dynamics of Simple Optical Systems with Instability

H. Hasegawa

Department of Physics, Kyoto University
Kyoto 606, Japan

1. Introduction

When we investigate the relaxation phenomena in nonlinear optical systems involving instabilities, we cannot disregard accompanying fluctuations: They give rise to a dissipation of energy of the system given up to its heat reservoir, but not in a simple fashion. For example, the threshold of a laser oscillation must be a phase transition associated with an unusual situation of such heat transfers. A natural way to approach the problems is the method of Langevin equations, and it is not strange that the method was adopted in the very beginning of the laser theory early in the decade of 1960 [1]. One would imagine that the Langevin theory of treating fluctuations must be a well-established chapter in the standard laser theory [2]. The recent development of nonlinear, nonequilibrium statistical mechanics [3], however, has revealed the necessity to revise the chapter so that a more satisfactory treatment of the nonlinear fluctuations can be achieved that is needed for a consistent macroscopic description of the instability.

Our aim is to establish a prototype of nonequilibrium thermodynamics in such a simple, ideal open system with instability for a test to wider classes of such systems. Indeed, if one looks at a recent monograph "Synergetics" [4], one finds that the laser is a real motivation of the author to the thermodynamic conception of challenge "self-organization". The present lecture outlines a contribution to this scope from our studies of the model optical system [5,6], where major efforts have been devoted to the bridge between statistical mechanics and thermodynamics.

A thermodynamic discussion about lasers and other related phenomena is a rather small part of the standard laser theory. The first interesting work, viz. to point out a remarkable similarity between the laser and ferromagnetism, was made by DEGIORGIO and SCULLY, which is summarized in [2] as a comparison between thermodynamic quantities (such as lasing mode amplitude vs. spontaneous magnetization etc.) belonging to both phenomena. A similar comparison is made in [4] between the laser and superconductivity. We note that such a comparison is merely an analogy argument that does not peneterate the real nonequilibrium characteristic of the laser. Later, GRAHAM [7] discussed this characteristic in an attempt to associate it with the GLANSDORFF-PRIGOGINE thermodynamics [8]; a really right direction. The difficulty in such an attempt, however, consists in the lack of reliable bases of how to introduce the entropy concept into such nonequilibrium open systems. (The information concept discussed in [4] only suggests a general direction.)

Here, we propose to recall a reasonable theoretical framework of open systems due to LEBOWITZ [9], who developed many years ago [10] a model by which the *entropy production* can be formulated with consistency between the thermodynamic and the information-theoretic contributions. This framework provides by no means an absolutely reliable solution to the above-mentioned difficulty, leaving the proof of a certain ansatz used which is still remaining unproved. Nevertheless, it is a promising framework, along which one can be motivated to explore many examples of nonequilibrium systems as well as to give further efforts for a general formulation of the statistical mechanics of open systems. What we have studied in [5,6] is that the standard simple model of nonlinear optical systems is indeed a fine example of the LEBOWITZ open system, which we will demonstrate in the subsequent sections. These are organized as follows:

Sections 2 and 5 are preparatory sections. A complete form of the starting Langevin equations is set up in the semiclassical framework of quantum optics, and an elementary introduction of the *stochastic calculus* is made for the noise treatment. Sections 3 and 4 are physical arguments of the deterministic part about the possible instabilities contained in the equations. We will not aim at a comprehensive treatment of them, confining ourselves to the three typical ones, namely, the *ordinary laser*, the *absorptive bistability* and the *superfluorescence*. The adiabatically reduced Langevin equations for these are presented in Sec.6, showing their adaptability to the LEBOWITZ model. Consequently, the rest sections are devoted to the thermodynamics at the stationary states and some transient processes to these states. Throughout the presentation, a number of improvements are included over the previous results [5,6].

2. The Model of Single Mode Operation

The optical system we consider is the familiar one; an assembly of N two-level atoms interacting with an active mode of the electro-magnetic field in a cavity, the Hamiltonian of which is given by

$$H=\hbar\omega b^*b + \hbar\omega_0 Z + \hbar g((b^*+\alpha_t^*)R + (b+\alpha_t)R^*) \,, \tag{1}$$

where the dynamical variables of the system are specified as follows:

b, b^* : complex amplitude of the em field active mode
R, R^* : complex atomic dipole collective mode
$Z=\frac{1}{2}(N_2-N_1)$: sum of two-level population difference (2, upper; 1, lower level).

The equations of motion pertaining to these variables consist of the hamiltonian part and the dissipative part, the former being derived from (1) by using the usual commutation relations

$$[b,b^*] = 1; \; [R,R^*] = -2Z, \; [Z,R] = -R, \; [Z,R^*] = R^* \,. \tag{2}$$

These are in the quantum mechanical version familiar from a quantum harmonic oscillator and angular momentum, but we put them into the semiclassical treatment: That is, these variables shall be regarded as continuously ranged, commutative variables in the equation of motion so that the complexity in the quantum formalism of dissipation there can be avoided. The largeness of N (the number of the atoms) will justify this treatment as long as the variables Z and $n=b^*b$ (photon number) are in the extensive range $O(N)$. Since, however, the treatment of the fluctuations about these variables pertains to the intensive range $O(1)$ of them, it requires a careful restoration of the noncommutativity in order not to miss the important quantum nature of the fluctuations.

Consequently, our treatment of the dissipative part of the equations of motion is in a phenomenological level, which we have called in [5] *Brownian motion of photons and spins*:

$$\dot{b}=-\kappa b+f,\quad \dot{b}^*=-\kappa b^*+f^*;\ \langle f^*(0)f(t)\rangle=2\kappa\bar{n}\delta(t) \tag{3}$$

$$\dot{R}=-\gamma_\perp R+F,\quad \dot{R}^*=-\gamma_\perp R^*+F^*;\ \langle F^*(0)F(t)\rangle=2\gamma_\perp\bar{M}\delta(t) \tag{4}$$

$$\dot{Z}=-\gamma_\parallel (Z-Z_0)+F_z;\ \langle F_z(0)F_z(t)\rangle=\gamma_\parallel \bar{M}\delta(t)\ . \tag{5}$$

Each of these equations is a familiar linear-Langevin equation with a fluctuating force satisfying the Gaussian white characteristic, as indicated by the δ-correlation in the end of each equation (3), (4) and (5) that assure the equilibrium distribution: $P^{eq}(b^*bR^*RZ)\propto\exp[-\frac{1}{\bar{n}}b^*b-\frac{1}{\bar{M}}(R^*R+(Z-Z_0)^2)]$. At this equilibrium every dynamical variable is normally distributed so that

$$\langle b^*b\rangle_{eq}(=\langle n\rangle_{eq})=\bar{n}=O(1)\ ; \tag{6}$$

$$\langle R^*R\rangle_{eq}=\bar{M}=O(N),\ \langle(Z-Z_0)^2\rangle_{eq}=\tfrac{1}{2}\bar{M}=O(N)\ . \tag{7}$$

Eq.(6) represents thermal photons (called black-body radiation) where the average number $\bar{n}$ is determined only by the temperature of the cavity wall for which quantum statistics tells us the well-known Planck formula. Eq.(7), on the other hand, has been introduced somewhat arbitrarily based on the consideration of the law of large numbers (i.e. a large number of equivalent, independent objects will become in a normal distribution). It will be shown later that the two-level structure of the atoms in the above normal distribution is compatible with the quantum nature of photons, disposing the variance parameter $\bar{M}=\frac{1}{2}N$ based on the condition of detailed balance and the operator symmetrization in the quantum version.

Let us now introduce two types of external driving forces into our system that may produce instability: The first type is a thermal constraint i.e. a pump of the two-level population which is expressed in the average of the fifth variable Z over the equilibrium distribution as

$$\langle Z\rangle_{eq}=Z_0=\tfrac{1}{2}(N_2^{eq}-N_1^{eq})\ . \tag{8}$$

This, which we call the pump parameter, is supposed to be externally controllable,[1] and to be changed from the ordinary negative value (corresponding to the cavity-wall temperature) to positive, inverted values of the population. The second type is a mechanical constraint i.e. an irradiation of the system by a coherent external field. This can be formulated by adding a complex amplitude α_t to the internal mode amplitude b in the atom-field interaction of the Hamiltonian, as indicated in (1). It is then more convenient to change the field variable from the internal field b to the total field a defined by

$$a=b+\alpha_t\ . \tag{9}$$

1. A more general scope of the laser theory would be such that the pumping action is included in the dynamics as in the consideration of 3- or 4-level lasers, for which Z_0 is not a controllable parameter. In the present discussion this complication is avoided, which enables us to treat the imput power of establishing a population inversion as a heat reservoir of the atoms (referred to as "atomic reservoir", see the model in [2] which gives a reduction of a 3-level to the present model).

It can be seen from the Hamiltonian (1) that this change results in a modification of the equations of motion in such a way that only the natural motion of the new amplitude a must be respected: In the absence of α_t the internal mode amplitude b oscillates such that $\dot{b}=-(i\omega+\kappa)b$, which is now replaced by $\dot{a}=-(i\omega+\kappa)(a-\alpha_t)+\dot{\alpha}_t$ and all the rest part of the equations is left unaltered (apart from the replacement of the b by a).

We are now prepared to write down the set of the equations of motion for our optical system. It is a simple sum of the hamiltonian part derived from (1) and the dissipative part based on (3)∿(5):

$$\begin{aligned}
&\dot{a}=-(\kappa+i\omega)(a-\alpha_t)+\dot{\alpha}_t-igR+f\\
&\dot{a}^*=-(\kappa-i\omega)(a^*-\dot{\alpha}_t^*)+\alpha_t^*+igR^*+f^*\\
&\dot{R}=-(\gamma_\perp+i\omega_0)R-2ig\ aZ+F\\
&\dot{R}^*=-(\gamma_\perp-i\omega_0)R^*+2ig\ a^*Z+F^*\\
&\dot{Z}=-\gamma_\parallel(Z-Z_0)+ig(a^*R-aR^*)+F_z\ .
\end{aligned} \tag{10}$$

This is a coupled set of nonlinear Langevin equations, its nonlinearity stemming, however, only from the purely mechanical interaction among the system (i.e. between the atoms and the total field) proportional to the coupling parameter g: In case of the vanishing limit of this coupling, (10) reduce to five independent, linear Langevin equations (their form equivalent to (3), (4) and (5) where κ is replaced by $\kappa\pm i\omega$ and $\gamma_\perp$ by $\gamma_\perp\pm i\omega_0$) subject, as we have seen already, to the second fluctuation dissipation theorem [11] (or, the Einstein relation) that guarantees the internal consistency between the stochastic process and the thermodynamics. We recapitulate it by

$$\begin{aligned}
&\langle f^*(t')f(t)\rangle=2\kappa\,\bar{n}\delta(t'-t),\ \langle F^*(t')F(t)\rangle=2\gamma_\perp\bar{M}\delta(t'-t)\ ,\\
&\langle F_z^*(t')F_z(t)\rangle=\gamma_\parallel\bar{M}\delta(t'-t)\quad \text{and all the rest vanishing correlations,}
\end{aligned} \tag{11}$$

where the two parameters representing the strength of the Langevin forces, $\bar{n}=\langle b^*b\rangle_{eq}$ and $\bar{M}=\langle R^*R\rangle_{eq}$, are yet to be determined later. The coupling parameter g is given (from the quantum theory of radiation [12]) in terms of the microscopic matrix element μ_{12} of the dipole by

$$g=\left(\frac{2\pi\omega\mu_{12}^2}{\hbar V}\right)^{1/2}=O(N^{-1/2})\ , \tag{12}$$

indicating that the interaction between the mode and the single atom is spread over the whole cavity volume V.

Thus, it is the most curious point of the laser action that the $O(N^{-1})$ fractions of the local interaction energies are accumulated from the N atoms that are randomly distributed into a single mode in a very narrow frequency range of the field, generating an extensive (O(N)) power of the energy that is emitted. This is the very self-organizing action which is caused from the alignment of the atomic dipoles, its phenomenological description of the formation being given in the next section. A satisfactory construction of the general theory including fluctuations requires a long space which we only discuss predictively in the last section. Also, it should be remarked in this connection that the use of the nonlinear Langevin equations (10), (11) with $g\neq 0$ depends upon the ansatz that the dissipative part of the system motion is unaltered by this coupling; an assumption its legitimacy being by no means obvious. It will be judged *a posteriori* from the overall consistency point of view of the entire outcome of the present approach.

3. Classification of the Instabilities

First, we discuss the possibility of a self oscillation in the absence of the external field, $\alpha_t=0$. This can be investigated by means of the phenomenological rate equation i.e. the equations of motion (10) with vanishing noise forces, $f=F=F_z=0$. We work with the rotating frame of the coordinates for the mode amplitude $a(=b)$ and the atomic dipole R to find the stationariness, $\dot{X}=0$ for all the five variables in the frame. Let us set $a=\tilde{a}e^{i\Omega t}$ and $R=\tilde{R}e^{i\Omega t}$ (with Ω undetermined) and consider $\dot{\tilde{a}}$ and $\dot{\tilde{R}}$. From 1st and 3rd eqs. of (10)

$$\dot{\tilde{a}} = -(\kappa + i(\omega - \Omega))\tilde{a} - ig\tilde{R} = 0$$

$$\dot{\tilde{R}} = -(\gamma_\perp + i(\omega_0 - \Omega))\tilde{R} + 2igZ\tilde{a} = 0,$$

which is considered as a linear, homogeneous equation for the two dimensional vector $(\tilde{a}, \tilde{R})$ with two undetermined parameters Ω and Z. A nontrivial solution other than $\tilde{a}=\tilde{R}=0$ is possible only from

$$\det\begin{vmatrix} \kappa + i(\omega - \Omega) & ig \\ 2igZ & \gamma_\perp + i(\omega_0 - \Omega) \end{vmatrix} = 0$$

i.e.,

$$(\Omega - \omega)(\Omega - \omega_0) + 2g^2Z - \kappa\gamma_\perp + i(\gamma_\perp(\Omega - \omega) + \kappa(\Omega - \omega_0)) = 0, \tag{13}$$

or,

$$\Omega=\frac{\gamma_\perp\omega+\kappa\omega_0}{\gamma_\perp+\kappa} \quad \left(\omega-\Omega=\frac{\kappa}{\gamma_\perp+\kappa}(\omega-\omega_0), \quad \omega_0-\Omega=\frac{-\gamma_\perp}{\gamma_\perp+\kappa}(\omega-\omega_0)\right) \tag{14}$$

$$Z=\frac{\kappa\gamma_\perp}{2g^2}\left(1+\frac{(\omega-\omega_0)^2}{(\kappa+\gamma_\perp)^2}\right)\equiv Z_{th}. \tag{15}$$

In this situation

$$\tilde{R}=\frac{2ig}{\gamma_\perp}\,\frac{Z_{th}\tilde{a}}{1-i(\omega-\omega_0)(\kappa+\gamma_\perp)^{-1}}, \quad \tilde{a}=\frac{-ig}{\kappa}\,\frac{\tilde{R}}{1+i(\omega-\omega_0)(\kappa+\gamma_\perp)^{-1}}, \tag{16}$$

both linear relations being identical. This means that under the special circumstance of the rotating frame and of the population of atoms predicted by (14) and (15), the self oscillation is possible. Its amplitude can be determined from the stationariness of Z, $\dot{Z}=0$ in (10), by substituting the linear relation (16), yielding

$$n(=a^*a=\tilde{a}^*\tilde{a})=\frac{\gamma_\parallel}{2\kappa}(Z_0-Z_{th})\equiv n_s \tag{17}$$

$$M(=R^*R=\tilde{R}^*\tilde{R})=\frac{4g^2}{\gamma_\perp^2}\,\frac{Z_{th}^2 n_s}{1+(\omega-\omega_0)^2(\kappa+\gamma_\perp)^{-2}}=\frac{\gamma_\parallel}{\gamma_\perp}Z_{th}(Z_0-Z_{th}) \tag{18}$$

which is conditioned by

$$Z_0 > Z_{th} \qquad \text{(the Schawlow-Townes threshold condition).} \tag{19}$$

Notice that the threshold value of Z, Z_{th} given by (15), is $O(N)$ from (12) and, since $\gamma_\parallel$, $\gamma_\perp$, $\kappa=O(1)$,

$$n_s = O(N), \qquad M = O(N^2) \; (|R| = O(N)). \tag{20}$$

Namely, a macroscopic dipole has resulted and, at the same time, an extensive photon energy condensed into the single mode has come out. This is what experimentalists call "coherence of dipoles" and may be considered as an elementary example of the self-organization [4].

We discuss such a coherent motion of the composite system by means of the method of adiabatic elimination [4].

Case $\gamma_\perp \gg \kappa$ $\quad (\Omega \doteq \omega,\ \omega_0 - \Omega \doteq -(\omega-\omega_0)$; see (14))

The first linear relation between $\tilde{R}$ and $\tilde{a}$ in (16) becomes

$$\tilde{R} = \frac{2igZ_{th}}{\gamma_\perp - i(\omega-\omega_0)}\tilde{a}, \text{ i.e. } \dot{\tilde{R}} = -(\gamma_\perp - i(\omega-\omega_0))\tilde{R} + 2igZ\tilde{a} = 0$$

in the rotating frame with angular frequency $\Omega=\omega$, for which

$$\dot{\tilde{a}} = -\kappa\tilde{a} - ig\tilde{R} = -\kappa\tilde{a} + \frac{2g^2 Z_{th}}{\gamma_\perp - i(\omega-\omega_0)}\tilde{a}. \tag{21}$$

It implies that one chooses the frame such that the rate of natural motion of the one variable is reduced as much as possible (that is for $\tilde{a}$), and then set the rate of the other variable equal to zero (that is for $\tilde{R}$) by which this variable is eliminated from the motion of the coupled system in the long time scale $t \gg \gamma_\perp^{-1}$. In this time scale the second variable for the fast subsystem is almost at rest, leaving a small fraction of it which moves "adiabatically following" the motion of the slow subsystem. Here, such an adiabatic motion of the N-independent dipoles following the single mode amplitude oscillation is the very origin of its organized motion provided $\tilde{a}^*\tilde{a}(=n)=O(N)$. This, in turn, can be assured from the adiabatically reduced equations of motion

$$\dot{n} = -2\kappa n + \frac{4g^2\gamma_\perp}{\gamma_\perp + (\omega-\omega_0)^2} Zn \tag{22}$$

$$\dot{Z} = -\gamma_\parallel (Z - Z_0) - \frac{4g^2\gamma_\perp}{\gamma_\perp^2 + (\omega-\omega_0)^2} Zn \tag{23}$$

which have under the threshold condition (19) (called "above threshold") the nontrivial fixed point

$$Z = \frac{\kappa\gamma_\perp}{2g^2}\left(1 + \frac{(\omega-\omega_0)^2}{\gamma_\perp^2}\right) \quad (= Z_{th} \text{ in the situation } \kappa \ll \gamma_\perp; \text{ see (15)}) \tag{15'}$$

$$n = \frac{\gamma_\parallel}{2\kappa}(Z_0 - Z_{th}) = n_s. \tag{17'}$$

Case $\gamma_\perp \ll \kappa$ $\quad (\Omega \doteq \omega_0,\ \omega - \Omega \doteq \omega - \omega_0$; see (14))

The second linear relation between $\tilde{a}$ and $\tilde{R}$ in (16) results from the adiabatic elimination of $\tilde{a}$ by

$$\dot{\tilde{a}} = -(\kappa + i(\omega-\omega_0))\tilde{a} - ig\tilde{R} = 0, \quad \dot{\tilde{R}} = -\gamma_\perp\tilde{R} + 2igZ\tilde{a} \tag{24}$$

in the rotating frame with angular frequency $\Omega=\omega_0$. Thus, the logic of the adiabatic elimination here is the same as before, the roles of $\tilde{a}$ and $\tilde{R}$ being just interchanged. However, not like the previous argument, one now has a difficulty of finding an intuitive reason why the randomness of the dipole phases be reduced by this adiabatic motion of the single mode to cause an organized motion of the dipoles. The possible reason may be found in the reduced equations of motion themselves:

$$\dot{M} = -2\gamma_1 M + \frac{4g2\kappa}{\kappa^2+(\omega-\omega_0)^2} ZM \tag{25}$$

$$\dot{Z} = -\gamma_{\parallel}(Z-Z_0) - \frac{2g^2\kappa}{\kappa^2+(\omega-\omega_0)^2} M. \tag{26}$$

The nontrivial fixed point of this set of equations is given by

$$Z = \frac{\kappa\gamma_\perp}{2g^2}\left(1 + \frac{(\omega-\omega_0)^2}{\kappa^2}\right) \quad (= Z_{th} \text{ in the situation } \kappa \gg \gamma_\perp\text{; see (14)}) \tag{15"}$$

$$M = \frac{\gamma_\parallel}{\gamma_\perp} Z_{th}(Z_0 - Z_{th}) = O(N^2) \quad (>0 \text{ above threshold}). \tag{18"}$$

The latter result can be seen to arise from the energy exchange between M and Z^2 rather than between M and Z due to the adiabatic reduction. Namely, the dipole energy gains from the thermal energy of the assembly in the form of Z^2 (instead of Z), as can be seen from $\dot{M}+2Z\dot{Z}$ in (25, 26).

$$\frac{d}{dt}(M + Z^2)_{ad} = 0 \qquad (\text{'ad' means the adiabatically reduced part}). \tag{27}$$

The ideal case of such is the "superfluorescence" [13] where $\gamma_\parallel$ and $\gamma_\perp$ can be entirely neglected. It is a special radiation from an assembly of atoms in cooperation predicted early by DICKE [14] by the name of "superradiance", and has been experimentally established during the decade of 1970 [15]. The difference of the two terminologies stems from the ways of preparating the initial atomic states; whether incoherently excited (superfluorescence) or coherently (superradiance). From the instability point of view both refer to purely transient phenomena. The dynamics in any case is simple enough because of the conservation (27) of the length of the Bloch vector (ReR, —ImR, Z). Thus, starting from a point of nearly complete inversion the Bloch vector undergoes an unstable motion towards the lower semisphere, emitting a coherent radiation. The radiation is of highest intensity when the vector passes the "equator zone" for which $M=O(N^2)$ with the dipole energy fully converted from the energy of the atoms initially stored.

Turning back from the special case $\gamma_\perp=\gamma_\parallel=0$, we see that the self oscillating stationary states just represent the laser action; in particular, the adiabatic state for $\kappa \ll \gamma_\perp$ the ordinary laser where the inequality is established e.g. by the Q-switching (a sudden increase of the $Q(=\omega/2\kappa)$ of a Fabry-Pérot cavity). The opposite case $\kappa \gg \gamma_\perp$, on the other hand, represents a different type of the laser action, and may be called a " mirrorless laser" that is realized as the stationary stage of an "amplified spontaneous emission (ASE)" [16]. These are classified in the first part of Table 1.

We now consider the effect of coherent external fields that produces another type of instability without the thermal pump. This is related to the recent topics of optical bistability predicted theoretically [17] for which an experimental test has been reported [18], stimulating a number of further theoretical investigations [19-25] as a typical first-order phase transition. Here, we aim to illustrate the simplified picture of BONIFACIO and LUGIATO

[19] in terms, again, of the adiabatic elimination in the equations of motion (10). We suppose that the external field α_t in (9) is monochromatic with frequency ω to which the internal mode b of the same frequency responds. Then, the natural rotating frame to describe the motion is the one with $\Omega=\omega$, and

$$\begin{aligned}
\dot{\tilde{a}} &= -\kappa(\tilde{a}-\alpha) - ig\tilde{R} \qquad (\alpha_t = \alpha e^{-i\omega t};\ \alpha = \text{real} > 0) \\
\dot{\tilde{R}} &= -(\gamma_\perp - i(\omega-\omega_0))\tilde{R} + 2igZ\tilde{a} \qquad (28) \\
\dot{Z} &= -\gamma_\parallel (Z-Z_0) + ig(\tilde{a}^*\tilde{R}-\tilde{a}\tilde{R}^*):
\end{aligned}$$

The adiabatic elimination of the dipole $\tilde{R}$ under the condition $\kappa \ll \gamma_\perp$ implies again the setting $\dot{\tilde{R}}=0$ in the second equation of (28) and substituting

$$\tilde{R} = \frac{2igZ}{\gamma_\perp - i(\omega-\omega_0)}\tilde{a} \qquad (\gamma_\perp \gg \kappa) \qquad (29)$$

so obtained into the other two equations. We write the resulting set of equations in terms of the two dimensional circular coordinate

$$\tilde{a} = re^{i\phi} \quad \text{or,} \quad r = (\tilde{a}^*\tilde{a})^{1/2}\ (=n^{1/2}) \ \text{and}\ \phi=\frac{1}{2i}\log\frac{\tilde{a}}{\tilde{a}^*} \qquad (30)$$

as follows:

$$\begin{aligned}
\dot{r} &= -\kappa(r-\alpha\cos\phi) + \frac{2g^2\gamma_\perp}{\gamma_\perp^2+(\omega-\omega_0)^2}rZ \\
\dot{\phi} &= -\frac{\kappa\alpha}{r}\sin\phi + \frac{2g^2(\omega-\omega_0)}{\gamma_\perp^2+(\omega-\omega_0)^2}Z \qquad (31) \\
\dot{Z} &= -\gamma_\parallel (Z-Z_0) - \frac{4g^2\gamma_\perp}{\gamma_\perp^2+(\omega-\omega_0)^2}r^2Z.
\end{aligned}$$

This is a fairly simple set of nonlinear deterministic equations in three dimension which might stimulate the specialists of "strange attractors" [26]. Recently, indeed, IKEDA has discovered a "chaotic" behavior in such a simple optical system under the mistuning condition, $\omega\neq\omega_0$ (but not with precisely the same model as (31), taking into account the propagation of the produced internal mode through several cavities chained in a ring shape (i.e. a ring cavity [27]), thereby treating the cavity-loss effect not phenomenologically). He has attributed the arising of the "chaos" to random phase shifts of the internal mode at the transmission point of the cavity for which the misstuning plays the role: thus it is to be categorized in the "dispersive bistability" [28]. This could open a new type of optical instability but must be left to future investigations.

Turning back to the simpler situation of (31), i.e. the exact resonance $\omega=\omega_0$, we make two further adiabatic eliminations: $\dot{\phi}=0$ and $\dot{Z}=0$. Since $\omega=\omega_0$, the former condition yields possible two fixed points $\phi=0,\ \pi$, and the latter

$$Z = \frac{Z_0}{1+sr^2}, \qquad \text{where} \qquad s = \frac{4g^2}{\gamma_\parallel\gamma_\perp}. \qquad (32)$$

Introducing the scaled variables

$$x = \sqrt{s}r, \quad y = \sqrt{s}\alpha \quad 0 \leq x,\ y < \infty, \qquad (33)$$

and a fixed parameter

$$c = \frac{g^2}{\gamma_\perp \kappa}(-Z_0) > 0 \quad (Z_0<0 \text{ without the population inversion}), \tag{34}$$

we obtain

$$\dot{x} = -\kappa\left(x + \frac{2cx}{1+x^2}\right) \pm \kappa y \quad (+ \text{ sign corresponds to } \phi=0). \tag{35}$$

This has a real fixed point, $\dot{x}=0$, only for $\phi=0$ for which

$$y = x + \frac{2cx}{1+x^2}, \tag{36}$$

and, in the circumstance $c > 4$, three branches of the real root $x(y)$ exist that are the origin of the bistable absorption [19]. We note that this bistability takes place also in the opposite adiabatic situation, $\kappa >> \gamma_\perp$, where interesting side peaks of the emitted light appear (the dynamical Stark shift (DSS)) in resonant fluorescence [29], [21, 23]). According to WALLS et al [23], there exists yet another type of instability with a threshold of the external field intensity above which the stationary state involves an irreversible limit cycle.

Table 1 Classification of the optical instabilities contained in the model (10)

adiabatic condition	$\kappa << \gamma_\perp, \gamma_\parallel$	$\kappa >> \gamma_\perp, \gamma_\parallel$	
dynamical variables	active mode of em field a, a^*	dipoles R, R^*	population difference Z
$\alpha = 0$ $Z_0 > Z_{th}$	ordinary laser	mirrorless laser (ASE) $\gamma_\perp, \gamma_\parallel \to 0$ superfluorescence	
$\alpha \neq 0$ $Z_0 < 0$	absorptive bistability dispersive bistability (chaos)	resonant fluorescence (DSS) $\gamma_\perp, \gamma_\parallel \to 0$ superradiance (limit cycle)	

4. Thermodynamic Significance of the Adiabatic Elimination

The adiabatic elimination of a fast relaxing variable from a set of equations of motion is not merely the matter of convenience in treating a composite dynamical system: It is based on an intuitive consideration that a rapidly relaxing subsystem of the total system can be settled into the reservoir with which this subsystem is in contact. If γ denotes the characteristic relaxation rate of the subsystem, the adiabatic description should be valid for the time range as long as $t >> \gamma^{-1}$.

Let S and S' denote two components of a closed dynamical system described by a Hamiltonian (consisting of the parts to describe each respective natural motions and the interaction between them) and R and R' their respective reservoirs each of which is characterized by a parameter, here a temperature. Roughly speaking, if the two characteristic rates γ_{S-R} and $\gamma_{S'-R'}$ with R and R', respectively, satisfy the condition

$$\gamma_{S-R} << \gamma_{S'-R'} \tag{37}$$

then the subsystem S' can be eliminated from the dynamics by incorporating it into its reservoir R', the reduced dynamics being S only but its thermal contact added to R by the second reservoir R' via the eliminated subsystem S' indirectly. Under the special circumstance, $\gamma_{S-R}=0$, i.e. in the absence of the thermal contact of S with R, the condition (37) must be replaced by certain other inequalities $\bar{\gamma}_{S-R} << \gamma_{S'-R'}$ with characteristic rates $\bar{\gamma}_{S-R}$ to be deduced from the interaction between S and S', but the thermal contact of S now is only through the indirect contact with R'. The effect of external fields acting on S can be considered in general as a special reservoir of the infinite temperature. The consideration is summarized in Table 2 for the three basic instabilities of our optical system, viz. the ordinary laser, the superfluorescence and the absorptive bistability. In each of these examples the adiabatically reduced equation of motion has a simple structure of manifesting the above mentioned contacts as follows:

the ordinary laser One more adiabatic elimination of the atomic degree of freedom, i.e. the variable Z, from (22) and (23) yields

$$\dot{n} = -Cn + \frac{A(\omega)n}{1+s(\omega)n}, \tag{38}$$

with constants

$$C = 2\kappa, \quad A(\omega) = \frac{4g^2\gamma_\perp\gamma_\parallel Z_0}{\gamma_\perp^2+(\omega-\omega_0)^2} = \gamma_\parallel s(\omega)Z_0 \tag{38a}$$

and

$$s(\omega) = \frac{4g^2\gamma_\perp\gamma_\parallel^{-1}}{\gamma_\perp^2+(\omega-\omega_0)^2}. \tag{38b}$$

Clearly, 1st and 2nd terms in (38) represent the thermal contact of the lasing mode with the cavity wall and the atomic reservoir, respectively.

the superfluorescence The special case $\gamma_\perp=\gamma_\parallel=0$ in (25) and (26) reduces to

$$\dot{Z} = -\gamma_r Z_0(1-\frac{Z^2}{Z_0^2}) \qquad (Z_0>0: \text{ constant of motion in (27)}) \tag{39}$$

with the superradiant rate

$$\gamma_r = \frac{2g^2\kappa Z_0}{\kappa^2 + (\omega-\omega_0)^2}. \tag{39a}$$

Table 2 Structure of thermodynamic contact with reservoirs in the three optical instabilities

system plus reservoir	S	R	S' + R'	
ordinary laser	active mode	cavity wall	(dipole, population diff.	atomic reservoir
superfluorescence	(dipole, population diff.	*	active mode	cavity wall
absorptive bistability	active mode	external field	(dipole, population diff.	(atomic reservoir, cavity wall

In terms of the Bloch angle θ (measured from the "north pole"), (39) is replaced by the simple, well-known pendulum equation

$$\dot{\theta} = \gamma_r \sin\theta \qquad 0 \le \theta \le \pi \; . \tag{39'}$$

The limit $\theta=\pi$ is the unique stable equilibrium implied by the thermal contact with the cavity wall via the active mode that is eliminated.
<u>the absorptive bistability</u> Recall the reduced equation (35):

$$\dot{x} = -\kappa\left(x + \frac{2cx}{1+x^2}\right) + \kappa y \qquad 0 \le x \; , \; y < \infty \; . \tag{40}$$

In the right-hand side, 1st term represents the thermal contact of the total field amplitude x with the two reservoirs unified into one, the cavity wall plus the atomic reservoir, and 2nd the driving by the external field y. It should be remarked that this simple structure results from the definition of the system variable x i.e. the total field amplitude rather than the pure internal field.

Thus, we see that the adiabatically reduced equation in the deterministic level has the structure that the total velocity is a simple sum of components, each representing the contact with a different reservoir. The argument should be completed, however, by succeeding the same structure in the level of including fluctuations, as will be seen in the next two sections.

5. Aspects of Stochastic Calculus for Nonlinear Fluctuations

Theory of Brownian Motion has a long history initiated by Einstein and Langevin. By its mathematical nature, substantial progresses of this theory had to be carried by mathematicians (see, a very physical account of the pertinent mathematics by NELSON [30]). But such progresses must certainly be rewarded for other fields; statistical physics, engineering and population biology etc. One of such progresses is the stochastic calculus on Gaussian white noises with a use of the *symmetric multiplication of Stratonovich* [31]. We give an account of this and demonstrate how it can be used conveniently combined with the Itô formula for our optical problem. The idea of STRATONOVICH first appeared in his paper of 1966 [32] and is summarized in a monograph [33].

In Table 3 the difference is illustrated in definition of the multiplication of dW by a (smooth) function of a stochastic process X_t, $g(X_t)$, which

<u>Table 3</u> Illustration of the Itô and Stratonovich Integrals

multiplication of a white noise by $g(X_t)$

(I)	$g(X_t)dW_{Ii}(t)$	$g(X_{ti})(W(t_{i+1}) - W(t_i))$ 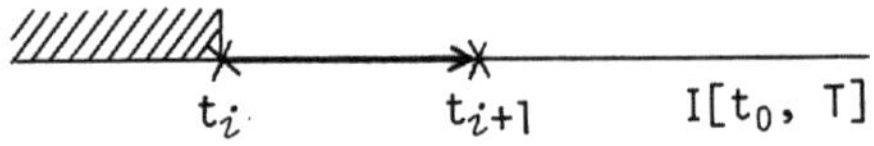
(S)	$g(X_t)\circ dW_{Ii}(t)$	$g\left(\frac{X_{ti}+X_{ti+1}}{2}\right)(W(t_{i+1}) - W(t_i))$ 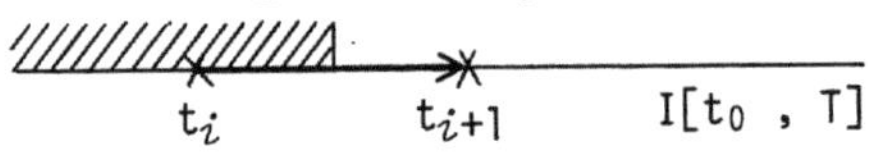

appears in the Itô integral and that in the Stratonovich integral, where $dW_{Ii}(t)=W(t_{i+1}) - W(t_i)$ i.e. the difference of the standard Wiener process in the subinterval $I(t_{i+1}, t_i)$ and X_t is assumed to correlate with this process in a probability space at each instant (i.e. both are sample functions of a common sample ω in the probability space; $X_t=X(t, \omega)$ and $W(t)=W(t, \omega)$). Note that the conventional writing the fluctuating force f(t) with unit strength, $< f(o)f(t) > = \delta(t)$, in a Langevin equation is based rigorously on the standard Wiener process that satisfies

$$< dW_{Ii}(t) > = 0, \quad < dW_{Ii}(t)dW_{Ii}(t) > = t_{i+1} - t_i, \tag{41}$$

i.e. being related to the latter as $\int_{t_0}^{t} f(t')dt = W(t)$; $t\varepsilon[t_0, T]$. The correlation of X_t with W(t) can be assumed to arise through an equation of the form

$$dX_t = a(X_t)dt + g(X_t)dW(t) \qquad t\varepsilon[t_0, T], \tag{42}$$

(with a possible exception at $t=t_0$ where X_t is independent of W(t)) that is a stochastic differential equation (SDE) of the Itô type, and the form $\Sigma g(X_t)dW(t)$ the Itô integral, provided $g(X_t)dW_{Ii}(t)$ is defined as in Table 3. For a more detail, the readers are referred to ARNOLD's book [34] which provides a plain account of the unique definition of the multiplication (I) in order for the integral to satisfy the *martingale* property: It assures the free expectation of the integral to vanish, $< g(X_t)dW(t) > = 0$. Any different choice of the multiplication gdW other than (I) in Table 3 does not satisfy this property: For example in (S), the Stratonovich symmetrized product,

$$\begin{aligned} g(\tfrac{1}{2}(X_{ti+1} + X_{ti}))dW(t) &= g(X_{ti} + \tfrac{1}{2}(X_{ti+1} - X_{ti}))dW(t) \\ &= g(X_{ti})dW(t) + \tfrac{1}{2}g'(X_{ti})\underbrace{(X_{ti+1} - X_{ti})(W(t_{i+1}) - W(t_i))}_{O(t_{i+1} - t_i)}, \end{aligned} \tag{43}$$

for which

$$< g(X_t)\circ dW(t) > = \frac{1}{2} g'(X_t)g(X_t)dt, \tag{44}$$

provided the process X_t is governed by the SDE (42).

ITÔ has shown [31] that the two definitions of the multiplication can be summarized by the most general relation

$$Y\circ dX = YdX + \tfrac{1}{2}dXdY \qquad X, Y\varepsilon Q(\text{local quasi-martingale}), \tag{45}$$

and any multiplications more than three dXs' may be discarded for any calculus within the infinitesimal duration dt(provided the Gaussian white characteristic is retained for X_t and Y_t). Consequently, the symmetric multiplication $Y\circ dX$ satisfies the same basic rules of algebra as usual:

$$(Y+Z)\circ dX = Y\circ dX+Z\circ dX, \qquad Y\circ(dX+dZ) = Y\circ dX+Y\circ dZ \tag{46}$$

Besides, $\quad Y\circ(Z\circ dX) = YZ\circ dX \quad$ (equivalent to $Y(ZdX) = (YZ)dX$). (47)

But one remerkable property of it is concerning differential calculus, given by

$$dg(X_1, \ldots, X_d) = \sum_{\mu}^{d} \frac{\partial g}{\partial X_\mu} \circ dX_\mu \qquad \text{(g: twice differentiable)} \tag{48}$$

$$= \sum_{\mu}^{d} \frac{\partial g}{\partial X_\mu} dX_\mu + \frac{1}{2} \sum_{\mu\nu}^{d} \frac{\partial^2 g}{\partial X_\mu \partial X_\nu} dX_\mu dX_\nu . \tag{48a}$$

The well-known Itô formula for the conditional expectation $F(t, X)(=\langle F(X_t)|X\rangle$ with $F(t_0, X)$ being given) and the Fokker-Planck equation for the probability distribution $P(t, X)$ to yield the expectation can be derived on this basis as

$$\frac{\partial F(t, X)}{\partial t} = a(X) \frac{\partial F}{\partial X} + \frac{1}{2} g^2(X) \frac{\partial^2 F}{\partial X^2} ,$$

$$\frac{\partial P(t, X)}{\partial t} = -\frac{\partial}{\partial X}(a(X)P) + \frac{1}{2} \frac{\partial^2}{\partial X^2}(g^2(X)P) \qquad \text{(associated with SDE (42))}. \tag{49}$$

The following is a concrete example applied to our optical problem [5]: transformation of the Brownian motion of photons to (n, ϕ) representation

$$db = -(\kappa + i\omega) b dt + dw, \qquad db^* = \text{c.c.} \qquad \langle dw^* dw \rangle = 2\kappa \bar{n} dt.$$

$$b = \sqrt{n} e^{i\phi}; \; n = b^* b, \qquad \phi = \frac{1}{2i} \log \frac{b}{b^*} . \tag{50}$$

Hence, by using (48) and (50),

$$dn = d(b^* b) = b^* \circ db + b \circ db^* = \mp 2\kappa n dt + b^* \circ dw + b \circ dw^*$$

to which the fundamental relation (45) is applied, resulting in

$$= -2\kappa n dt + b^* dw + b dw^* + \frac{1}{2}(db^* dw + db dw^*)$$

$$= -2\kappa(n - \bar{n}) dt + (b^* dw + b dw^*) \qquad \text{(Itô SDE)}. \tag{51}$$

It says that the deterministic equation for the intensity is subject to a real noise correction such that n tends to $\bar{n}$ expected from the Einstein relation. As to the phase part ϕ, such noise correction is absent, because

$$d\phi = -\omega dt + \frac{1}{2i}(b^{-1} \circ dw - b^{*-1} \circ dw^*) = -\omega dt + \frac{1}{2i}(b^{-1} dw - b^{*-1} dw^*). \tag{52}$$

Note that the extra drift $2\kappa\bar{n}$ in the Itô formula disappears again by the symmetric representation of the diffusion part: We see this in the F-P eq.,

$$\frac{\partial P}{\partial t} = 2\kappa\bar{n} \frac{\partial}{\partial n}\left(n\left(\frac{1}{\bar{n}} + \frac{\partial}{\partial n}\right)P\right) + \left(\omega \frac{\partial}{\partial \phi} + \frac{\kappa\bar{n}}{2n} \frac{\partial^2}{\partial \phi^2}\right)P \tag{53}$$

which allows the equilibrium distribution $P^{st}(n) = \bar{n}^{-1} e^{-n/\bar{n}}$ consistent with $\langle n \rangle_{eq} = \bar{n}$.

Our aim to extend the adiabatic elimination of fast relaxing variables to the Langevin equations including Gaussian white noises will be aided by the above stochastic calculus: We set up the following ansatz [5].

Set the rate of the fast-relaxing stochastic variable equal to zero just as the case in the deterministic equations, and substitute the variable thus obtained into the rest of the equations, where the multiplication of the white noise by a factor (generally, a smooth function of other stochastic variables) is carried out in accordance with the symmetrized multiplication (43).

We note that this prescription is seemingly adapted to the original STRATONOVICH's idea of symmetry restoration of time [34]. However, one must have a precise condition of its applicability for a given set of Langevin equations, which we hope to clarify near future. The Langevin equations (10), in fact, contain several subtle points in this respect which suggest that the ansatz is not entirely correct for (10) (see Footnote 2 and sec. 10), but let us proceed.

By way of example, let us apply it to the elimination of the dipole variables R and R* from (10) under the adiabatic condition $\kappa \ll \gamma_\perp$ by writing conventionally $\dot{R}=0$ (in the rotating frame with $\Omega=\omega$), yielding

$$\tilde{R} = \frac{-2ig}{\gamma_\perp - i(\omega-\omega_0)}\,\tilde{a}Z + \frac{1}{\gamma_\perp - i(\omega-\omega_0)}\,F, \qquad \tilde{R}^* = \text{c.c.}$$

and substituting them into the other three equations. The results are given by

$$\dot{\tilde{a}} = -\kappa(\tilde{a}-\alpha) + \frac{2g^2 Z}{\gamma_\perp - i(\omega-\omega_0)}\,\tilde{a} + \left(f + \frac{-ig}{\gamma_\perp - i(\omega-\omega_0)}F\right), \qquad \dot{\tilde{a}}^* = \text{c.c.}$$

$$\dot{Z} = -\gamma_\parallel(Z-Z_0) - \frac{4g^2\gamma_\perp n}{\gamma_\perp^2+(\omega-\omega_0)^2}\,Z + \left(F_z + \frac{ig\tilde{a}^*\circ}{\gamma_\perp - i(\omega-\omega_0)}\,F + \text{c.c.}\right),$$

where the circle $\circ$ in the bracket of *r.h.s* of the second equation indicates that the multiplication of the white noise F by $\tilde{a}^*$ is in the Stratonovich sense. That is, if $Fdt=dW$ and multiplied by $\tilde{a}^*$, then $\tilde{a}^*dW$ must be understood as $\tilde{a}^*\circ dW$. Let this noise term be reexpressed in terms of the product in the Itô sense by the fundamental relation (45):

$$\tilde{a}^*\circ dW = \tilde{a}^*dW + \frac{1}{2}\,d\tilde{a}^*dW = \tilde{a}^*dW + \frac{1}{2}\,\frac{ig}{\gamma_\perp + i(\omega-\omega_0)}\,dWdW^*$$

where

$$d\tilde{a}^* = \dot{\tilde{a}}^*dt = \frac{ig}{\gamma_\perp + i(\omega-\omega_0)}\,dW^*$$

has been inserted. Therefore,

$$\frac{ig\tilde{a}^*\circ}{\gamma_\perp - i(\omega-\omega_0)}\,F = \frac{ig}{\gamma_\perp - i(\omega-\omega_0)}\,\tilde{a}^*F - \frac{1}{2}\,\frac{2g^2\gamma\bar{M}}{\gamma_\perp^2+(\omega-\omega_0)^2}$$

which yields

$$\dot{Z} = -\gamma_\parallel(Z-Z_0) - \frac{4g^2\gamma_\perp}{\gamma_\perp^2+(\omega-\omega_0)^2}\left(nZ+\frac{1}{2}\bar{M}\right) + (\text{noise in the Itô sense}).$$

This is compared with $\dot{n}$ to be obtained from $\tilde{a}^*\circ d\tilde{a} + \tilde{a}\circ d\tilde{a}^*$, i.e.

$$\dot{n} = -2\kappa(n - \text{Re}\,\alpha\tilde{a}^* - \bar{n}) + \frac{4g^2\gamma_\perp}{\gamma_\perp^2+(\omega-\omega_0)^2}\left(nZ+\frac{1}{2}\bar{M}\right) + (\text{noise in the Itô sense}).$$

Here, the noise correction $\propto \bar{M}$ has come out as a supplement of the term nZ just as the correction $\propto \bar{n}$ of n. Notice that the two corrections for $\dot{n}$ and $\dot{Z}$ are totally cancelling, reflecting its internal nature of the noise force.

6. The Adiabatically Reduced Langevin Equations

When $\alpha_t=0$, the phase angle ϕ of the active mode and that of the dipole is separable from the rest of the equations, and the adiabatic eliminations for the two cases are performed (as shown before) with the following results:

Case $\gamma_\perp \gg \kappa$

$$\dot{n} = -2\kappa(n-\bar{n}) + \frac{4g^2\gamma_\perp}{\gamma_\perp^2+(\omega-\omega_0)^2}(nZ+\frac{1}{2}\bar{M}) + f_n \tag{54}$$

$$\dot{Z} = -\gamma_\parallel(Z-Z_0) - \frac{4g^2\gamma_\perp}{\gamma_\perp^2+(\omega-\omega_0)^2}(nZ+\frac{1}{2}\bar{M}) + f_z , \tag{55}$$

where the Gaussian white noises f_n and f_z are related to F, F_z in (10) by

$$f_n = (a^*f + \frac{-iga^*F}{\gamma_\perp - i(\omega-\omega_0)}) + c.c. , \qquad f_z + (\frac{iga^*F}{\gamma_\perp - i(\omega-\omega_0)} + c.c.) \tag{56}$$

with the multiplication defined in the Itô sense so that

$$< f_n(t) > = < f_z(t) > = 0. \tag{57}$$

Case $\gamma_\perp \ll \kappa$

$$\dot{M} = -2\gamma_\perp(M-\bar{M}) + \frac{4g^2\kappa}{\kappa^2+(\omega-\omega_0)}(MZ + \bar{n}(2Z^2-M)) + f_M \tag{58}$$

$$\dot{Z}^2 = -\gamma_\parallel(2Z(Z-Z_0) - \bar{M}) - \frac{4g^2\kappa}{\kappa^2+(\omega-\omega_0)^2}(MZ + \bar{n}(2Z^2-M)) + f_{Z2} , \tag{59}$$

where

$$f_M = (R^*F + \frac{igZR^*f}{\gamma_\perp + i(\omega-\omega_0)}) + c.c. , \qquad f_{Z2} = 2ZF + (\frac{-igZR^*f}{\gamma_\perp + i(\omega-\omega_0)} + c.c.) \tag{60}$$

with vanishing expectation as before.

In the former case, a further adiabatic elimination of the atomic freedom, the variable Z, is to be made under the condition $\gamma_\parallel \gg \kappa$. This is included in the following treatment for $\alpha \neq 0$ (in the presence of the external field):

$$\dot{n} = v_n^\alpha - 2\kappa(n-\bar{n}) + \frac{4g^2\gamma_\perp}{\gamma_\perp^2+(\omega-\omega_0)^2}(nZ+\frac{1}{2}\bar{M}) + f_n \tag{61}$$

$$\dot{\phi} = v_\phi^\alpha + \frac{2g^2(\omega-\omega_0)}{\gamma_\perp^2+(\omega-\omega_0)^2} Z + \tilde{f}_\phi \tag{62}$$

with

$$v_n^\alpha = 2\kappa\alpha n^{1/2}\cos\phi, \qquad v_\phi^\alpha = -\kappa\alpha n^{1/2}\sin\phi \tag{63}$$

and

$$f_\phi = \frac{1}{2i}(a^{-1}f + \frac{ig}{\gamma_\perp i(\omega-\omega_0)} a^{-1}F) + \text{c.c.} \tag{64}$$

We introduce the following notations (C an A refer to 'cavity' and 'atoms', respectively):

$$n_C = \bar{n}\,, \qquad n_A = \frac{\bar{M}}{-2Z_0} = \frac{\bar{M}}{N_1^{eq}-N_2^{eq}}\,. \tag{65}$$

Also using the notations previously introduced in (38a, b) i.e. $C=2\kappa$, $A(\omega)=\gamma_\parallel s(\omega)Z_0$, we have[2]

$$\dot{n} = v_n^\alpha + \{-C(n-n_C) + (a^*f+\text{c.c.})\}$$

$$+ \left\{ \frac{A(\omega)}{1+s(\omega)n}(n-n_A) + \frac{s(\omega)n}{1+s(\omega)n}F_z + \left(\frac{-ig}{1+s(\omega)n}\,\frac{a^*F}{\gamma_\perp - i(\omega-\omega_0)} + \text{c.c.}\right) \right\} \tag{66}$$

and a similar equation for $\dot{\phi}$. The full expressions for the pair of these variables, r and ϕ where $r^2=n$, are listed below for the special case of the resonance i.e. $\omega=\omega_0$:

$$\dot{r} = \kappa\alpha\cos\phi + \{-\kappa(r-\frac{n_C}{2r}) + \mathrm{Re}\; a^*f\}$$

$$+ \{\frac{2g^2\gamma_\perp^{-1}Z_0}{1+sr^2}\,(r - \frac{n_A}{2r}) + \mathrm{Jm}\;(\frac{g\gamma_\perp^{-1}}{1+sr^2}e^{-i\phi}F) + \frac{sr^2}{1+sr^2}\,\frac{F_z}{2}\} \tag{67}$$

$$\dot{\phi} = -\frac{\kappa\alpha}{r}\sin\phi + \mathrm{Jm}\;(a^{-1}f + ig\gamma_\perp^{-1}a^{-1}F). \tag{68}$$

Each curly bracket in the above expressions, (66) and (67), consisting of a drift term and fluctuating forces represents the contact with a unit thermal reservoir, here, the cavity wall and the atomic reservoir. Each fluctuating force, Gaussian white but with nonconstant multiplicative factors, is represented by the Itô definition so that the above Langevin equations are all Itô SDE. The physical significance of the noise correction thereof being included fully into the systematic part is that the variable n is ensured to tend to its expectation value at the respective equilibrium. Namely,

---

2. The adiabatic elimination of Z from (54) by setting $\dot{Z}=0$ according to the prescription results in the fluctuating force terms in 2nd bracket of (66) given by $\frac{s(\omega)n}{1+s(\omega)n}\circ F_z + (\frac{-ig}{\gamma_\perp - i(\omega-\omega_0)}\,\frac{1}{1+s(\omega)n}\circ a^*F + \text{c.c.})$. It turns out that this expression produces vanishing noise correction (an exact cancellation between the two corrections of 1st and 2nd terms), so that it is just equal to the same form under the Itô definition, as in (66). This is a subtle point of (66), which differs from the result by the simultaneous elimination [6]; an indication that our ansatz is not entirely correct. (See Remark in Sec. 10.) Accordingly, a correction is necessary for (66): n_A in the curly bracket should be replaced by $n_A(1+s(\omega)n)^{-1}$. Similarly for (67): n_A by $n_A(1-sr^2)(1+sr^2)^{-1}$.

$$n \longrightarrow n_C \quad \text{and} \quad n \longrightarrow n_A \quad (>0; \text{ for the ordinary population } Z_0<0).$$

While the former behavior of "tending" is exact (based on (51) and (53)), the latter involves the problem of nonlinear relaxation (exact only in the weak coupling limit $s(\omega)\propto g^2\to 0$) and hence it is understood as in the decoupled version of the nonlinear saturation factor $(1+s(\omega)n)^{-1}$. We also note that in (67) where the variable is changed into $r=n^{1/2}$, the noise correction arises in the form $(r-(2r)^{-1}n_C)$ and $(r-(2r)^{-1}n_A)$ rather than $(r-r^{-1}n_C)$, and $(r-r^{-1}n_A)$, respectively. This is due to the additional correction arising from the transformation (the rule subject to the stochastic calculus), and corresponds to the Jacobian prefactor r of the distribution function. (cf. Footnote 2)

Besides, in the presence of the external field, the reduced Langevin equations contain an extra drift velocity $v^\alpha(v_n^\alpha, v_n^\alpha)$ given by (63), or given by

$$v_r^\alpha = \kappa\alpha\cos\phi, \qquad v_\phi^\alpha = -\frac{\kappa\alpha}{r}\sin\phi. \tag{63'}$$

No fluctuating force is associated with this velocity, and it satisfies the condition of vanishing divergence (called the reversible drift $v^{(r)}$ [7]), i.e.

$$\operatorname{div} v^\alpha = \frac{\partial v_n^\alpha}{\partial n} + \frac{\partial v_\phi^\alpha}{\partial \phi} = \frac{1}{r}\frac{\partial}{\partial r}(r v_r^\alpha) + \frac{\partial v_\phi^\alpha}{\partial \phi} = 0. \tag{69}$$

This is the thermodynamic condition that the external field supplies energy with the system only in the form of *work* without the entropy production: For any probability density $P(x)$, $\int(\partial_\mu v_\mu P)\log P dx=-\int v_\mu(\partial_\mu P)dx=\int \operatorname{div} v P dx=0$.

Considering the drift velocity v^α in the reduced Langevin equation (66) or (67) as one independent part from the rest bracketed parts, we see that the fluctuating forces belonging to different parts are of no correlations such that $\langle f_i(o)f_j(t)\rangle = 0$ ($i\neq j$, and for the external field part, i=0, f=0). This is the most important requirement of the LEBOWITZ model, namely, the corresponding Fokker-Planck operator is of the structure that it is decomposed into a sum

$$\frac{\partial P}{\partial t}=LP, \quad L=L_0+L_1+L_2+\cdots; \quad L_0=-\frac{\partial}{\partial x_\mu}(v_\mu^{(r)}\cdot), \quad \operatorname{div} v^{(r)}=0, \tag{70}$$

and the other L_i an elementary Fokker-Planck operator representing the contact with i-th reservoir. (For the explicit presentation of these operators, see Table 4.)

7. Thermodynamic Characteristic of the Ordinary Laser

The laser instability has been compared with ferromagnetism [2]; the lasing state above threshold vs. the ordered state with spontaneous magnetization, and the simple model of the ordinary laser vs. the well-known Weiss mean-field theory of ferromagnetism. For this reason it was suggested to study the partition function of the Hamiltonian (1) with an indication of a critical point behavior, which is almost meaningless in view of the openness of the system. Therefore, one looks for some other way of illustration of the comparison. This is given by Fig. 1 to show the macroscopic laser light intensity to arise spontaneously with an increase of the pump. The extensive photon number n_s corresponds to the stable fixed point of the nonlinear rate equation (38) (thus exhibiting a *bifurcation* nature of the laser instability); i.e. the point of balance between the two rates, the power emission from the atoms and the power dissipation through the cavity wall. The nonequilibrium characteristic of this phase transition vs. the static phase transition in ferromagnetism consists in the above dynamic nature of the balanced quantities, and is summarized as follows:

(a) The threshold Z_{th} is the critical value of the pump Z_0 above which both powers become extensive quantities: Below threshold, the powers are $O(1)$, the determination of which is possible only from the associated fluctuation i.e. by taking into account the noise corrections, n_C and n_A ($=O(1)$) in the rate equation. Above threshold, the extensive powers represent the heat flows; the one produced from the atomic reservoir and the other that flow poured onto the cavity wall (which is actually partly taken outside of the cavity and partly dissipated). The former flow is given by

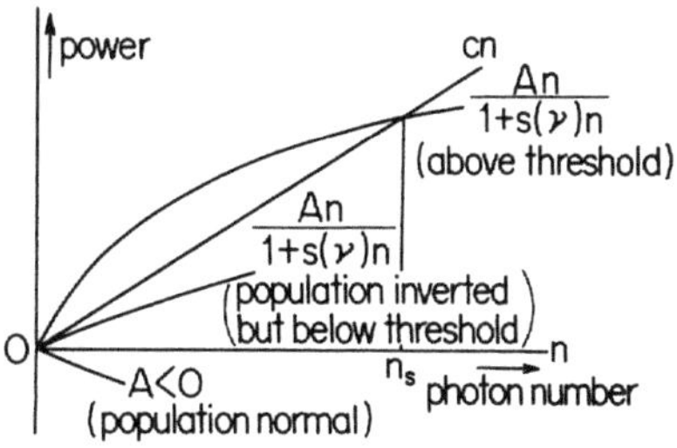

Fig. 1 The formation of the laser instability based on (38)

$$J_A^{(h)} = \left\langle \frac{A(n-n_A)}{1+s(\omega)n} \right\rangle \doteq \frac{An_s}{1+s(\omega)n_s} > 0, \tag{71}$$

where the average is over the stationary distribution of the lasing photons.
(b) The reason for the occurrence of the threshold is that the dynamical system (the lasing mode) is attracted by the two different thermal reservoirs, and the attraction by the atomic reservoir is driven far and far away from the other reservoir of the cavity wall to get a breakdown of this attraction; a picture fit for the Lebowitz description [9, 10] (see also [6]) of 2-reservoir open systems.
(c) The special circumstance of the dynamical system gaining heat from the atomic reservoir raises a thermodynamic question about the Second Law; if any alteration of this law. The question is linked also with the population inversion expressed by the negative absolute temperatures: Consider the rate equation (38) with the noise correction included ((66) with $v^{\alpha}=0$, and averaged)

$$\dot{n} = -C(n-n_C) + \frac{A}{1+s(\omega)n}\left(n - \frac{n_A}{1+s(\omega)n}\right) \qquad \text{(cf. Footnote 2)}$$

where by (65)

$$n_A = \frac{\overline{M}}{N_1^{-eq} - N_2^{eq}} < 0, \; Z_0 > 0. \tag{72}$$

However, an application of the Lebowitz formulation to the present situation assures no alteration of the Second Law, yielding the following expression:

$$\dot{S} = \sigma(P) + \frac{1}{n_C}J_C^{(h)} + \frac{1}{n_A}J_A^{(h)} \geq \frac{1}{n_C}J_C^{(h)} + \frac{1}{n_A}J_A^{(h)}, \tag{73}$$

where $\dot{S}$ is the rate of the information entropy of the dynamical system, $S=-\mathrm{Tr}\rho\log\rho$ (ρ: the probability distribution P in the classical sense by which the stochastic process (66) is described, i.e. a solution of the resulting Fokker-Planck equation), and $\sigma(P)$ a positive functional of the P called the entropy production [9]. At the stationary state of lasing $P=P^{st}$, $\dot{S}=0$, and $J_C^{(h)}=-J_A^{(h)}$ in the absence of the external work so that

$$\left(\frac{1}{n_C} - \frac{1}{n_A}\right)J_A^{(h)} = \sigma(P^{st})(>0) = O(1) \qquad \text{below threshold} \tag{74}$$

$= O(N)$ above threshold.

In the above threshold, the stationary entropy production is of the form

$(\frac{1}{n_C} + \frac{1}{|n_A|})J_A^{(h)}$, the positive sign of which necessitates $J_A^{(h)}>0$

that is the consequence of the Second Law. Nevertheless, the historical heat engine-theoretic formulation of the Second Law has a limitation (the Kelvin-Planck statement restricts the absolute temperatures to be positive), for which efforts of satisfactory reformulations have been made [35, 36] (see, a recent best formulation by NAKAGOMI [37], which is made consistent with (73)).

(d) The effective negative temperature above threshold has come out, (72), in a natural manner by virtue of the stochastic calculus (such was intuitively considered in the early laser equation by RISKEN [38]). In conjunction with the Planck formula and the quantum statistics of Pauli spins, it can be summarized by the *quantum Nyquist-noise* formula:

$$n_C = <\frac{1}{2}(b^*b + bb^*)> = \frac{1}{2}\coth\frac{\hbar\omega}{2kT_C},$$
$$n_A = \frac{\bar{M}}{N_1^{eq} - N_2^{eq}} = \frac{1}{2}\coth\frac{\hbar\omega_0}{2kT_A} \qquad (\bar{M} = <\frac{1}{2}(R^*R + RR^*)> = \frac{N}{2}). \tag{75}$$

Perhaps this 2-temperature description provides the best answer to the question of generalizing the temperature concept concerning optical pumps raised by INGARDEN [39].

8. Thermodynamic Characteristic of the Absorptive Bistability

The S-shaped curve of the total (transmitted)-vs. incident-field amplitude (x vs. y in (36) [19]) is shown in Fig. 2. A major theoretical effort seen in the recent investigations [20~25] is to find out a reasonable fluctuating force term that supplements the rate equation in order to fix the point of discontinuity on this curve: There must be a point y_0 on the y-axis (in Fig. 2) at which the lower branch of the total field x undergoes a transition to the upper branch, or vice versa. This point should be identified with the coincidence point of the double minimum of the potential function $U_y(x)=-\log P_y^{st}(x)$ (the double maximum of the stationary distribution $P_y^{st}(x)$), for which an adequate Fokker-Planck equation must be set up associated with the rate equation (40). It has been expected that the diffusion coefficient D would be not constant:

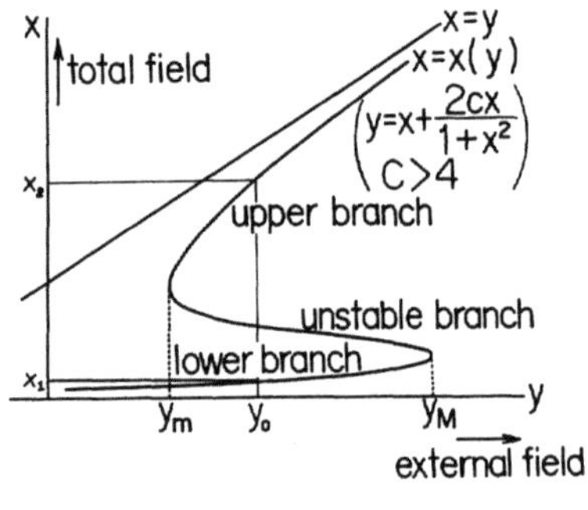

Fig. 2 The absorptive bistability based on (40) and (78)

$$U_y(x) = -\int^x \frac{v_y(x')}{D(x')}dx', \qquad v_y(x) = -\kappa\left(x + \frac{2cx}{1+x^2} - y\right). \tag{76}$$

This is the general structure of a one dimensional diffusion process and its stationary state whose maxima of the probability coincide with

$$\frac{\partial}{\partial x} U_y(x) = 0 \qquad \text{i.e.} \qquad y = x + \frac{2cx}{1+x^2}, \tag{77}$$

and therefore the point of discontinuity is given by

$$\int_{x_1(y_0)}^{x_2(y_0)} \frac{v_{y_0}(x)}{D(x)}\, dx = 0. \tag{78}$$

Several proposals to determine the unknown function $D(x)$ have been made in the cited literatures, but our solution in [6] is believed as most satisfactory, which we now summarize:

(a) The reduced Langevin equation which substitutes the rate equation (40) can be obtained by setting $\phi=0$ in (67, 68), and may be written in the scaled coordinates $x(=\sqrt{s}\,r)$ and $y(=\sqrt{s}\,\alpha)$ as

$$\dot{x} = \kappa y - \kappa\left(x - \frac{n_C s}{2x} + \frac{2c}{1+x^2}\left(x - \frac{n_A s}{2x}\frac{1-x^2}{1+x^2}\right)\right) + F_x\,, \quad \text{(cf. Footnote 2)} \tag{79}$$

where

$$F_x = \frac{\sqrt{s}}{2}(f+f^*) + \frac{\sqrt{s}g\gamma_\perp^{-1}}{1+x^2}\frac{1}{2i}(F-F^*) + \frac{sx}{1+x^2}\frac{1}{2}F_Z\,, \tag{80}$$

for which

$$\langle F_x(0)F_x(t)\rangle = \kappa s\left(n_C + \frac{2cn_A}{1+x^2}\right)\delta(t) \tag{81}$$

(cf. (11), (32) and (34)).

As we have discussed in the previous section, the system implied in (79) is in contact with three reservoirs; the cavity wall whose temperature is n_C, the atomic reservoir (the temperature $n_A>0$), and the external field (the temperature ∞), but in the thermal equilibrium condition

$$n_C = n_A = \bar{n} \quad \text{(condition of detailed balance in (79)).} \tag{82}$$

This means that the two thermal reservoirs are identical so that the system is attracted by one thermal reservoir and the external field, effectively. The diffusion coefficient is then from (81)

$$D(x) = \frac{\kappa s\bar{n}}{2}\left(1 + \frac{2c}{1+x^2}\right). \tag{83}$$

(b) One remarkable point of the form (83) is that for $y\to 0$

$$U_y'(x) = -\frac{v_y(x)}{D(x)} = -\frac{2}{s\bar{n}}x \qquad \left(P_0^{st}(x) \propto r e^{-\frac{1}{\bar{n}}r^2}\right), \tag{84}$$

which is consistent with the black-body radiation in the absence of the external field. (The proposed forms in [20-25] are either constant or only the part due to the atomic noise: Both do not satisfy the above condition.) For an arbitrary (positive) y

$$U_y(x) = -\frac{1}{s\bar{n}}\left\{(x-y)^2 + \frac{4cy}{\sqrt{2c+1}}\tan^{-1}\frac{x}{\sqrt{2c+1}}\right\}. \tag{85}$$

(c) The necessity of the use of the radial coordinate r (the amplitude of the total field) rather than n(the intensity) stems from the fact that it enables one to eliminate the phase ϕ without violating the divergenceless character of the external field ($v_r^\alpha=\kappa\alpha$=const., and hence $\frac{d}{dr}v_r^\alpha$(div v^α)=0). Here, the electric current is produced (besides the heat flow) as a response to the external field expressed as a constant current $J=\kappa\alpha$, the response being nonlinear. Hence, the question arises about a test of extendability of the elementary linear-irreversible thermodynamics in terms of entropy production.

(d) The linkage between the potental function $U_y(x)$ and the diffusion coefficient $D(x)$ indicated in (76) and explicitly given by (83, 85) is of the unique structure for one-dimensional systems such that the response of the total field against the external current is given by the average of the inverse of $D(x)$ over the stationary state $P_y^{st}(x) \propto r\exp(-U_y(x))$, and that by this nonlinear relation the entropy production is expressed as [flux] x [force]:

$$< r > = \frac{s\bar{n}}{2} < D(x)^{-1} > J \qquad \text{(a nonlinear function of } J = \kappa\alpha) \tag{86}$$

$$\sigma(P^{st}) = \frac{2}{\bar{n}} J < r > = s < D(x)^{-1} > J^2. \tag{87}$$

Table 4 Fokker-Planck operator and the stationary distribution associated with the Langevin equation for each reduced process (the phase part separated or omitted)

	operation on prob. distribution P	Stationary distribution P^{st}
ordinary laser	$L=L_C + L_A$	(far from thermal equilibrium)
	$L_C=C_{n_C}\frac{\partial}{\partial n}(n(\frac{1}{n_C}+\frac{\partial}{\partial n})$.	const. e^{-n/n_C}
	$L_A=\lvert A(\omega)n_A\rvert\frac{\partial}{\partial n}(\frac{n}{1+s(\omega)n}(\frac{1}{n_A}+\frac{\partial}{\partial n})$.	const. e^{-n/n_A} †
absorptive bistability	$L=L_0 + L_{C+A}$	(far from thermal equilibrium)
	$L_0=\kappa\frac{\partial}{\partial x}(y$.	
	$L_{C+A}=\kappa\frac{\partial}{\partial x}(\frac{2c}{1+x^2}(x-\frac{s\bar{n}}{2x}+\frac{s\bar{n}}{2}\frac{\partial}{\partial x})$.	const. $xe^{-x^2/s\bar{n}}$ †
superfluorescence	$L=L_\theta$(or L_Z;$z=\cos\theta$)	(thermal equilibrium)
	$L_\theta=\gamma_r\frac{\bar{n}}{Z_0}\frac{\partial}{\partial\theta}(-\frac{Z_0}{\bar{n}}\sin\theta+\frac{\partial}{\partial\theta}\frac{1}{\sin\theta}$.	const. $\sin\theta\, e^{(-Z_0/\bar{n})\cos\theta}$
	$L_Z=\gamma_r\frac{\bar{n}}{Z_0}\frac{\partial}{\partial z}((1-z^2)(\frac{Z_0}{\bar{n}}+\frac{\partial}{\partial z})$.	const. $e^{(-Z_0/\bar{n})z}$

† See Remark in the end of Sec.10.

So far discussions are restricted to stationary states; both the lasing state and the bistable state, classified into the so-called far from thermal equilibrium state driven by external means (see Table 4). The rest part will be devoted to a feature of transient fluctuations associated with unstable motions; typically with a superfluorescent dipole motion.

9. Transient Characteristic of the Superfluorescence

Under the special condition $\gamma_\perp = \gamma_\parallel = 0$, the reduced Langevin eqs. (58, 59) are simplified to be written as

$$\dot{M} = \frac{4g^2\kappa}{\kappa^2+(\omega-\omega_0)^2}\,(MZ + \bar{n}(2Z^2-M)) + f_M\,, \qquad f_M = 2\mathrm{Re}\,\frac{igZR^*f}{\gamma_1+i(\omega-\omega_0)}$$
$$\dot{Z} = -\frac{4g^2\kappa}{\kappa^2+(\omega-\omega_0)^2}\,(MZ + \bar{n}(2Z^2-M)) - f_M\,, \tag{88}$$

ensuring the Bloch-vector length conservation including the noise. Hence, by setting $M=Z_0^2\sin^2\theta$ and $Z=Z_0\cos\theta$, we get (by the stochastic calculus [5])

$$\dot{\theta} = \gamma_r(\sin\theta + \varepsilon\cot\theta) + f_\theta. \qquad < f(0)f(t) > = \gamma_r 2\varepsilon\delta(t);\ \varepsilon = \frac{\bar{n}}{Z_0} \tag{89}$$

The resulting Fokker-Planck equation is listed in Table 4.

In Sec. 3 we made a brief discussion on the unstable motion of the pendulum in the upper semisphere where the drift velocity, $\gamma_r(\sin\theta+\varepsilon\cot\theta)$, is positive and has positive derivatives in $0<\theta_0<\theta\le\frac{\pi}{2}$ so that the Bloch angle is accelerated. It has been argued that a great enhancement of fluctuations occurs associated with such an unstable motion, named "anomalous fluctuations" [40]. A rough feature of this can be analyzed by the time development of the *variance* of a (normalized) macrovariable following the system-size expansion theory [41], [42]:

$$\dot{y} = C_1(y), \qquad \dot{\sigma} = 2C_1'(y(t))\sigma + C_2(y(t)), \tag{90}$$

where $\sigma(t)=<(x-y(t))^2>$ represents the variance of the actual stochastic processes around the deterministic path $y(t)$ subject to 1st eq. of (90). It can be seen from 2nd eq. of (90) that the variance σ increases exponentially in the range $C_1'(y)>0$ that is the case for the Bloch angle θ ($\theta_0 < \theta \le \frac{\pi}{2}$; $\theta_0=O(N^{-1/2})$). Eqs. (90) refer to the stochastic process described by

$$\frac{\partial}{\partial t}P(t,x) = -\frac{\partial}{\partial x}(C_1(x)P) + \frac{1}{2}\frac{\partial^2}{\partial x^2}(C_2(x)P) \qquad C_2(x) > 0 \tag{91}$$

We will not go into the problem raised by VAN KAMPEN that (91) should be replaced by the equation proposed by him [41], but will note a counter proposal that (90) be replaced by

$$\dot{y} = C_1(y) + \frac{\varepsilon}{2}C_1''(y)\sigma, \qquad \dot{\sigma} = (2C_1'(y) + \frac{\varepsilon}{2}C_2''(y))\sigma + C_2(y), \tag{90'}$$

to determine an approximate solution to (91) in the form

$$P(t,x) = \frac{1}{\sqrt{2\pi\varepsilon\sigma(t)}}\exp[-\frac{1}{2\varepsilon\sigma(t)}(x-y(t))^2]. \tag{92}$$

(This is one step higher than the stated system-size expansion scheme, and can be deduced from (91) correctly by the variational method [43].) Our aim is to find a characterization of this enhancement of fluctuations in terms of the entropy production, motivated by the Glansdorff-Prigogine concept of "evolution criterion" [44].

We shall be based on the stability argument of the Liapunov function theory [45]. Namely, if a scalar function $V(x)$ of a vector x exists satisfying

$$V(x(t) \geq 0, \quad \frac{d}{dt}V(x(t)) \leq 0 \quad t_0 \leq t < \infty, \tag{93}$$

then, the vector-valued solution $x(t)$ of a set of evolution equations is stable. For a Markov process a universal Liapunov functional of the probability density P is known to exist, i.e. the *relative entropy*, $S(P|P^{st})=\int(\log P-\log P^{st})P dx$ which satisfies $S(P|P^{st})\geq 0$, $\frac{d}{dt}S(P|P^{st})\leq 0$. (See, the latest theory in the quantum dynamical semigroup version [46].) In view of its universal nature, this functional cannot provide a criterion of instability and so one looks at its time derivative, that is the entropy production in the Lebowitz theory: $\sigma(P)=-\frac{d}{dt}S(P|P^{st})\geq 0$. The motion is stable as far as $\frac{d}{dt}\sigma(P)\leq 0$ from (93), which is shown to be the case for sufficiently near the stationary state; i.e. for t>T with sufficiently large T. Therefore, it is interesting to see the time behavior of the $\sigma(P)$ for the superfluorescence whether the instability could be correctly predicted by this quantity. This has been done by using the time-dependent variational solutions of the superfluorescence Fokker-Planck equation (for the Gaussian trial distribution of the type (92) with $x=z=\cos\theta$), exhibiting a clear, affirmative answer [47]. (Fig. 3 shows this: $\sigma(P)$ once decreases and then increases attaining a maximum near τ_m, the maximum variance point, and finally decreases steadily.)

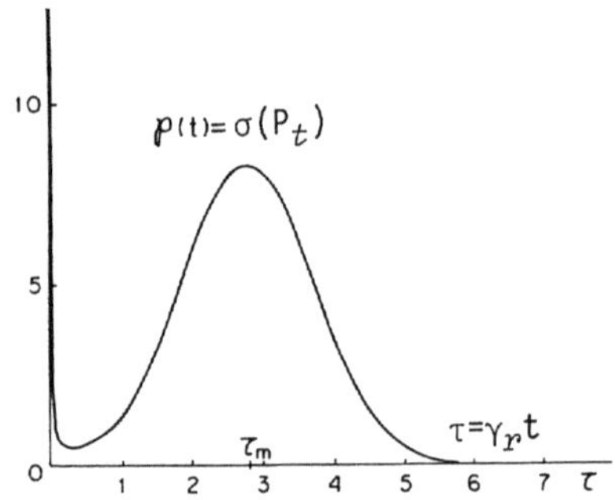

Fig. 3 The entropy production $\sigma(P)$ for the superfluorescence showing an unstable range

10. Towards the Theory of Self-Organization

A remarkable advantage of the Lebowitz model of open systems is that it provides a clear structure of the far-from-thermal equilibrium state: It is by referring it to a decomposition into several thermal equilibrium states so that every thermodynamic quantities can be expressed by this reference. Whenever more than two such reference states for a dynamical system are far distant from each other, the resulting stationary state of the system could be non-thermal, which is to be determined by solving an equation of the form

$$LP^{st}=0, \quad L=L_0+\Sigma_{i=1}^{r}L_i \text{ (discussed in connection with (70)).}$$

For the present example of the nonlinear optical system, we have shown how such decomposition can be made by means of the adiabatic elimination procedure.

Among various thermodynamic quantities the entropy production is of a special importance, for which LEBOWITZ [9, 10] prescribes the decomposition

$$\sigma(P) = \sum_{i=1}^{r} \sigma_i(P), \quad \sigma_i(P) = \mathrm{Tr}(L_i P)(\log P_i^{eq}-\log P) \geq 0. \tag{94}$$

This is a natural generalization of the argument in Sec. 9, and it is easy to see, by noting $\mathrm{Tr}(L_0P)\log P=0$, that the $\sigma(P)$ can be reexpressed as

$$\sigma(P) = -\frac{d}{dt}\,\mathrm{Tr}P\log P + \sum_{i=1}^{r}\mathrm{Tr}(L_iP)\log P_i^{eq}$$
$$= \frac{d}{dt}(-\mathrm{Tr}P\log P) - \sum_{i=1}^{r}\beta_i\mathrm{Tr}(L_iP)H \qquad (\log P_i^{eq} = -\beta_i H), \tag{95}$$

i.e. 1st term the total time-derivative of the information entropy and 2nd the sum of [inverse temperature] x [heat flux]. At the stationary state, the total time-derivative vanishes so that $\sigma(P)=-\Sigma\beta_iJ_i$ $(J_i=\mathrm{Tr}(L_iP^{st})H)$, thus, it becomes equal to the thermodynamic entropy production. It does not vanish unless $\beta_1=\beta_2=\cdots$ and the external field is absent: Therefore, the entropy production provides a measure of deviation from the absolute thermal equilibrium for the system in the stationary state, and furthermore from a transient dynamics point of view it would be a Liapunov functional to predict the presence of instability. (The latter is only a guess based on our experience.) It is a natural physical intuition that the anomalous enhancement of transient fluctuations seen in such unstable dynamics is a reflection of the counter ordering-process executed by the fast relaxing subsystem. Namely, in the process of lasing oscillation, for example, the dipole alignment enhances necessarily such an anomalous fluctuation, and in the superfluorescence the dipole fluctuation we have obtained must reflect a kind of ordered state in the superfluorescent light, perhaps its *coherence*. This is an expected subject of self-organization in our model optical system by means of the entropy production [48]. It is noted that the thermodynamic principle of *minimal entropy production* has a statistical significance that it characterizes the stationary state in the presence of *flows* [49], to which the JAYNES principle [50] is not applicable.

It is an important theoretical task to examine and establish the basis of the simple formula (94). For this purpose one must have a more powerful method of approach than the conventional adiabatic elimination, and it should be provided by the *projector method* [51] [52].

Remark During writing this article, the author has been noticed by prof. H. Mori that the adiabatic elimination of fast relaxing variables in a set of Langevin equations in the conventional way (the ansatz stated in Sec. 5) is not generally correct and that the correct result should be provided by means of the projector method (his paper is to appear in Prog. Theor. Phys. [53]). This has stimulated the author to investigate and establish the precise condition of applicability of the conventional elimination: Consider a set of real Langevin equations of the form

$$\dot{x}_\mu = v_\mu(x) + \Sigma_i a_{\mu i}(x)u_i + f_\mu \qquad \mu = 1, 2 \cdots d$$

$$\dot{u}_i = -\Sigma_j(\gamma_{ij}(x)u_j + b_i(x)) + f_i \qquad i = d+1, d+2 \ldots d+m,$$

where f_μ and f_i are δ-correlated white noises defined in the Itô sense and the $m \times m$ regression matrix $\Gamma=(\gamma_{ij}(x))$ is positive $(\Sigma\gamma_{ij}u_iu_j>0,\ \forall u(\neq 0)\in m$-dimensional real vectors). For simplicity, we assume the stationary covariance matrix V of u_i's to be independent of x. Then, the elimination of u_i's from the above set of equations for a long-time scale can be correctly performed as stated in the ansatz (i.e. $\dot{u}_i = 0$ to obtain u_i's which are inserted into the rest equations) for arbitrary $a_{\mu i}(x)$'s, if and only if the following conditions are fulfilled, namely [54]

(1) the potential condition, $\Gamma V = V\Gamma'$ (Γ': the transpose of Γ)

(2) no correlation between f_μ's and f_i's, $< f_\mu(o) f_i(t) >=0$, $\forall \mu, i$.

It can be generalized to complex Langevin equations of the similar form (for which Γ' is replaced by the hermitian conjugate Γ^+ and $V^+=V$). Consequently, the starting laser Langevin equations (10) can be reexamined by this criterion, and it can be seen that both (1) and (2) are fulfilled when the dipole R and R* or the mode amplitude a and a* are regarded as the u-variables, which ensures the validity of (54, 55) or (58, 59). For the ordinary laser ($\kappa \ll \gamma_\perp$, $\gamma_\parallel$), the further elimination of Z in going from (54, 55) to (66) has a problem that (2) is violated, and also for the simultaneous elimination of R, R* and Z that are considered as the u-variables (1) is violated. Thus, the reduced Langevin equations (54, 55) and (58, 59) are valid, but corrections are necessary for the results (66) and (67). These corrections avoid an ambiguous argument employed in [5, 6] to deduce the Fokker-Planck operators for the ordinary laser and the absorptive bistability which are listed in Table 4.

References

1 Lecture Note given by H. Haken and by M. Lax in Dynamical Processes in Solid-State Optics, ed. by R. Kubo and H. Kamimura, Syokabo (1966).
2 M. Sargent, M. O. Scully and W. E. Lamb, Laser Physics, Addison-Wesley Pub. Comp. London (1974).
3 For example, Proc. Int. Symp. on Nonlinear Nonequilibrium Statistical Mechanics, Prog. Theor. Phys. Suppl. No. 64 (1978).
4 H. Haken, Synergetics-An Introduction (Springer-Verlag Berlin Heidelberg New York, 1978).
5 H. Hasegawa and T. Nakagomi, J. Stat. Phys. 21, 191 (1979).
6 H. Hasegawa, T. Nakagomi, K. Kondo and M. Mabuchi, Nonequilibrium Thermodynamics of Lasing and Bistable Optical Systems, J. Stat. Phys. to be published. See also Opt. Comm. 32, 136 (1980).
7 R. Graham, in Coherence and Quantum Optics, ed. by L. Mandel and E. Wolf, Plenum Press, New York, (1973) p. 851: in Springer Tracts in Modern Physics, No. 66 (1973).
8 P. Glansdorff and I. Prigogine, Thermodynamic Theory of Structure, Stability and Fluctuations, Wiley-Interscience, London (1971).
9 H. Spohn and J. L. Lebowitz, Adv. Chem. Phys. 38, 109 (1978).
10 J. L. Lebowitz, Phys. Rev. 114, 1192 (1959).
11 R. Kubo, Rept. Progr. Phys. 24 Part 1, 255 (1966).
12 W. H. Louisell, Radiation and Noise in Quantum Electronics, McGraw-Hill, New York (1964).
13 R. Bonifacio and L. A. Lugiato, Phys. Rev. A11, 1507 (1975); A12, 587 (1975).
14 R. H. Dicke, Phys. Rev. 93, 99 (1954).
15 Cooperative Effects in Matter and Radiation, ed. by C. M. Bowden, D. W. Howgate and H. R. Robl (Plenum, New York, 1977).
16 J. Okada, K. Ikeda and M. Matsuoka, Opt. Comm. 26, 189 (1978); See also Theory of Cooperative Cascade Emission I, II to be published in J. Phys. Soc. Japan.
17 S. L. McCall, Phys. Rev. A9, 1515 (1974).
18 H. M. Gibbs, S. L. McCall and T. N. C. Venkatesan, Phys. Rev. Lett. 36, 1135 (1976).
19 R. Bonifacio and L. A. Lugiato, Opt. Comm. 19, 172 (1976); Phys. Rev. Lett. 40, 1023 (1978).
20 R. Bonifacio, M. Gronchi and L. A. Lugiato, Phys. Rev. A18, 2266 (1978).
21 G. S. Agarwal, L. M. Nardacci, R. Gilmore, and Da Hsuan Feng, Phys. Rev. A18, 620 (1978).
22 C. R. Willis, Opt. Comm. 23, 151 (1977); 26, 62 (1978).

23 D. F. Walls, P. D. Drummond, S. S. Hassan and H. J. Carmichael, Prog. Theor. Phys. Suppl. No. 64, 307 (1978). See also Opt. Comm. 27, 160 (1978); 27, 480 (1978).
24 A. Schenzle and H. Brand, Opt. Comm. 27, 485 (1979).
25 F. T. Arecchi and A. Politi, Opt. Comm. 29, 361 (1979).
26 D. Ruelle, Prog. Theor. Phys. suppl. No. 64, 339 (1978).
27 R. Bonifacio and L. A. Lugiato, Lett. Nuovo Cimento 21, 505 (1978); 510 (1978).
28 K. Ikeda, Opt. Comm. 30, 257 (1979).
29 C. Cohen-Tannoudgi and S. Reymond, J. Phys. B 10, 365 (1977).
30 E. Nelson, Dynamical Theories of Brownian Motion (Princeton University Press, 1967).
31 K. Itô, Applied Math. Optimization 1, 374 (1975); Lecture Notes in Physics 39, 218 (1975).
32 R. L. Stratonovich, J. SIAM Control 4, 362 (1966).
33 R. L. Stratonovich, Conditional Markov Processes and Their Application to Optimal Control, Elsevier, N. Y. (1968).
34 L. Arnold, Stochastic Differential Equations, Wiley-Interscience Pub. N. Y. (1973).
35 The necessity of revise of the heat-engine theoretic formulation of the Second Law was first discussed by N. F. Ramsey, Phys. Rev. 103, 20 (1956).
36 P. T. Landsberg, J. Phys. A 10, 1773 (1977); See also Thermodynamics Interscience Pub. N. Y. and London (1961).
37 T. Nakagomi, J. Phys. A 13, 291 (1980).
38 H. Risken, Z. Physik 186, 85 (1965); See also Prog. in Optics 8, 239 (1970).
39 R. S. Ingarden, Notion de Temperature et Pompage Optique, Panstwowe Wydawnictwo Naukowe, Warszawa (1968).
40 M. Suzuki, Physica 86A, 622 (1977); 90A, 303 (1978); Prog. Theor. Phys. Suppl. No. 64, 402 (1978).
41 N. G. van Kampen, Adv. Chem. Phys. 34, 245 (1976); Phys. Lett. 62A, 383 (1977).
42 R. Kubo, K. Matsuo and K. Kitahara, J. Stat. Phys. 9, 51 (1973).
43 H. Hasegawa, Prog. Theor. Phys. 58, 128 (1977); also work with M. Mabuchi (unpublished).
44 P. Glansdorff and I. Prigogine, Physica 30, 351 (1964); See also 8.
45 For example, J. La Salle and S. Lefschetz, Stability by Liapunov's Direct Methods and Applications, New York, Academic Press (1967).
46 H. spohn, J. Math. Phys. 19, 1277 (1978).
47 H. Hasegawa, S. Sawada and M. Mabuchi, in Coherence and Quantum Optics IV, ed. by L. Mandel and E. Wolf, Plenum Press, New York, (1978) p. 667.
48 H. Hasegawa and T. Nakagomi, Suppl. Prog. Theor. Phys. 64, 321 (1978).
49 H. Hasegawa and T. Nakagomi, On the Characterization of the Stationary State for a Class of Dynamical Semigroups, to be published in J. Stat Phys.
50 E. T. Jaynes, Phys. Rev. 106, 620 (1957).
51 R. Zwanzig, Phys. Res. 124, 783 (1961); Suppl. Prog. Theor. Phys. 64, 74 (1978).
52 H. Mori, Prog. Theor. Phys. 33, 423 (1965).
53 H. Mori, T. Morita and K. Mashiyama, Contraction of State Variables in Nonequilibrium Open Systems I, II, to be published in Prog. Theor. Phys.
54 H. Hasegawa and M. Mabuchi, Phys. Lett. to appear.

New Photon-Echo Phenomena in a Localized-Electron Phonon System

M. Aihara

Department of Physics, Faculty of Liberal Arts, Yamaguchi University
Yamaguchi 753, Japan

ABSTRACT

It is predicted that a new type of photon echoes appears in a strongly coupled localized-electron phonon system with a broad spectrum. These echoes arise because of the reversibility of the decay of the phonon wave-packet, and their formation is independent of the inhomogeneous broadening of spectral line due to the spatial crystalline field inhomogeneity.

1. Introduction

The problem of resonant light scattering and luminescence in solids has been the subject of a number of theoretical and experimental investigations. Recently, many publications concerning the time-dependent phenomena have provided essential information for a better understanding of resonant light scattering and luminescence. These works have, however, been restricted to the limiting situation of a weak excitation, that is, a linear response. It seems to me that a nonlinear resonance scattering in solids is one of the most interesting problems demanding solutions, since nonlinear effects caused by intense incident radiation become observable in the resonance condition. The characteristic feature of the nonlinearity is a violation of the principle of superposition. One of the most fundamental problems associated with the nonlinear transient response is the two-pulse excitation. That is, the two-pulse excitation is expected to induce an interesting transient response, which cannot be explained by a simple superposition of two independent responses due to a single-pulse excitation.

In the field of the coherent nonlinear optics, it is well-known that the two-pulse excitation gives rise to a photon echo [1]. This phenomenon is essentially due to the reversibility of the dephasing process of many oscillating dipoles with different frequencies. The purpose of this paper is, however, to show that a new photon-echo phenomenon arises in a single localized-electron which is strongly coupled with the phonon system, and to investigate how echo characteristics reflect the dynamics of the localized-electron phonon system.

2. Conventional Echo Theory

First of all, let us briefly review how conventional photon echoes [1] arise in an inhomogeneously broadened system. In order to confine our attention to the essential aspects of echo formation, we consider here a simple nondegenerate two-level system. The Hamiltonian for a two-level atom in a classical optical field is expressed in terms of the Pauli operators σ_z and σ_x;

$$H = (\Omega/2)\sigma_z - \mu\sigma_x E \cos \omega t . \qquad (1)$$

Here, Ω is the atomic transition frequency, μ is the matrix element of the electric-dipole moment operator, E is the amplitude of the electric field, and ω is the optical frequency.

In describing the echo phenomena, it is convenient to make use of the density operator $\rho(t)$ in the rotating-frame representation:

$$\rho(t) = e^{(i/2)\omega\sigma_z t} \rho_s(t)\, e^{-(i/2)\omega\sigma_z t} , \qquad (2)$$

where $\rho_s(t)$ is the density operator in the Schrödinger picture. It is easy to show that $\rho(t)$ satisfies the following equation of motion:

$$i \frac{d\rho(t)}{dt} = [(\Delta\omega/2)\sigma_z - (\mu E/2)\sigma_x, \rho(t)] . \qquad (3)$$

Here, the rotating-frame approximation has been used, and $\Delta\omega=\Omega-\omega$ is the detuning parameter. The advantage of this rotating-frame picture over the usual interaction picture is that the Hamiltonian becomes time independent. Therefore, we can immediately obtain the solution for $\rho(t)$ which has the form

$$\rho(t) = U(t, t_0)\, \rho(t_0)\, U^{\dagger}(t, t_0) , \qquad (4)$$

where the time-evolution operator $U(t, t_0)$ is simply expressed by

$$U(t, t_0) = \exp[-(i/2)(\Delta\Omega\sigma_z - \mu E\sigma_x)(t-t_0)] . \qquad (5)$$

When the two-level atom is excited by two optical pulses applied at time zero and τ_s as shown in Fig. 1, the density operator $\rho(t)$ takes the form

$$\rho(t) = e^{-(i/2)\Delta\Omega(t-\tau_s)\sigma_z} U_2\, e^{-(i/2)\Delta\Omega\tau_s\sigma_z} U_1\, \rho(0) \times U_1^{\dagger}\, e^{(i/2)\Delta\Omega\tau_s\sigma_z} U_2^{\dagger} e^{(i/2)\Delta\Omega(t-\tau_s)\sigma_z} , \qquad (6)$$

where, $\rho(0)$ is the initial density operator. The operators $U_j=\exp[i\theta_j\sigma_x/2]$, (j=1,2), in the above equation describe the change in the system due to the first and second pulses, respectively. Here, $\theta_j=\mu E_j\Delta t$ is the pulse area, and the pulse duration Δt is assumed to be much shorter than the inverse

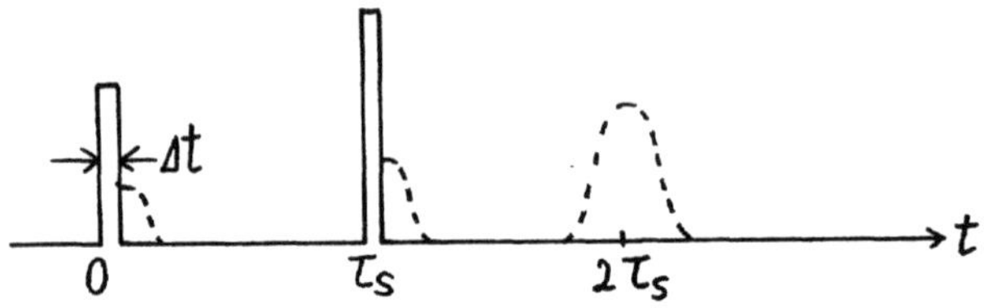

Fig. 1 Schematic representation of a sequence of two pulse excitation. The dashed curves indicate emitted radiation.

of the detuning parameter $\Delta\Omega$. The exponential operator $\exp[-(i/2)\Delta\Omega\tau_s\sigma_z]$ in (6) describes the phase change of the system caused by the off-resonance during the interval between the two pulses, and $\exp[-(i/2)\Delta\Omega(t-\tau_s)\sigma_z]$ describes the phase change of the system after the second pulse.

The induced electric-dipole moment is obtained from the off-diagonal element of the density operator $\rho(t)$, and is expressed by

$$\mu(t) = \mu <-|\rho(t)|+> + \text{c.c.}$$

$$= \mu \sum_{\alpha,\beta=\pm 1}\sum \exp[i\Delta\omega(t-\frac{2+\alpha-\beta}{2}\tau_s)] <-|U_2|\alpha><\alpha|U_1|-><+|U_2|\beta>^*<\beta|U_1|->^* + \text{c.c.} \tag{7}$$

Here, the minus and plus signs denote the ground and excited states, respectively. Kets $|\alpha>$ and $|\beta>$ are the states after the first pulse for forward and backward propagations, respectively. If we set α at +1 (excited state) and β at -1 (ground state), then we obtain the factor $t-2\tau_s$ which suggests that the echo is formed at time $2\tau_s$. Other combinations of α and β are not related to echo formation but to emissions occurring immediately after the excitation pulses, which are called the optical free induction decay.

Restricting our attention to the one term related to the echo formation, the induced electric-dipole moment has the form

$$\mu_n(t) = \mu\, e^{i\Delta\omega_n(t-2\tau_s)}\, e^{-t/T_2} <-|U_2|+><+|U_1|-><+|U_2|->^*<-|U_1|->^* . \tag{8}$$

If we take the ion-phonon interaction into account, and make use of the Markov approximation, we find that the magnitude of the induced electric-dipole moment decreases exponentially with the transverse relaxation time T_2. In (8), the suffix n is introduced in order to explicitly indicate that each impurity ion in a solid has a different transition frequency Ω due to crystalline field inhomogeneity. As is evident from (8), we immediatly find that, at time $2\tau_s$, every dipole moment oscillates in the same phase. Therefore, radiation emitted from all the dipoles is coherently superposed, so that a burst of radiation appears at time $2\tau_s$. Squaring the total dipole moment which is obtained by summing up (8) over n, we derive the echo intensity which has the maximum at $t=2\tau_s$:

$$I_{echo}(t) \propto \sin^2\theta_1 \sin^4(\theta_2/2)\, |F(t-2\tau_s)|^2 \exp(-2t/T_2) \tag{9}$$

Here, the function F is the Fourier transform of the inhomogeneous line-shape function f:

$$F(t) = \sum_n \exp(i\Delta\Omega_n t) \simeq \int d(\Delta\Omega)\; f(\Delta\Omega)\; \exp(i\Delta\Omega t) \; . \tag{10}$$

The essential point of the echo formation is that the second pulse exchanges the probability amplitudes for the excited and ground states as is indicated by (8), so that it transforms the off-diagonal element of the density operator to its complex conjugate. Therefore, the second pulse reverses the sense of time. As is seen from the exponential term $\exp(-t/T_2)$ in (8), the transverse relaxation process, which results mainly from the ion-phonon interaction, reduces the echo intensity. That is, if the ion-phonon interaction is so strong that T_2 becomes shorter than the inverse of the inhomogeneous spectral width, then the photon echoes disappear.

However, we should note that this result is based on the Markov approximation. In the ultra-short time region, where the Markov approximation is no longer valid, the echo decay cannot be expressed by a simple exponential function with the transverse relaxation time T_2. Furthermore, if the electron-phonon coupling is sufficiently strong, the frequency broadening due to the electron-phonon coupling not only causes the echo decay, but also is expected to play the essential role in the echo formation because of the inhomogeneous character of the spectrum. That is, a photon-echo phenomenon is expected to arise without the spatial crystalline field inhomogeneity due to the lattice imperfection, such as dislocation. In the next section, we will show that such a phenomenon actually occurs, and that echo characteristics reflect the dynamics of the electron-phonon coupling.

3. Formulation

We consider a two-level localized-electron coupled with a phonon system. This system resonantly interacts with the radiation field which is supposed to be so coherent and intense that the semiclassical treatment is applicable. The Hamiltonian for this system is expressed by

$$H = H_o + H_{ep} + H_{er} \; , \tag{11}$$

$$H_o = E_g|g\rangle\langle g| + E_e|e\rangle\langle e| + \sum_k \omega_k\, b_k^{\dagger}\, b_k \; , \tag{12}$$

$$H_{ep} = V(\{b_k + b_k^{\dagger}\})|e\rangle\langle e| \; , \tag{13}$$

$$H_{er} = -(1/2)\mu E\, e^{-i\omega t}\; |e\rangle\langle g| + \text{h.a.} \; , \tag{14}$$

where $|g\rangle$ and $|e\rangle$ denote the ground and excited electronic states, ε_k is the phonon energy with the momentum k, and b_k and $b_k^{\dagger}$ are the associated annihilation and creation operators. The electron-phonon coupling is denoted by H_{ep}, and an operator V is a function of phonon amplitudes. The electron-radiation interaction is denoted by H_{er}, and the electric-dipole interaction is assumed.

Since the dynamic behavior of this coupled system plays an essential role in the echo formation in the present problem, the exact calculation should be carried out without making use of the Markovian type approximation. Time-evolution operator in the interaction picture, during the absence of an excitation pulse is given by

$$U(t, t_o) = e^{iH_g t} e^{-iH_e(t-t_o)} e^{-iH_g t_o} |e><e| + |g><g| , \qquad (15)$$

where H_g and H_e are defined by

$$H_g = \sum_k \omega_k b_k^\dagger b_k , \qquad H_e = H_g + V . \qquad (16)$$

On the other hand, the change in the system caused by the excitation pulse, in the case of a short pulse limit, is described by

$$U_j = \cos(\theta_j/2) (|e><e| + |g><g|) + i \sin(\theta_j/2) (|e><g| + |g><e|), \qquad (17)$$

where the subscripts j=1,2 indicate the first and second pulses, respectively, and θ_j is the pulse area defined by $E_j \mu \Delta t$ as before.

The density operator in the interaction picture after the two-pulse excitation as shown in Fig.1 is given by

$$\rho(t) = U(t, \tau_s) U_2 U(\tau_s, 0) U_1 \rho(0) U_1^\dagger U^\dagger(\tau_s, 0) U_2^\dagger U^\dagger(t, \tau_s) . \qquad (18)$$

Here, the operators U_1 and U_2 describe the changes in the system caused by the two excitation pulses, and $U(\tau_s, 0)$ and $U(t, \tau_s)$ describe the time-evolution of the phonon system during the absence of the excitation pulses. Assuming that the electron is initially in the ground state and that phonon system is in the thermal equilibrium described by $\rho_p = \exp(-\beta H_g)/z$, we obtain the off-diagonal element of $\rho(t)$ with the form

$$\begin{aligned} <g|\rho(t)|e>_{echo} &= <g|U(t, \tau_s)|g><g|U_2|e> <e|U(\tau_s, 0)|e> <e|U_1|g> \\ &\times \rho_p <g|U_1^\dagger|g><g|U^\dagger(\tau_s, 0)|g><g|U_2^\dagger|e> <e|U^\dagger(t, \tau_s)|e> , \end{aligned} \qquad (19)$$

A subscript echo indicates that only the term contributing to the echo formation is considered here; that is, as mentioned before, the electronic state after the first pulse is set at the excited state $|e>$ in the forward propagation, and is set at the ground state $|g>$ in the backward propagation. Substituting the specific forms (15) and (17) for the time-evolution operators in (19), we obtain

$$\begin{aligned} <g|\rho(t)|e>_{echo} &= \frac{i}{2} \sin\theta_1 \sin^2(\theta_2/2) e^{iH_g\tau_s} e^{-iH_e\tau_s} \rho_p \\ &\times e^{iH_g\tau_s} e^{iH_e(t-\tau_s)} e^{-iH_g t} \end{aligned} \qquad (20)$$

It should be noted that this matrix element of $\rho(t)$ is not a classical number but the operator in the phonon space. From expression (20), we obtain the amplitude of the induced electric-dipole moment:

$$\mu(t)_{echo} = Tr[(\mu|e\rangle\langle g| + \mu^*|g\rangle\langle e|)\, \rho(t)]$$

$$= \frac{i}{2}\mu \sin\theta_1 \sin^2\frac{\theta_2}{2} \left\langle e^{iH_e(t-\tau_s)} e^{-iH_g(t-\tau_s)} e^{-iH_e\tau_s} e^{iH_g\tau_s} \right\rangle + c.c.. \quad (21)$$

Here we have rearranged the order of the operators in the trace operation, and $\langle \cdots \rangle$ is an abbreviation of $Tr[\rho_p \cdots]$. The operator $\exp(-iH_e\tau_s)\exp(iH_g \times \tau_s)$ in the above equation describes the time-evolution of the phonon system during the interval between the two excitation pulses, and the operator $\exp[iH_e(t-\tau_s)]\exp[-iH_g(t-\tau_s)]$ describes the time-evolution of the phonon system after the second pulse; we should here note that signs of the time argument of the former exponential operators are opposite to those of the latter exponential operators. Therefore, the second excitation pulse partially reverses the time evolution of the phonon system, and gives rise to the echo phenomenon.

4. Model Calculation

In order to facilitate more explicit discussion on this point, we assume the electron-phonon coupling is linear in the phonon amplitude. In this case, the operator V takes the specific form as follows;

$$H_g = \sum_k \omega_k b_k^\dagger b_k ,$$

$$H_e = H_g + V = \sum_k \omega_k b_k^\dagger b_k + \sum_k h_k \omega_k (b_k + b_k^\dagger) , \quad (22)$$

where h_k is the dimensionless coupling constant for the mode k. The assumption that the electron-phonon coupling is linear makes exact evaluation of the average in (21) possible. That is, the product of exponential operators in (21) can be rewritten in terms of the generator G of the phonon coherent state in the interaction picture;

$$e^{iH_e(t-\tau_s)} e^{-iH_g(t-\tau_s)} e^{-iH_e\tau_s} e^{iH_g\tau_s}$$

$$= G^\dagger(0)\, G(t-\tau_s)\, G^\dagger(0)\, G(-\tau_s) \exp(-i \sum_k \omega_k h_k^2 t) , \quad (23)$$

$$G(t) = \exp[\sum_k h_k(b_k^\dagger e^{i\omega_k t} - b_k e^{-i\omega_k t})] . \quad (24)$$

The intensity of the coherent radiation is proportional to the squared amplitude of the total dipole moment. Since the spatial inhomogeneity of the crystalline field is not considered in the present problem, there is no phase difference among the oscillating dipoles, so that the intensity is expressed by

$$I(t) = \frac{4\Omega^3}{3c^3} \left| \sum_{n=1}^{N} \mu_n(t) \right|^2 = \frac{4\Omega^3 N^2}{3c^3} |\mu(t)|^2 . \quad (25)$$

This situation should be compared with the case of the conventional echo phenomenon mentioned before, in which the phase differences among the oscillating dipoles play the essential role in the echo formation. In the present problem the magnitude of a single dipole moment changes with time to form an echo. After some straightforward calculations, we obtain the echo intensity:

$$I(t)_{echo} = \frac{N^2 I_0}{4} \sin^2\theta_1 \sin^4\frac{\theta_2}{2} \exp\{-2[3S+S(t)-2S(t-\tau_s)-2S(\tau_s)]\} , \tag{26}$$

where, $S(t)$ is defined by

$$S(t) = \int d\varepsilon\ h(\varepsilon)[1+2n(\varepsilon)] \cos \varepsilon t . \tag{27}$$

In (27), $h(\varepsilon)$ is the coupling density:

$$h(\varepsilon) = \sum_k h_k^2\ \delta(\varepsilon - \omega_k) , \tag{28}$$

and $n(\varepsilon)$ is the phonon distribution function. The parameter S is an abbreviation of $S(t)$ at time zero, and is a measure of the coupling strength.

From this equation, it is found that the photon echo phenomenon arises if the electron-phonon coupling is sufficiently strong. In order to see this echo formation more directly, we first consider the case of the strong coupling limit, where the exponential function in (26) has a significant value when t is much shorter than the inverse of the average phonon frequency. Considering the lowest order term we obtain the approximate expression of (26) which has the form

$$I(t)_{echo} \simeq \frac{N^2 I_0}{4} \sin^2\theta_1 \sin^4\frac{\theta_2}{2} \exp[-D^2 (t - 2\tau_s)] , \tag{29}$$

Here, D is defined by

$$D^2 = \int d\varepsilon\ h(\varepsilon) [1 + 2n(\varepsilon)] \varepsilon^2 . \tag{30}$$

This parameter D is a measure of the spectral width of the strongly-coupled localized-electron phonon system, and is approximated to $\sqrt{S}\,\omega_p$, where ω_p is the average phonon frequency. The expression (29) clearly shows that an echo with a Gaussian profile is formed at time $2\tau_s$. It should be noted that the echo width depends on temperature through the phonon distribution function $n(\varepsilon)$. This result should be compared with the fact that a conventional photon echo formed in solids has a temperature-independent width due to the inhomogeneous broadening of the spectral line caused by the spatial crystalline field inhomogeneity. The next highest order term in the argument of the exponential function in (26) is related to the echo decay, as expressed by

$$I(2\tau_s)_{echo} \propto \exp(-D^{4} \tau_s^{4}) . \tag{31}$$

where D' is defined by

$$D'^{4} = \int d\varepsilon\ h(\varepsilon)\ [\ 1 + 2n(\varepsilon)\ ]\ \varepsilon^{4}\ , \tag{32}$$

and is roughly approximated to $S\omega_p{}^4$. This result should be also compared with the fact that the τ_s dependence of conventional photon echo is expressed by the simple exponential decay with the transverse relaxation time.

In order to see how echo intensity for an arbitrary coupling strength changes with τ_s in a larger time scale where these equations are no longer valid, we assume that the coupling density $h(\varepsilon)$ defined by (28) has the following Gaussian form with a maximum value at ω_p, and with a width of γ_p:

$$h(\varepsilon) = (S/\sqrt{\pi}\,\gamma_p)\exp[-(\varepsilon-\omega_p)^2/\gamma_p{}^2]\ . \tag{33}$$

Using this expression, and assuming for simplicity the absolute zero temperature, we derive the specific form of $S(t)$ defined by (27);

$$S(t) = \exp(\ -\frac{1}{4}\gamma_p{}^2\ t^2\ )\cos\omega_p t \tag{34}$$

This damped oscillation is directly related to the relaxation of the interaction mode [2] in the upper adiabatic potential. Substituting this equation into (26), we obtain

$$I(t)_{echo} = \frac{N^2 I_o}{4}\sin^2\theta_1\ \sin^4\frac{\theta_2}{2}\exp[-2S(3 + e^{-\gamma_p{}^2 t^2/4}\cos\omega_p t - 2e^{-\gamma_p{}^2 (t-\tau_s)^2/4}\cos\omega_p(t-\tau_s) - 2e^{-\gamma_p{}^2 \tau_s{}^2/4}\cos\omega_p\tau_s\ . \tag{35}$$

Fig.2 graphically shows the solution (35) for several values of the pulse separation τ_s. Here we set the coupling strength S at 50, and $\gamma_p = 0.1\,\omega_p$. In Fig.2, the vertical straight lines indicate the positions of the second pulses, and the free induction decay caused by a first pulse applied at time 0 is indicated by the dotted line. In the present example, the coupling is sufficiently strong (S=50), so the echo decay as a function of τ_s is approximately expressed by (31). We should also note that when the echo decay becomes significant the position of the echo peak shifts away from time $2\tau_s$. This point will be discussed in the next section. Fig.3 shows the case of S=5, taking several values of τ_s nearly equal to $2\pi\omega_p{}^{-1}$. In this case, the optical free induction decay after the first excitation pulse, described by the dotted line, shows the damped oscillation which reflects the relaxation of the interaction mode in the upper adiabatic potential. As shown in this figure, in the case of electron-phonon coupling of intermediate strength, the τ_s dependence of echo intensity cannot be represented by a monotonically decreasing function. That is, when the position of the second pulse approaches the average lattice vibration period $2\pi\omega_p{}^{-1}$, an echo reappears.

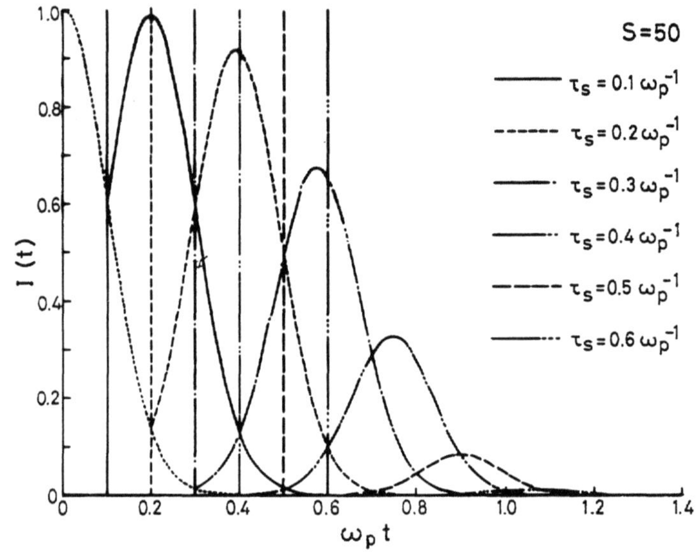

Fig. 2 Photon echo profiles in the case of strong electron-phonon coupling.

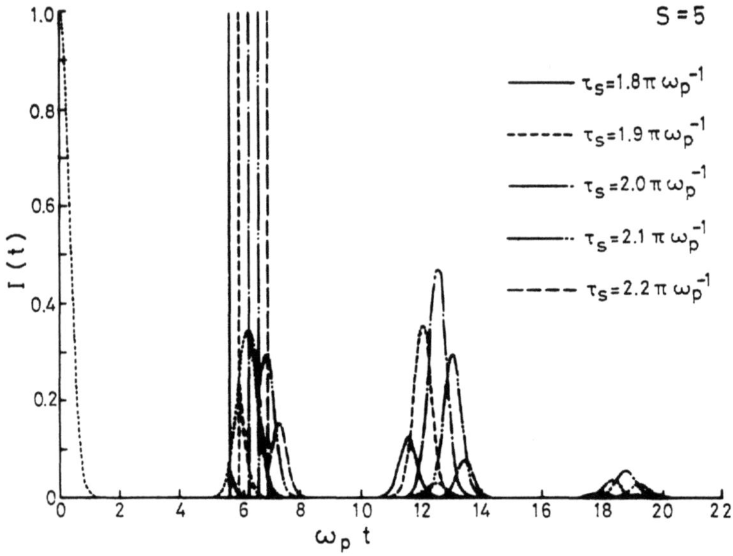

Fig. 3 Photon echo profiles in the case of the intermediate electron-phonon coupling.

5. Discussion

The predicted result that a photon echo is formed even in a single* localized-electron seems to be puzzling at first, since it cannot be understood by the well-known explanation that the second pulse reverses the dephasing process of many oscillating dipoles with different frequencies. However, I should point out that there exists the common physical basis between the conventional echo phenomenon and the predicted one. As shown in the preceding analysis, the echo amplitude is proportional to the thermal average of the product of the exponential operators which can be rewritten in terms of the time-ordered exponential operator as

$$\left\langle e^{iH_e(t-\tau_s)}\, e^{-iH_g(t-\tau_s)}\, e^{-iH_e\tau_s}\, e^{iH_g\tau_s}\right\rangle = \left\langle \exp_-\Big[\, i\int_0^{t-\tau_s} dt'\; V(t')\Big]\exp_-\Big[i\int_0^{-\tau_s} dt''\; V(t'')\Big]\right\rangle . \quad (36)$$

In order to discuss the relationship between the conventional and predicted echo phenomena, let us restrict our attention to the time region much shorter than the lattice vibration period. In this case, we can ignore the time-dependence of $V(t)$, and (36) is rewritten, in the configuration coordinate representation, as

$$\left\langle e^{i(t-2\tau_s)V}\right\rangle \propto \prod_k \int dq_k\; \exp[-\omega_k\; \tanh(\omega_k q_k{}^2/2kT)] \times \exp[-i\; h_k\; (2\omega_k{}^3)^{1/2}\; (t-2\tau_s)\; q_k] \; . \quad (37)$$

Here, we should note that this expression is formally the same as (10) in the case of the coventional photon echo, and the distribution function of the phonon amplitude q_k indicated by the lower dashed curve in Fig.4 plays a role of the inhomogeneous line shape function. That is, the ensemble of continuously distributed vertical transitions in the configuration coordinate space plays the role of the inhomogeneous broadening. However, such an analogy is valid only within a short-time region where the dynamical motion of the phonon system in the upper adiabatic potertial can be neglected. In general, the phonon wave-packet in the upper adiabatic potential relaxes toward the new equilibrium position, as schematically shown in Fig.4. Therefore, the vertical transitions for different q_k are no longer independent with each other, so that the system cannot be regarded as a simple inhomogeneously broadened system, and the lattice relaxation is reflected to the decay of the predicted photon echo.

In discussing how the dynamical motion of the phonon system is related to the echo characteristics, it is convenient to express the echo intensity in terms of the coherent state [4];

---

*Strictly speaking, in order to obtain information on the induced electric-dipole moment, we need at least two localized electrons [3]. However, the difference between the one- and two-electron problems is not important in the present problem, since a great number of localized electrons are necessary in observing the conventional echo phenomenon.

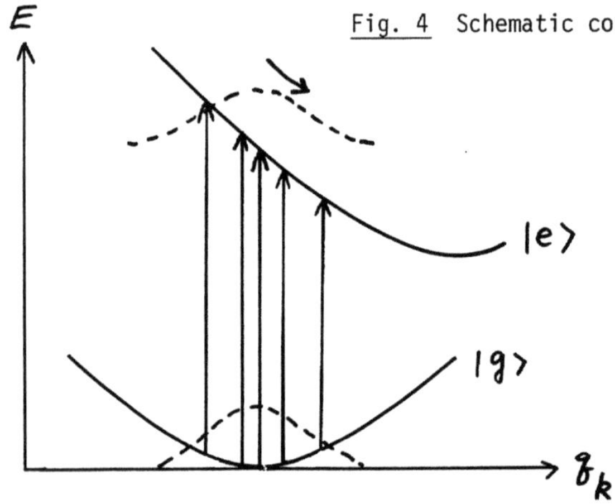

Fig. 4 Schematic configuration coordinate diagram.

$$I(t)_{echo} \propto \left| \left\langle e^{iH_e(t-\tau_s)}\, e^{-iH_g(t-\tau_s)}\, e^{-iH_e\tau_s}\, e^{iH_g\tau_s} \right\rangle \right|^2$$

$$\propto \prod_k \left| \langle 0 \,|\, \alpha_k \rangle \right|^2 = \prod_k \exp(\, -|\alpha_k|^2 \,) \;, \tag{38}$$

$$\alpha_k = (1+2n_k)^{1/2}\, h_k \,[\, 1 - e^{i\omega_k\tau_s} + 1 - e^{-i\omega_k(t-\tau_s)}] \,. \tag{39}$$

The coherent state $|\alpha_k\rangle$ in the above expression describes the phonon wave-packet moving in the upper adiabatic potential, and the vacuum state describes the phonon wave-packet located at the minimum point of the lower adiabatic potential. The overlap between these two states, $\langle 0|\alpha_k\rangle$, is

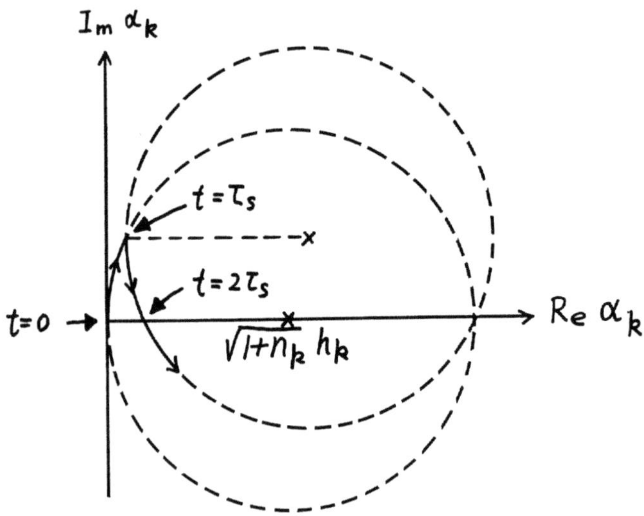

Fig. 5 Motion of α_k in the complex plane, which indicates the dynamical motion of the phonon wave-packet in the phase space.

proportional to the induced electric-dipole moment. The time dependence of α_k which is directly related to the echo intensity is graphically shown in the complex α_k plane (Fig.5). Here we should note that the well-known expression:

$$\mathrm{Re}\,\alpha_k = (\omega_k/2)^{1/2} \langle\alpha_k|q_k|\alpha_k\rangle \quad ,$$
$$\mathrm{Im}\,\alpha_k = (2\omega_k)^{-1/2} \langle\alpha_k|p_k|\alpha_k\rangle \quad . \tag{40}$$

After the first excitation pulse, α_k moves along the lower circle in Fig.5 with center at $(\sqrt{1+n_k}h_k, 0)$. This circular motion with the angular frequency ω_k describes the oscillation of the phonon wave packet in the upper potential. When the second excitation pulse is applied at time τ_s, α_k changes its direction, and follows the upper circle in Fig.5. If $\omega_k\tau_s$ is much less than unity, and the radius of the circles is sufficiently large, that is, the coupling is sufficiently strong, then the overlap $\langle 0|\alpha_k\rangle$ is almost recovered at time $2\tau_s$. This movement of α_k represented by the solid curve, clearly illustrates the process of echo formation. As we see in Fig.5, the deviation of the path after τ_s from the path before τ_s due to the finite curvature causes the decrease of echo intensity with τ_s. Furthermore, the absolute value of α_k takes the smallest value when α_k intersects the line connecting the origin and the center of the upper circle, so that the echo has its maximum befor $2\tau_s$. This shift of the echo formation time increases with τ_s (see Fig.2).

As shown in Fig.3, in the case of the intermediate coupling, a photon echo arises when τ_s is nearly equal to $2\pi\omega_p^{-1}$. In this case, we should take into consideration the distribution of the phonon frequency ω_k due to the phonon dispersion. This frequency distribution, or equivalently, the distribution of α_k gives rise to the damping of the oscillation of the phonon wave-packet. We do not consider here k dependence of h_k which causes the additional distribution of α_k. Figure 6 shows the path of α_k in the complex plane in the case of $\tau_s=2\pi\omega_p^{-1}$. For the mode with the frequency ω_k larger than the average phonon frequency ω_p, the path of α_k is shown in Fig.6(a); for the mode with ω_k less than α_p, the path of α_k is shown in

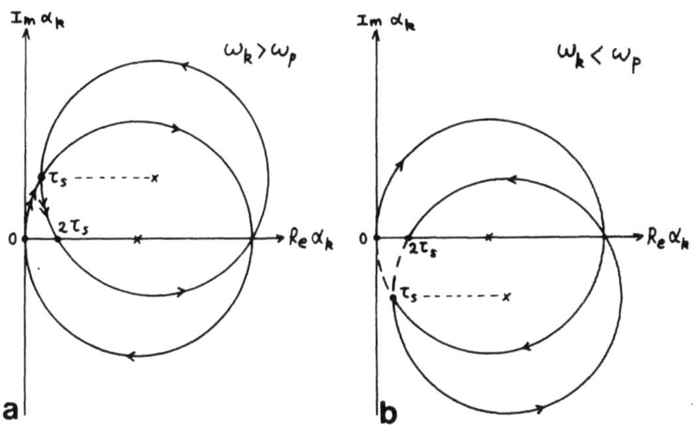

Fig. 6 Motion of α_k in the complex plane in the case of $\tau_s=2\pi\omega_p^{-1}$.

Fig.6(b). In this way, the distribution of α_k along the circle at τ_s is converted at time $2\tau_s$ into the narrower distribution near the origin on the real axis. Therefore, the free induction decay damped at τ_s is recovered at time $2\tau_s$; this illustrates the process of echo formation for $\tau_s \simeq 2\pi\omega_p^{-1}$.

6. Summary

Finally, let us summarize the characteristic features of predicted photon echoes. (i) The dependence of the echo intensity on the pulse separation τ_s is expressed by (31) for small τ_s. This result should be compared with the fact that the τ_s dependence of the conventional photon echo intensity is expressed by the simple exponential decay with the transverse relaxation time T_2. (ii) The width of echo depends on temperature through the average phonon number (see (30)). This result should be also compared with the fact that the conventional photon echo formed in solids has a temperature-independent width due to the inhomogeneous broadening of the spectral line caused by the spatial crystalline field inhomogeneity, which arises from the lattice imperfection, such as dislocations. It should be emphasized here that, in the present problem, the phonon system not merely causes the echo decay, but plays the essential role in the echo formation itself. (iii) Echoes are formed when τ_s is nearly equal to $2\pi\omega_p^{-1}$, which seems to be a powerful tool to investigate the relaxation of the interaction mode. In conclusion, we have investigated the new photon echo phenomenon associated with a localized-electron phonon system. The predicted results seems to be one of the various interesting phenomena associated with the coherent nonlinear resonance scattering.

Acknowledgements

The author would like to thank Prof. Y. Toyozawa, Prof. E. Hanamura, Prof. T. Yajima, and Dr. K. Nasu for helpful conversations. He is also indebted to Prof. E.L. Hahn for his valuable comments. This work is supported in part by the Grant-in-Aid for Scientific Research from the Ministry of Education.

References

1 I.D. Abella, N.A. Kurnit and S.R. Hartmann: Phys. Rev. 141, 391 (1966).
2 Y. Toyozawa and M. Inoue: J. Phys. Soc. Jpn. 21, 1663 (1966).
3 See for example, G.S. Agarwal: Quantum Optics (Springer Berlin, Heidelberg, New York, 1974), chap. 10 and 11.
4 R.J. Glauber: Phys. Rev. 131, 2766 (1963).

Part IV

Molecular Crystal and Biological System

Relaxation of Electronic Excitation in Biology: The Initial Processing of Photon Energy and Information*

R.S. Knox†

Department of Physics, Faculty of Science, University of Kyoto
Kyoto, Japan

ABSTRACT

A review of the qualitative features of electronic excitation relaxation processes important to biological physics is given, with special emphasis on photosynthesis.

1. Introduction

Under the conditions prevailing at the earth's surface, optical excitation in the 1-10 eV range places any complex system in a highly specific nonequilibrium state. Prior to re-emission of a photon, if that occurs at all, the system passes through a sequence of states involving high vibrational excitation, may make transitions between electronic excited states, and sometimes undergoes a drastic alteration of the original molecular structure. In terms of scale, of composition, and of interaction strengths, there is little difference among solids, liquids, and biological material. Therefore studies of the initial photoexcitation of these systems are intimately related.

In a field of vast as biology, or as wide as photobiology, specialization of some kind is particularly essential. The transient nature of a photoexcitation provides a rather interesting way to specialize, namely, by time domain. The phenomena observed occupy a very small fraction (perhaps 10^{-10} to 10^{-14}) of the life cycle of an organism, but the measurability of the unique events associated with the special nonequilibrium state is rapidly increasing as laser technology develops. The relative importance of that state is therefore much greater, we may hope, than its relative lifetime.

Specialization will also be required in this article. As mentioned in the title, only electronic state relaxation will be treated. The original electronic excitation will be considered completely relaxed as soon as pronounced physical or chemical changes (for example, conformational change or charge transfer, respectively) have occurred, or when the system has reached the ground electronic state. Furthermore, we shall not go into the physical aspects of the relaxation processes. As we have implied above, all biosystems

*Research supported in part by National Science Foundation grant PCM-77-20230. The author thanks the Japan Society for Promotion of Science for partial support in the preparation and presentation of this paper.

†Visiting Scholar. Permanent address: Department of Physics and Astronomy, University of Rochester, Rochester, New York 14627, USA.

are physicochemical and their fundamental processes, especially at these short times, are presumably identical to those covered by the rest of this Symposium. We can therefore examine qualitative issues. Finally, in concentrating directly on the visual and photosynthetic primary processes we will ignore excitations of the damaging or ionizing type, excitations designed to study biomolecules which are not normally photoexcited, and excitations which are normally produced chemically.

2. Survey of Temporal Regimes

2.1 1 fs to 1 ps

To the extent that individual molecules are excited, early events in photobiology can be characterized on the basis of the conventional picture for molecules in solution. The chromophore, such as the porphyrin ring of a chlorophyll molecule, is placed in a state which is initially a linear combination of vibronic states whose phases are determined by the polarization of the absorbed photon and the system geometry. The early photoexcited state is also strongly affected by the structure of the excitation pulse, as emphasized by RHODES [1]. This state dephases rapidly... roughly speaking, the dephasing time is $\sim$ 10 fs, the uncertainty time of a state involved in an optical transition 20 nm wide at 650 nm and also the order of magnitude of the vibrational period of typical modes. One might prepare a particular biomolecule in a matrix at low temperature and excite it to a similar state which could be well defined, in principle, for times approaching vibrational relaxation times. However, at present such procedures are probably more relevant to the general discipline of chemical physics than to biology. It is not known what kind of information derived from subpicosecond studies will be useful for gaining biological insights. A qualitative review of the early dephasing events is given by FREED [2].

To the extent that more than one molecule is involved in a photo-excitation, very little more can be said about the initial excitation except that it is "delocalized" over some set of equivalent molecules and an intermolecular phase relationship is set up by the light wave. In a crystal at low temperatures we expect polariton formation and coherent motion of the excitation over relatively large distances (see, e.g., KNOX [3]). However, the processes which dephase the ensemble of individual excited molecules also remove phase relations between molecules initially coherently excited, leading rather quickly to a motion which is describable as exciton diffusion over the set of equivalent molecules. Thus even if the initial excitation can be specified precisely, the subsequent behavior of the system is very complicated, as may be seen in studies of the case of two molecules (BURLAND and ZEWAIL [4], RAHMAN *et al*. [5]). Low-temperature spectra of chlorophyllins, chlorophyll-protein complexes about which more will be said in part 3, demonstrate the existence of delocalized electronic states through their characteristic circular dichroism and energy level splittings (PHILIPSON and SAUER [6], SCOTT and GREGORY [7], VAN METTER [8]). However, any persistent spatial coherence associated with the delocalization is probably restricted to the chlorophyllin itself, or a region of the order of 3-4 nm in extent and containing 5-10 chromophores. There is no indication that any further spatial coherence of such states has diagnostic or functional importance *in vivo* at the moment.

2.2 1 ps to 10 ps

Most of the functionally important electronic relaxation in photobiology occurs in this domain. Consider the case of rhodopsin, which is a primary absorber in the vision process (e.g., WALD [9]; HONIG [10]; YOSHIZAWA and TOKUNAGA [11]). On the basis of induced absorbance changes it has been found that excitation of animal rhodopsin relaxes to a new state in less than 6 ps, a state which then lasts for 30 ns (BUSCH *et al.* [12]). The rhodopsin is excited at 530 nm (2.34 eV) and the new state is detected by interrogation at 561 nm (2.21 eV). From the strict relaxation point of view adopted here, a most important question is whether the new absorption is simply that of an excited state of rhodopsin in essentially its original conformation. This is not the case. A new ground state species appears to have been formed since the rhodopsin ground state repopoulation kinetics differ from the kinetics of the decay of the transient species. By use of an analogy with known states in a similar artificial chromophore, it has long been argued that a certain specific *cis-trans* isomerization has taken place. At the least, a substantial change of some kind occurs in rhodopsin after a very brief electronic relaxation. HONIG *et al.* [13] review these arguments and suggest that the intermediate state, whatever it is, involves a charge separation at very short times.

Most chlorophyll and bacteriochlorophyll molecules produced by nature for the purpose of energy conversion are organized by complexing with proteins in rather rigid local environments, according to current thinking (see THORNBER *et al.* [14]). The chlorophylls which behave passively, i.e., those which relax only by transferring excitations on to other chlorophylls, are regarded as part of an "antenna" or "light-harvesting" system. Two broad categories of singlet relaxation other than simple intersystem crossing and radiationless transitions to the ground state may therefore be distinguished. The first is the transfer of excitation energy from one antenna molecule to another in a random walk which ends with highest probability at a special site or region called the reaction center. The fast process involved here is neighbor-to-neighbor transfer, whose rate can be estimated by FÖRSTER's theory [15] as $w = (1/\tau)(\bar{R}_0/R)^6$, where $(1/\tau)$ is the monomolecular decay rate of a singlet excitation, R the distance between donor and acceptor, and $\bar{R}_0$ a parameter whose value is about 5.8 nm for two chlorophyll *a* molecules (see KNOX [16], Table II). Taking a typically quoted interchromophore distance of 1.5 nm, and using τ = 5.2 ns, one finds $w = (1.6\ \text{ps})^{-1}$. With an equally resonable choice R = 2.0 nm, the predicted pairwise rate is $w = (8.7\ \text{ps})^{-1}$.

The second nontrivial type of relaxation in the case of energy conversion is the production of charge-transfer states within the reaction center. These states are the precursors of complete charge separation. Bacteria have been the most fully characterized in this respect because of successes in preparing reaction center complexes containing no antenna chlorophyll (CLAYTON and WANG [17], FEHER [18]). They do contain four bacteriochlorophylls (BChl) and two bacteriopheophytins (BPh), among other molecules involved in the charge-transfer chemistry (STRALEY *et al.* [19]). In these complexes excitation reaches the special site by direct absorption or by at most one or two transfers. Then an intermediate species appears in less than 10 ps (KAUFMANN *et al.* [20]). In the same frame, $(BChl_2)^+$ is seen to appear. It is now well established that the intermediate is $(BChl_2)^+BPh^-$. These studies are reviewed by HOLTEN and WINDSOR [22] and by SAUER [23].

2.3 10 ps and later

In the visual pigments, events in this time domain are dominated by photochemical and thermally induced transitions among a variety of intermediate rhodopsin structures. Each structure is characterized by a spectral signature in the visible region, and the structures are identified by making time-resolved studies at different temperatures where different thermal reactions are frozen out (e.g., SHICHIDA *et al.* [24]). Because relaxation in the simplest sense is clearly not involved here, we shall not continue the discussion of visual pigments into this era.

In photosynthesis, it is useful in this time domain to view the combined antenna complexes as one large excited system in which the overall relaxation is accomplished by a sequence of rapid pairwise transfers over at least one unit which consists of N antenna molecules and a trap (reaction center). In the generally two-to-three-dimensional topology of the antenna in the membranes, the number of steps in the random walk before the first visit to the trap will be of the order of magnitude of N (MONTROLL [25]). The effective pairwise rate in an aggregate of coordination number c is cw, where w has been estimated above. (Concerning the factor c, refer to KNOX [26], page 56 and Fig. 4.7.) Therefore, if we assume that the first visit to the trap is fatal to the excitation, the relaxation rate by trapping is of the order of $(N/cw)^{-1}$. This leads to rates $(0.39\ N\ ps)^{-1}$ and $(2.2\ N\ ps)^{-1}$ for the respective choices of 1.5 and 2.0 nm for typical pairwise separations and an arbitrarily chosen coordination number of 4. The bacterial photosynthetic unit is known to contain N = 20 to 50 bacteriochlorophylls and the green plant unit N = 100 to 300 (see, e.g., SAUER [27]). Therefore the overall trapping times may range from 10 ps to 650 ps. Under conditions in which reaction centers are "busy" and cannot utilize the energy in the antenna, fluorescence yields may increase because the centers return or reflect the excitation to the antenna chlorophylls. In green plants one of the two types of reaction centers (system I) disposes of trapped energy without this reradiation regardless of its oxidation state (see BUTLER [28]). Any return of the excitation to antenna increases the chances for triplet formation, about which we will comment later. The triplet state of the reaction center $BChl_2$ in the case of bacteria is involved explicitly in relaxation when chemistry is blocked (see PARSON and MONGER [29]).

3. Further Remarks on Relaxation Processes Involving Chlorophyll

3.1 Chlorophyll in vivo

In the previous section we reviewed the normal operation of the antenna-trap system after photoexcitation. Although the picture presented there is generally accepted, the excitation diffusion process, particularly as it predicts fluorescence as a function of time, is in a state of animated discussion (see the various comments in the Ciba Foundation volume whose reference is given under KNOX and VAN METTER [30]).

An alternative to trapping when saturating conditions exist is intersystem crossing to the triplet state. In ether solutions, chlorophyll singlets cross to triplets with a quantum yield of 2/3, implying a rate of $(7.6\ ns)^{-1}$ (BOWERS and PORTER [31]). Therefore, when the antenna lifetime is lengthened by failure to trap, intersystem crossing becomes a competing channel. Since triplet chlorophyll is easily destroyed by oxidation, the resulting triplet is quenched *in vivo*, generally by carotenoids (see RENGER and WOLFF [32]).

In solution the chlorophyll triplet lives for 1 ms, but *in vivo* appears to vanish in about 40 ns (BRETON and MATHIS [33]). [The carotenoid is also able to act as an accessory pigment for supplying singlet excitation to the chlorophyll and is therefore a remarkable adjunct to photosynthesis.]

At very high intensities, 10^{14} to 10^{16} photons cm^{-2} per laser pulse of about 20 ps, the fluorescence yields of green plants and of bacterial both begin to drop (representative papers, respectively are GEACINTOV *et al.* [34] and CAMPILLO *et al.* [35]). This effect is reversible and not due to damage, rather, it has been attributed quantitatively to singlet-singlet annihilation of antenna excitations. This unique relaxation process has been used to infer topologies of photosynthetic units and to infer the range of migration of excitations before trapping (PAILLOTIN *et al.* [36]). An equally effective annihilation process is singlet-triplet fusion, which has a high biexcitonic rate constant (RAHMAN and KNOX [37]) and which is operative during pulse-train experiments when triplet populations build up appreciably (e.g., MONGER and PARSON [38]).

Exciton fission, a relaxation process which is the inverse of annihilation (SWENBERG and STACY [39]), may be possible in bacteriochlorophyll where the lowest triplet has about half the singlet energy (CONNOLLY *et al.* [40]). To our knowledge, this process has not yet been observed in photosynthesis.

3.2 Chlorophyllins

Optical properties of fragments of the antenna have been studied extensively. In the green bacterium *Prosthecocloris aestuarii* the fragments are water soluble and crystallize sufficiently well for accurate structural determination by X rays (MATTHEWS *et al.* [41]). Even when not water soluble, antenna fragments are apparently quite uniform and consist of proteins complexed with various small numbers ($\lesssim$ 10) of chlorophyll molecules. THORNBER *et al.* [14] recommend the use of the term *chlorophyllins*.

In trap-containing chlorophyllins, which are usually somewhat larger, relaxation processes are expected to be qualitatively identical to those of the entire antenna-trap system with a correspondingly smaller number N (see part 2). Relaxation processes in pure-antenna chlorophyllins are essentially those of the individual chromophores included, with the following important exceptions: if there exists an inhomogeneous distribution of chromophore excitation energies, the higher states will be relaxed by transfer to the lower ones; and there may be accessory pigments acting as triplet quenchers which would function mainly in the event that antenna triplets build up.

An example of an inhomogeneous chlorophyllin is the light-harvesting Chl-*a*/*b* complex (THORNBER *et al.* [42], VAN METTER [43]). In this case three chlorophyll *b* molecules, whose excitation energies are 0.06 eV higher than that of chlorophyll *a*, coexist with at least three chlorophyll *a*'s, one of which may be shifted by 0.019 eV lower than the other two. Absorption-emission analysis shows that the excited states of these six molecules relax into very good metastable thermal equilibrium with each other and the environment prior to emission [30].

Every antenna chlorophyllin may be spectrally inhomogeneous in the following sense: exciton interactions lead to a distribution of eigenvalues, even if all first-order monomer excitation energies are identical. This is clearly a reality in complexes from the bacterium *Chloropseudomonas ethylica*

[6] and from P. aestuarii [41, 44], where exciton interactions are strong enough to produce observable splitting.

3.3 Chlorophyll in Solution

Despite its long history as an object of study, the monomeric chlorophyll molecule in solution is not completely characterized, especially with respect to its singlet relaxation. We call attention to the fact that absorption-emission analyses reveal anomalies when expressed in terms of a "Stepanov temperature" T*. Although this temperature parameter (STEPANOV [45]; VAN Metter and KNOX [46]) should be equal to the ambient temperature, it is found to be considerably higher at both room temperature (SZALAY *et* al. [47]) and liquid nitrogen temperature (HEVESI and SINGHAL [48]), where it is 417K and 126K, respectively, in very dilute chlorophyll solutions.

An explanation of the anomalies was given by VAN METTER and the author [46], including an explanation of an observed excitation-wavelength dependence of T*. Our model was based on the assumption of an inhomogeneous distribution of excitation energies in solution. Unfortunately, more recent data (VAN METTER [49]) indicate no dependence of emission spectra of Chl *a* on excitation wavelength in the same solvents for which the Stepanov anomaly exists. Such a dependence is expected for a sufficiently long-lived inhomogeneous distribution.

3.4 General Remarks

The general tool of temperature variation, so important to condensed matter physics in nonbiological materials, may be used in the biological case with appropriate precautions. Certain species of chlorophyll which fluoresce only at low temperature have proven invaluable in the monitoring of energy distribution between photosynthetic reaction centers [28]. However, in the general case one must keep in mind that because of irreversible phase transitions in membranes the system at low temperature may be poorly related to the "real" one. Many low-temperature results on monomeric chlorophylls have been similar to that on other organic molecules (see, e.g., FÜNFSCHILLING and WILLIAMS [50]). For these reasons but primarily because of time and space we will not review low-temperature studies.

4. Information Considerations

As is plainly evident from the development of fiber optic communications, light may contain information as well as energy. The processing of light information by the visual and photosynthetic systems differs greatly, as summarized in Table 1. Intensity and color are sensed at the molecular level in vision by photon counting (see RIPPS and WEALE [51]) and by wavelength-selective pigments, respectively [9,10]. For obvious reasons no energy transfer is desirable between different color sensors and there is none, nor is there transfer between molecules of the same color type (HAGINS and JENNINGS [52]). Color is not sensed by photosynthetic pigments in general, although, in green plants, subsystems absorbing selectively in the red are connected with two different types of reaction centers [28]. The general objective in photosynthesis seems to be complete sharing of the energy absorbed by chlorophyll and accessory pigments across the 400-700 nm region. This energy relaxes to the lowest excited states of the antenna and reaction centers and undergoes a distribution that is optimally efficient for the complete primary

Table 1. Features of continuous light which may in principle contain information. Columns 2 and 3 contain the response to that information by two important photobiological systems. Light's directionality is, of course, acknowledged at macroscopic levels by most light-sensitive biological systems

Feature of Light	Photosynthesis	Vision
Intensity	Requires $\gtrsim$ 8 photons per O_2 evolved; operates at low intensity; saturates physiologically at about 5 mW cm^{-2}.	Photon counter, especially at low intensity
Frequency (wavelength)	Broadly unspecific in 440-700 nm; energy shared by all pigments, excess over $E(S_1)$ probably lost.	Specific spectral regions sensed by three "color pigments," do not share energy.
Polarization, coherence	No evidence that either is functionally utilized at this stage of evolution.	Eye may possibly respond as photon counter to sense degree of coherence, no functional use known.
Directionality	Lost at molecular level; this reduces available free energy from sunlight.	Lost at molecular level.

photochemical system [28]. Polarization and coherence have no identifiable effect at the molecular level, at least to the knowledge of this author. At the macroscopic level it has been found in one case (KARLANDER and KRAUSS [53]) that coherent light has some effect on cell growth, but this interesting experiment appears not to have been followed up systematically.

Directionality is a special feature of light. At the macroscopic level it is recognized explicitly by most organisms. In the visual case there is eye focus and pattern recognition; in the photosynthetic case there is a variety of light-seeking dispositions for the plant or cell. In addition, photokinesis and phototaxis are apparent in various organisms which neither "see" nor photosynthesize (e.g., THOMAS [54]). Apart from the macroscopic effects, however, the information contained in the collimated rays from the sun is completely lost. The excited state produced by absorption of photons from the sun's beam of solid angle $\Delta\Omega \sim 6.8 \times 10^{-5}$ sr contains virtually the same information as that produced by light of the same intensity incident from 4π sr. Therefore an overall entropy increase of at least $k_B ln(4\pi/\Delta\Omega) = k_B ln(1.85 \times 10^5) \approx 12\ k_B$ accompanies the absorption of each photon from direct sunlight. At 295 K, the corresponding loss of free energy is 0.31 eV per photon, a thermodynamic fact which restricts the amount of work derivable from unfocussed solar energy to about 70% in the red region of the spectrum. This factor was first found by DUYSENS [55]. More detailed versions of these considereations can be found in papers of ROSS [56], KNOX [57] and PARSON [58].

The "directional relaxation" just described seems inevitable. To prevent the loss of directional information in the absorption process, photon wave vector must be conserved as well as photon energy. Therefore a polariton state might at first be considered suitable as a required primary excitation of a system with increased conversion efficiency. However, directionality must eventually be lost. As the polariton ensemble relaxes by scattering, its kinetic temperature drops and its entropy increases. Even if this were somehow avoided in a device, sooner or later the energy would have to be coupled into a three-dimensional machine ... or a one-dimensional machine which would do work on and give up heat to a three-dimensional world.

5. Concluding Remarks

One omission due to this article's specialization which deserves at least a mention is the bacteriorhodopsin system of the purple membrane from Halobium halobacterium (OESTERHELT and STOCKENIUS [59]). Phenomena relevant to the electronic relaxation time domain are reviewed by CAMPILLO and SHAPIRO [60]. An interesting relaxation problem relating to localization and delocalization of excitation energy in the purple membrane is discussed by EBREY et al. ([61], pp. 395-6).

The need for close examination of the most fundamental exciton ideas is nowhere more apparent than in the case of P. aestuarii, where straightforward theory has been shown to fail in calculating the optical properties of a complex whose structure is known [44]. Time-resolved spectroscopy may help resolve the problem through dynamic studies of the interactions analogous to those in the purple membrane case. Without some such resolution, all detailed exciton studies in photosynthesis may be forced to rest on an uncertain foundation.

The variety of singlet excitation relaxation mechanisms in biological systems is rich. Ultrafast optical techniques are being used to probe these mechanisms on an increasingly large scale. In the resulting study of the early stages of primary photobiological processes, the distinction between biological and chemical physics is becoming vanishingly small. As a result, new techniques and objects of investigation are opening up for both disciplines.

References

1 W. Rhodes: Chem. Phys. 22, 95-103 (1977)
2 K. F. Freed: Accts. Chem. Res. 11, 74-80 (1978)
3 R. S. Knox: Theory of Excitons (suppl. 5 of Solid State Physics, ed. by F. Seitz and D. Turnbull, Academic Press, N.Y., 1963), §§8-12
4 D. M. Burland, A. H. Zewail: Adv. in Chem. Phys. 40, 369-484 (1979)
5 T. S. Rahman, R. S. Knox, V. M. Kenkre: Chem. Phys. 44, 197-211 (1979)
6 K. D. Philipson, K. Sauer: Biochemistry 11, 1880-1885 (1972)
7 B. Scott, R. P. F. Gregory: Biochem. J. 149, 341-347 (1975)
8 R. L. Van Metter: Biochim. Biophys. Acta 462, 642-658 (1977)
9 G. Wald: Science 162, 230-239 (1968)
10 B. Honig: Ann. Rev. Phys. Chem. 29, 31-57 (1978)
11 T. Yoshizawa, F. Tokunaga: Photochem. Photobiol. 29, 197-202 (1979)
12 G. Busch, M. Applebury, A. Lamola, P. Rentzepis: Proc. Nat. Acad. Sci. (USA) 69, 2802-2806 (1972)
13 B. Honig, T. Ebrey, R. H. Callender, U. Dinur, M. Ottolenghi: Proc. Nat. Acad. Sci. (USA) 76, 2503-2507 (1979)

14 J. P. Thornber, J. P. Markwell, S. Reinman: Photochem. Photobiol. 29, 1205-1216 (1979)
15 Th. Förster: Ann. Physik 2, 55-75 (1948)
16 R. S. Knox: Bioenergetics of Photosynthesis, ed. Govindjee (Academic Press, New York, 1975), Chap. 4, pp. 183-221
17 R. K. Clayton, R. T. Wang: Methods Enzymol. 23, 696-704 (1971)
18 G. Feher: Photochem. Photobiol. 14, 373-387 (1971)
19 S. C. Straley, W. W. Parson, D. C. Mauzerall, R. K. Clayton: Biochem. Biophys. Acta 305, 597-609 (1973)
20 K. J. Kaufmann, P. L. Dutton, T. L. Netzel, J. S. Leigh, P. M. Rentzepis: Science 188, 1301-1304 (1975)
21 M. G. Rockley, M. W. Windsor, R. J. Cogdell, W. W. Parson: Proc. Nat. Acad. Sci. (USA) 72, 2251-2255 (1975)
22 D. Holten, M. W. Windsor: Ann. Rev. Biophys. Bioeng. 7, 189-227 (1978)
23 K. Sauer: Ann. Rev. Phys. Chem. 30, 155-178 (1979)
24 Y. Shichida, T. Kobayashi, H. Ohtani, T. Yoshizawa, S. Nagakura: Photochem. Photobiol. 27, 335-341 (1978)
25 E. W. Montroll: J. Math. Phys. 10, 753-765 (1969)
26 R. S. Knox: in Primary Molecular Events in Photobiology (NATO Advanced Study Institute held at the Badia Fiesolana, 4-16 Sep 1972), ed. A. Checcucci and R. A. Weale (Elseiver Sci. Publ. Co., Amsterdam, 1973), pp. 45-78.
27 K. Sauer: in Bioenergetics of Photosynthesis, ed. Govindjee (Academic Press, New York, 1975), Chap. 3, pp. 115-181.
28 W. L. Butler: Ann. Rev. Plant Physiol. 29, 345-378 (1978).
29 W. W. Parson, T. G. Monger: in Chlorophyll-proteins, Reaction Centers, and Photosynthetic Membranes, Brookhaven Symposium in Biology no. 28, ed. J. M. Olson and G. Hind (Brookhaven National Laboratory, Upton, N.Y., 1977), pp. 195-212
30 R. S. Knox, R. L. Van Metter: in Chlorophyll Organization and Energy Transfer in Photosynthesis, ed. Ciba Foundation, Ciba Foundation Symposium 61 (new series) (Excerpta Medica, Amsterdam, 1979), pp. 177-190
31 P. G. Bowers, G. Porter: Proc. Roy. Soc. (London) A296, 435-441 (1967)
32 G. Renger, Ch. Wolff: Biochim. Biophys. Acta 460, 47-57 (1977)
33 J. Breton, P. Mathis: Compt. Rend. Acad. Sci (Paris) 271, 1094-1096 (1970)
34 N. E. Geacintov, J. Breton, C. E. Swenberg, G. Paillotin: Photochem. Photobiol. 26, 629-638 (1977)
35 A. J. Campillo, R. C. Hyer, T. G. Monger, W. W. Parson, S. L. Shapiro: Proc. Nat. Acad. Sci. (USA) 74, 1997-2001 (1977)
36 G. Paillotin, C. E. Swenberg, J. Breton, N. E. Geacintov: Biophys. J. 25, 513-533 (1979)
37 T. S. Rahman, R. S. Knox: Phys. Stat. Solidi (b)58, 715-720 (1973)
38 T. G. Monger, W. W. Parson: Biochim. Biophys. Acta 460, 393-407 (1977)
39 C. E. Swenberg, W. T. Stacy: Chem. Phys. Lett. 2, 327-328 (1968)
40 J. S. Connolly, D. S. Gorman, G. R. Seely: Ann. N.Y. Acad. Sci. 206, 649-669 (1973)
41 B. W. Matthews, R. E. Fenna, M. C. Bolognesi, M. F. Schmid, J. M. Olson: J. Mol. Biol. 6, 391-396 (1967)
42 J. P. Thornber, R. P. F. Gregory, C. A. Smith, J. L. Bailey: Biochemistry 6, 391-396 (1967)
43 R. L. Van Metter: Biochim. Biophys. Acta 462, 642-658 (1977)
44 R. M. Pearlstein, R. P. Hemenger: Proc. Nat. Acad. Sci. (USA) 75, 4920-4924 (1978)
45 B. I. Stepanov: Dokl. Akad. Nauk SSSR 112, 839-841 (1957) [English translation Sov. Phys.-Doklady 2, 81-84 (1957)]
46 R. L. Van Metter, R. S. Knox: Chem. Phys. 12, 333-340 (1976)

47 L. Szalay, G. S. Singhal, E. Tombacz, L. Kozma: Acta Phys. Acad. Sci. Hungaricae 34, 341-350 (1973)
48 J. Hevesi, G. Singhal: Spectrochim. Acta 25A, 1751-1758 (1969)
49 R. L. Van Metter: Ph.D. thesis, University of Rochester, Rochester, N.Y. (1977)
50 J. Fünfschilling, D. F. Williams: Photochem. Photobiol. 26, 109-113 (1977)
51 H. Ripps, R. A. Weale: in The Eye, Vol. 2A, ed. H. Davson (Academic Press, New York, 2nd ed., 1976), Chap. 2
52 W. A. Hagins, W. H. Jennings: Discuss. Faraday Soc. 27, 180-185 (1960)
53 E. P. Karlander, R. W. Krauss: Biochim. Biophys. Acta 153, 312-313 (1968)
54 J. B. Thomas: Primary Photoprocesses in Biology (North Holland Publ. Co., Amsterdam, 1965), Chaps. 7-9
55 L. N. M. Duysens: in The Photochemical Apparatus, Its Structure and Function, Brookhaven Symposium in Biology no. 11 (Brookhaven National Laboratory, Upton, N.Y., 1959), pp. 10-25
56 R. T. Ross: J. Chem. Phys. 45, 1-7 (1966)
57 R. S. Knox: Biophys. J. 9, 1351-1362 (1969); in Light-Induced Charge Separation in Biology and Chemistry, ed. H. Gerischer and J. J. Katz (Berlin, Dahlem Konferenzen 1979), pp. 45-59
58 W. W. Parson: Photochem. Photobiol. 28, 389-394 (1978)
59 D. Oesterhelt, W. Stoeckenius: Nature (London) 233, 149-152 (1971)
60 A. J. Campillo, S. L. Shapiro: in Ultrashort Light Pulses: Picosecond Techniques and Applications, ed. S. L. Shapiro (Topics in Applied Physics, v. 18, Springer-Verlag, Heidelberg, 1977), Chap. 7, pp. 317-376.
61 T. G. Ebrey, B. Becher, B. Mao, P. Kilbride, B. Honig: J. Mol. Biol. 112, 377-397 (1977)

The Origin of the J Band

J. Tanaka, M. Tanaka, and N. Kanamaru

Department of Chemistry, Faculty of Science, Nagoya University
Chikusa, Nagoya, 464, Japan

ABSTRACT

The J band is a very sharp absorption band which has been found in the aggregate of some cyanine dyes. It is characterized by a red shifted unusual sharp absorption band and zero Stokes shift of fluorescence. The condition for the appearance of the J band is investigated in order to find the mechanism of molecular interaction involved in the excited state of dye aggregates. Electronic spectra of single crystals of several dyes are presented and discussed from a view of specific polarization character of the J band. Comparison of spectra of many dyes suggests that strong π electron overlap and large exciton band splitting are important requisite for the appearance of the J band. It implies that the charge resonance state mixes with molecular exciton state when favourable conditions are satisfied, and this will enable fast excitation transfer among molecules in the aggregate.

1. Introduction

The J band was found in 1936 by JELLY [1] and SCHEIBE [2] on 1,1'-diethyl-2,2'-cyanine aggregate in aqueous solution and its origin has been discussed by many people from different aspect. The spectral change found with the J aggregation is quite unusual as compared with other cases of dye association. The aggregate is very effective for photographic sensitization of silver halide crystals.

The characteristic of the J band may be summarized as follows. First the absorption band intensity is centered on a single 0-0 vibronic band and the band width is sharpened as compared to the monomer dye. Secondly the fluorescence band is also sharp and it appears without Stokes shift. According to Franck-Condon principle for the absorption and fluorescence processes, these characteristics may be explained in terms of fast excitation transfer among molecules in the aggregate that they have not enough time to distort before the excitation moves to other sites.

In the molecular exciton theory the strong and weak coupling concept has been used by SIMPSON and PETERSON [3] to describe the coupling strength of the excitation transfer. The J aggregate has been referred as a case of the strongest excitation transfer. BRIGGS and HERZENBERG [4] and SUMI [5] used coherent potential approximation to explain the absorption band shape in terms of the exciton band width B and the electron phonon coupling parameter S. Following these authors the J band may appear when $B > S\hbar\omega$ where ω is the characteristic frequency of the coupled molecular vibration, hence the J band should appear when B is appreciably larger than the molecular vibrational energies.

We have been studying the origin of the J band by measuring the crystalline absorption spectra of several cyanine dyes. MARCHETTI et al. [6] presented a similar study on 1,1'-diethyl-2,2'-cyanine dyes. The merit of single crystal measurement of different kinds of crystals is several fold; the polarization character of the J type absorption may be found along particular crystalline direction and the J band may be found only for limited kind of crystals. The arrangement of molecules necessary for the J band may give an idea of important molecular interaction for the J band. The molecular exciton splitting is critically dependent on both the arrangement of the dye molecules in the crystal and the direction of the propagation vector of light with the crystalline surfaces [7,8], therefore the measurement with different crystalline faces will provide valuable information on the exciton interaction. Then we will be able to compare the observed band splitting with the calculated ones and confirm our theoretical basis of spectral analysis. Also the comparison of crystalline spectra of different exciton band width will give a test of above-mentioned mechanism of band sharpening.

The J aggregate in aqueous solution is presented at the begining in order to explain basic facts and properties of the J aggregate.

2. The J Band of Dyes in Solution

Figs. 1 and 2 illustrate typical J band in solution. The solid line in Fig.1 is the monomer absorption spectra of 1,1'-diethyl-2,2'-cyanine iodide in ethyleneglycol water (EGW) solution and when its concentrated solution is cooled to 77 K the spectrum change to a dotted curve by aggregation. The intensity of the sharp J band at 17300 cm^{-1} is dependent on experimental conditions and other bands at higher energy region 19000 - 20000 cm^{-1} are accompanied with the aggregate. The association number in solution is suggested to be at least 7-25 molecules by DALTROZZO et al.[9].

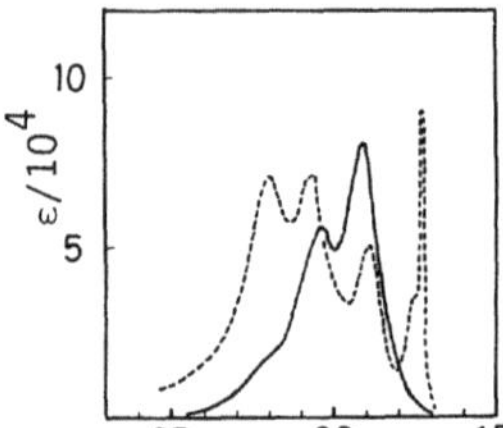

Fig.1 Solution spectra of 1,1'-diethyl-2,2'-cyanine iodide in EGW solution at 198 K ($2x10^{-5}$M ———) and at 77 K ($2x10^{-3}$M ------------)

The structure of the dye aggregate in solution is not known a priori, and the comparison with crystalline spectra discussed below showed that the aggregate of 1,1'-diethyl-2,2'-cyanine might have a similar structure with that of the iodide crystal [6]. The size of the aggregate mentioned above indicates that the J band is a spectral phenomenon associated with at least 7-25 molecules and it does not necessarily require a large number of molecules. Although the similarity between the crystalline spectra with the solution spectra is fairly good but not perfect, still the mode of interaction in the crystalline spectra may reflect the necessary condition for the appearance of the J band. In particular the J type band could be found in crystalline spectra along specified direction,and other accompanied bands found in aggregate solution are ascribed to the transitions to other directions of microcrystalline aggregate. Actually the different polarization

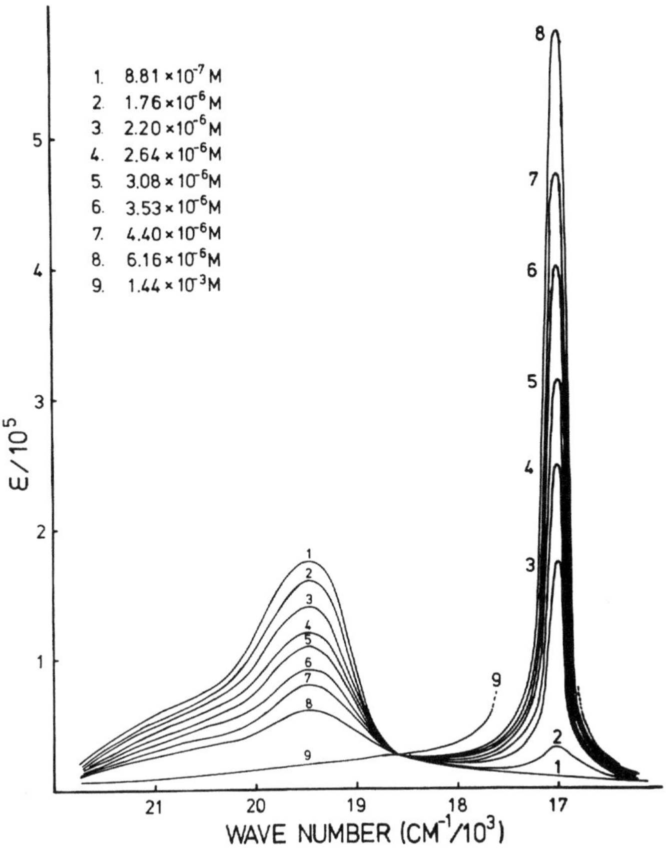

Fig.2 Change of electronic absorption spectra with concentration of 5,5'-6,6'-tetrachloro-1,1'-diethyl-3,3'-disulfobutylbenzimidazolocarbocyanine sodium in a dilute alkaline aqueous solution (pH=11.0)

character of the J band and other higher energy transitions were confirmed by the fluorescence polarization measurement that they are nearly perpendicular to each other [9].

In Fig.2 the spectra of aggregate of 5,5'-6,6'-tetrachloro-1,1'-diethyl-3,3'-disulfobutylbenzimidazolocarbocyanine sodium in alkaline aqueous solution is shown, where the association number is estimated to be 8.0 from equilibrium measurement [10]. The large red shift of 2460 cm^{-1} from monomer peak is explained by an exciton type interaction of strong coupling type within the octamer assuming that the molecular arrangement is the same with that of the crystal containing two methanol molecules [11].

In contrast to the case of 1,1'-diethyl-2,2'-cyanine iodide, a single peak was observed in solution spectra of Fig.2 implying that a better condition for the J band is fulfiled in solution as compared to crystalline

Table 1 Fluorescence data of the J aggregate of 5,5',6,6'-tetrachlor-1,1'-diethyl-3,3'-disulfobutylbenzimidazolocarbocyanine sodium in alkaline aqueous solution

	ϕ	τ_f [ps]	k_r [s^{-1}]	k_{nr} [s^{-1}]
monomer	0.03	120 ± 30	2.5×10^8	8.1×10^9
J aggregate	0.25	62 ± 10	4.0×10^9	1.2×10^{10}

case shown below. The half band widths of the absorption and fluorescence bands of J type bands are 245 and 115 cm^{-1}, respectively, while those of the monomer first band are 1295 and 900 cm^{-1}, respectively; thus the sharpening is observed to reduce the band width even less the single vibronic band width.

The fluorescence spectra of the dye aggregate is discussed in detail in a separate paper [10], but the quantum yield ϕ, lifetime τ_f and rate of radiative (k_r) and non-radiative processes (k_{nr}) are listed in Table 1. It is interesting that the radiative lifetime is increased 16 times in the aggregate as compared to the monomer, indicating that the octamer aggregate may dimerize in the excited state and the coherent exciton state has 16 times radiative probability as compared to the monomer. This is consistent with the earlier studies [12] on the lifetime of the J aggregate fluorescence that it is very short to be measured. Contrast to k_r it is of interest that k_{nr} is not so much different in the aggregate as compared to the monomer. Thus a coherent exciton state is shown to be really important in the decaying process of the excited state of the J aggregate.

3. Crystalline Spectra of 1,1'-Diethyl-2,2'-Cyanine Halides

The crystalline spectra of the title dye iodide (DYE·I) are obtained from Kramers-Kronig tranform of reflectivity measured with the (100) and (011) faces as shown in Fig.3. The arrangement of molecules in the crystal is presented in Fig.4 according to X-ray crystal structural result of Nakatsu et al.[13]. The transition moment of the molecular π electron transition of the dye is along the long-axis of the molecule, and by the oriented gas model the intensity along the b-axis is expected to be much smaller as compared with the a- and c-axes. The J type absorption was found only along the b-axis, along which the molecular plane stacks extensively. Along the a- and c-axes the intense molecular bands are shifted to blue by the molecular exciton splitting. In addition several weak humps are observed at 17000 - 19000 cm^{-1} region besides molecular exciton bands. They are ascribed to the charge resonance (CR) bands between the stacked molecules. The exciton band splitting on these faces are well explained by the calculation discussed below.

The crystalline spectra of the dye bromide (DYE·Br) are measured by the same method with the (100), (101) and (10$\bar{1}$) faces as shown in Fig.6 The arrangement of molecules in the crystal is also presented in Fig.7 according to the X-ray crystal analysis [14]. The J type absorption was found

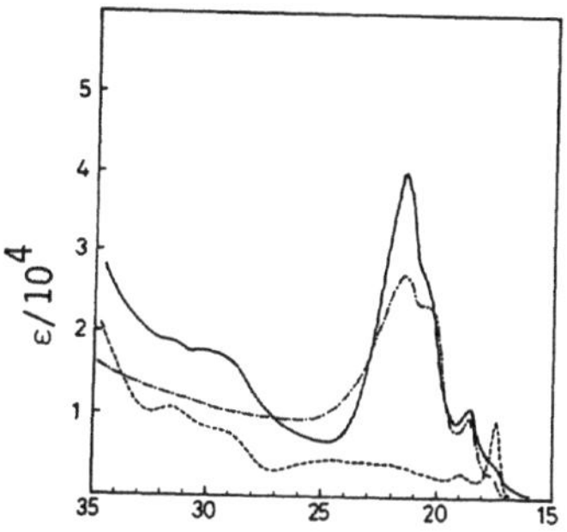

Fig.3 Crystalline spectra of 1,1'diethyl-2,2'-cyanine iodide calculated from Kramers-Kronig transformation of reflectivity. Polarized spectra along the a-axis (——·——·——) is measured with the (011) face, along the b-(············) and c-axis (————————) are obtained from the (100) face

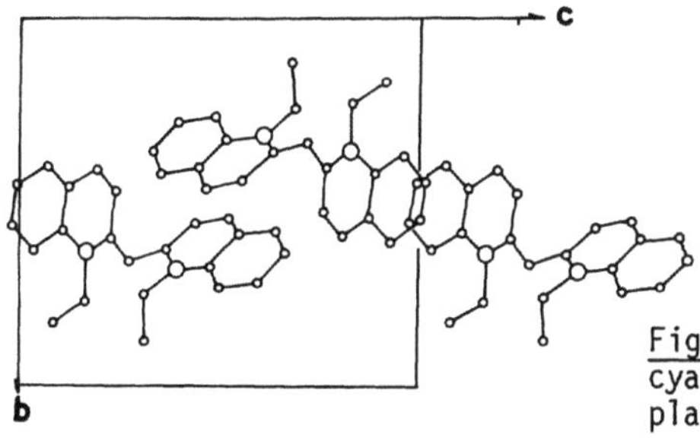

Fig.4 Projection of 1,1'-diethyl-2,2'-cyanine iodide crystal onto the (100) plane

only along the b-axis where the π electron overlaps most strongly. A subtle J look band was observed along the [10$\bar{1}$] direction which may be due to the CR band mixed with the MX band. Although the MX band is very strong and largely red shifted along the c-axis as well as b-axis, it does not show the J type band shape and this result is not in accord with earlier theories [4,5] that the same B will give the same spectra if S is not different along these axes. Along the [101] direction the molecular bands are shifted to blue and weak humps are found around 18000 - 19000 cm^{-1} region which may be ascribed to the CR band.

In monomer spectrum of solution (Fig.1), a vibrational mode of λ = 1300 cm^{-1} is resolved, and the same vibrational progressions are observed in the crystalline spectra of Figs. 3 and 6, hence a weak coupling model may be applicable in which each vibronic state interacts individually. We may designate these progressions as the 0-0, 0-1 and 0-2 bands.

On the basis of weak coupling model the crystalline spectra are analyzed as follows. First, the monomer absorption curve of the dye may be simulated by a formula

$$g(E) = \sum_{n=0}^{2} Im \frac{|p_n|^2}{E - E_0 - n\lambda - i\gamma_n} \qquad (1)$$

where p_n is proportional to the transition moment and we take it from the observed monomer vibronic intensities, E_0 is the transition energy of the monomer dye, λ is the vibrational energy of the coupled normal vibration and γ_n is the half band width of each vibronic band. The band width shows the coupling of each vibronic level with other low frequency intramolecular and solvents modes. Numerical values are taken as E_0 = 19000 cm^{-1}, λ = 1300 cm^{-1} and γ_n = 500 cm^{-1}, and a reasonable fit is obtained.

These crystals belong to a monoclinic system and includes four molecules in a unit cell. The surface exciton band splitting is calculated for each face with a dipole-dipole approximation over 100 - 200 A radius and 10-20 A depth [6,7] . By using values of exciton band shift B_n for each vibronic level the crystalline absorption band may be simulated by

$$G(E) = \sum_{n=0}^{2} Im \frac{|p_n|^2}{E - E_0 - n\lambda - B_n - i\Gamma_n} \qquad (2)$$

where Γ_n is the half band width for the n-th vibronic level. The calculated simulation spectra is shown in Fig.5, where each vibronic bands are assumed to have same intensity with the solution spectra. The calculated shifts are in good agreement with observed spectra as regard to band splitting. The results of calculation may be briefly expressed as follows.

(a) When B_n is negative and large, the red shift of the splitted band is observed with increasing separation of the vibrational progressions.

(b) When B_n is positive and large, the blue shifted vibronic bands are overlapped and converged into a single peak. These tendencies were clearly found in the spectra shown in Figs. 3 and 6.

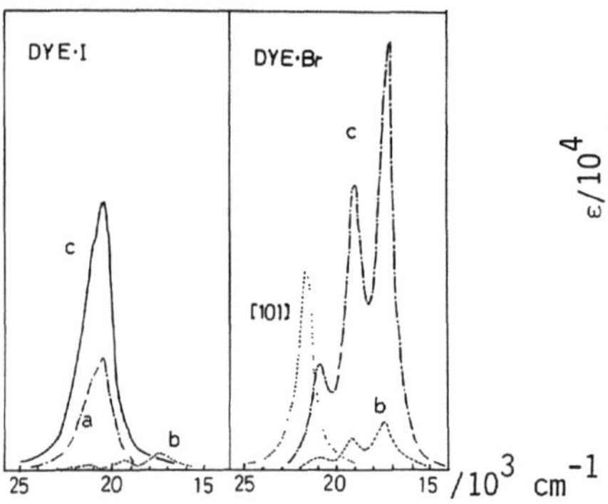

Fig.5 Simulation spectra of 1,1'-diethyl-2,2'-cyanine iodide and bromide crystals. The polarization directions are shown in the figure

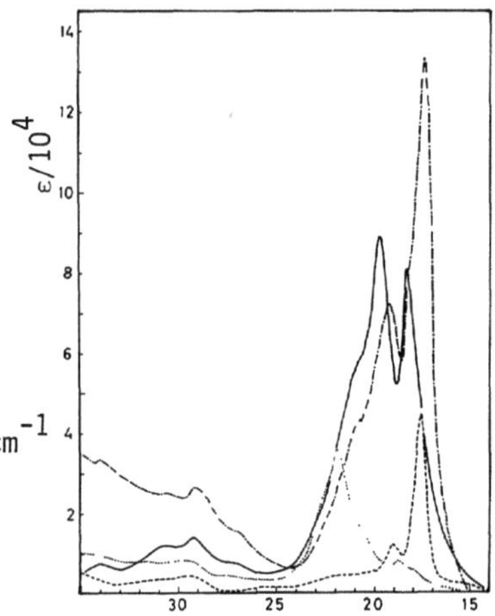

Fig.6 Crystalline spectra of 1,1'-diethyl-2,2'-cyanine bromide calculated from Kramers-Kronig transform of reflectivity. Polarized spectra along the b- (------------) and c-axes (—— · —— · ——) are obtained from the (100) face, the [101] direction (··············) from (10$\bar{1}$) face and [10$\bar{1}$] direction (————————————) from the (101) face

The comparison of the observed spectra with the simulated spectra gives us a clear insight into the origin of the J band. First of all the sharp J type absorption bands are observed along the b-axis of both the iodide and bromide crystals, where only the 0-0 band is enhanced and sharpened and the 0-1 and 0-2 bands are broadened. Since the dye molecules are stacked so closely to each other in this direction with unusually short intermolecular contact of 3.6 to 3.3 A, the CR interaction should occur naturally along this direction, and the CR absorption is expected to occur along the stacking direction of molecules. Usually charge transfer absorption of the electron donor acceptor complex is broad because its energy depends on the interplanar separation and vibration. The mixing of the CR and MX bands is very conceivable when their energies coincide and the matrix element is large by the effective π electron overlap.

The location of the CR level is estimated from the ionization potential, electron affinity, Coulomb and polarization energies of the dyes in the crystal, and it is reasonably regarded at 17000 - 19000 cm^{-1} region. The MX 0-0 band seems to be sharpened and intensified by mixing with the CR band and the 0-1 and 0-2 bands are broadened along the b-axis. In the a-and c-axes of the iodide crystal the MX bands are shifted to blue, therefore the mixing of the MX and CR states is not so extensive, and the CR bands are observed separately.

In the bromide crystal, the MX band is strong along the c-axis, but it does not mix with the CR band because the electron overlap is less effective along the c-axis than the b-axis. The absorption band along the [101] direction seems to be mixed CR and MX bands, the first band may have much CR character and the second and third ones are much MX nature. The absorption along the [10$\bar{1}$] direction shows broad CR band in the 17000 - 19000 cm^{-1} region and blue shifted MX vibronic bands at 22000 cm^{-1}; they appeared separately because they are not mixed strongly.

Thus the spectral changes occurred from monomer spectra in solution to that found in crystalline state are well explained by assuming the strong mixing of the CR and MX bands. Neither the simulation of MX spectra shown

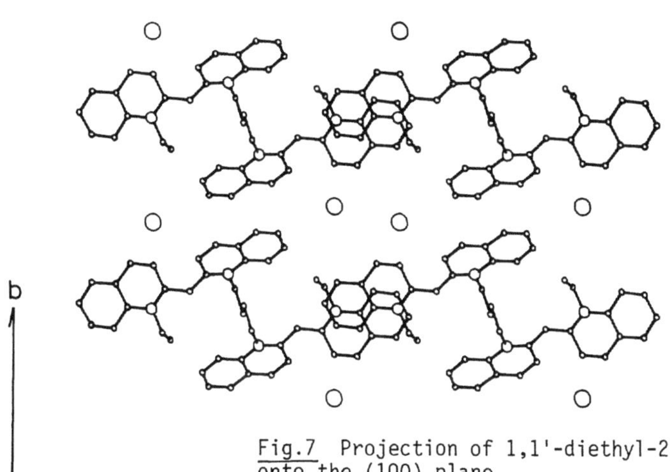

Fig.7 Projection of 1,1'-diethyl-2,2'-cyanine bromide onto the (100) plane

in Fig.5 alone nor the earlier theories [4,5] can explain the complicated spectra found in actual crystals unless we consider the effect of CR interaction seriously.

4. Crystalline Spectra of 5,5',6,6'-Tetrachloro-1,1',3,3'-Tetraethylbenzimidazolocarbocyanine Iodide

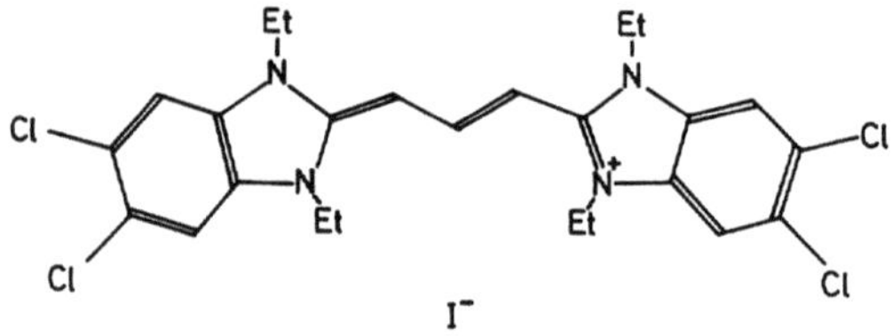

The dye molecule shown above is one of the most effective sensitizer for silver halides and has been used in photographic industry. The dye forms a typical J aggregate on the surface of silver halides, therefore the aggregate has an intrinsic importance for the mechanism of the photographic sensitization. The derivative of this dye forms a beautiful J aggregate in alkaline aqueous solution as shown in Fig.2.

The dye forms more than five crystals from several solvents and comparative studies of the crystalline spectra elucidate the mechanism of appearance of the J band. In this article some representatives, one is 2:1 methanol dye solvate (DYE·2M) and another is 1:1 methanol dye solvate (DYE·1M) and 2:1 dichloromethane dye solvate (DYE·DC) are presented.

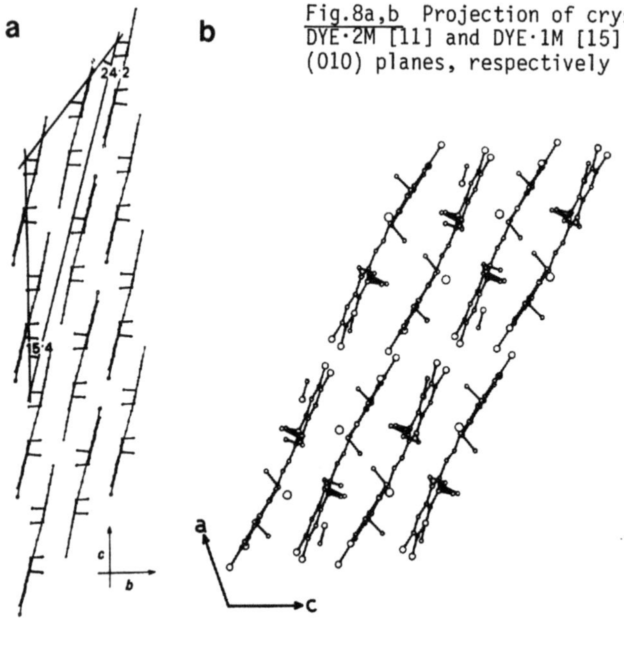

Fig.8a,b Projection of crystalline structures of DYE·2M [11] and DYE·1M [15] onto the (100) and (010) planes, respectively

The crystalline structures of DYE·2M and the dye acetonitrile solvate (DYE·A) were analyzed by SMITH and LUSS [11] and that of DYE·DC, the ethanol dye solvate (DYE·E) and DYE·1M were analyzed by ourselves [15].

The crystalline structure of DYE·2M is monoclinic and a layer structure is shown in Fig.8(a) where the dye molecules are stacked extensively and the interplanar separation is only 3.45 A in spite of electrostatic repulsion of same charge. The mode of dye arrangement in the crystal may be specified by lateral displacement and tilt angle between the line connecting centers of molecules and the molecular long-axis. It has been believed that for the stacked dimer the red shift may be found when the tilt angle is less than 32° , and actually they are 15° and 24° in the pairs of dyes in DYE·2M, 17° and 23° in DYE·A and 16° in DYE·DC. On the other hand a blue shifted H band is anticipated when the slip angle is more than 32°, and they are 55° for DYE·E and 54° for DYE·1M (Fig.8(b)).

The extent of molecular interaction through electron exchange is quantitatively evaluated from the overlap integrals S between the highest occupied MO (HOMO, 16) and the lowest unoccupied MO (LUMO, 17) as shown in Table 2. The electron transfer integral t is approximately given by

$$t = kS \tag{3}$$

where k is a constant which is estimated to be -15 eV. From Table 2 it is seen that the matrix elements are nearly 0.1 eV for DYE·2M and 1M, but

Table 2 Overlap integrals between HOMO (16) and LUMO(17) in DYE crystals. Overlap types are shown in Fig.9

	OVERLAP FORM	MO	SINGLE TERM AO 16	17
DYE·2M	TYPE C	16	0.0015	0.0074
		17		0.0043
	TYPE B	16	0.0011	-0.0036
		17		0.0026
	TYPE A	16	-0.0043	-0.0003
		17		0.0000
DYE·A	TYPE B	16	-0.0032	-0.0024
		17		0.0022
	TYPE A	16	0.0001	0.0001
		17		0.0000
DYE·DC	TYPE B	16	-0.0008	0.0012
		17		0.0001
	TYPE A	16	-0.0011	0.0011
		17		0.0001
DYE·E	FACED STACK	16	0.0041	-0.0018
		17		-0.0031
DYE·1M	FACED STACK	16	0.0053	-0.0024
		17		-0.0037

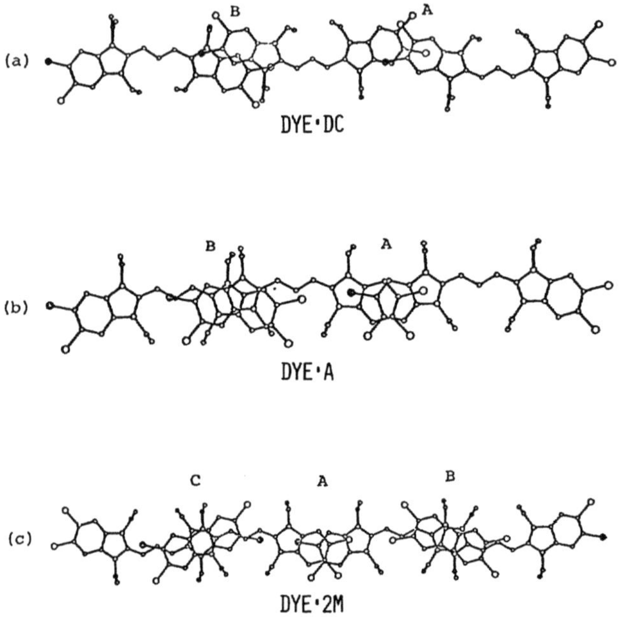

Fig.9 Overlap types of DYE solvate crystals

for DYE·DC they are negligibly small. Actually solvent molecules are inserted between cations in DYE·DC crystal and the molecular overlap is greatly hindered. The t integrals in DYE·2M and 1M are same magnitude with those of TTF·TCNQ, although it is smaller than the exciton band width B. The extent of molecular overlap is shown in Fig.9, where that of the DYE·2M is found to be largest. The calculation of overlap integrals is performed with a single exponent Slater orbitals.

The crystalline spectra of DYE·2M is presented in Fig.10 which is measured with the (100) face. The molecular long-axis is closely tilted to the c-axis, therefore absorption intensity is much larger along the c-axis than the b-axis. The J type absorption spectra is found only along the b-axis, while the metamorphoric sharp absorption is found along the c-axis accompanied with broad vibronic bands. The red shift of the b-and c-axes spectra from the center of the crystalline excited level is predicted by the calculation of dipole-dipole interaction.

Although the broad CR band is not seen clearly in Fig.10 from reflectivity transformed spectra, but a single crystal absorption spectrum on thin crystal shows a broad absorption starting at 12500 cm^{-1} and an anti-resonance dip at 16500 cm^{-1} and a sharp peak at 16500 cm^{-1}. This result suggests that the interaction of the CR and MX states takes place and the MX 0-0 band is sharpened. Along the b-axis the electron overlap is largest and only a single 0-0 peak is observed, which means that higher vibronic bands are blurred off by the interaction with the CR state.

The crystalline spectra of the DYE·DC (Fig.11) is a contrast to that of DYE·2M, namely only broad vibrational progressions with wide band width are found with the (01$\bar{1}$) face along the a-axis. With the (100) face along the c-axis, the superposed blue shifted vibronic bands are observed, and the tendency of the band shift is well explained by the calculation of surface

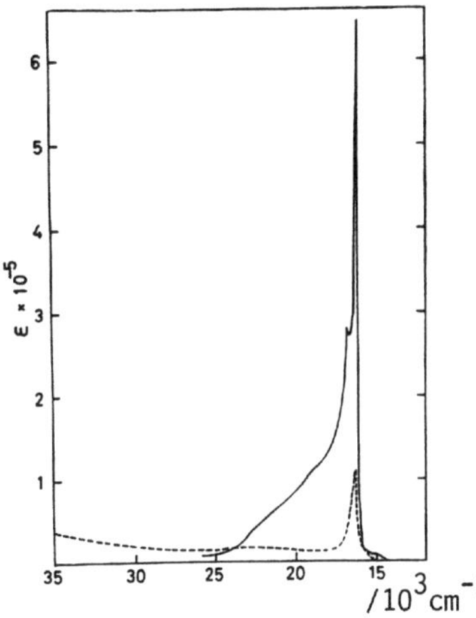

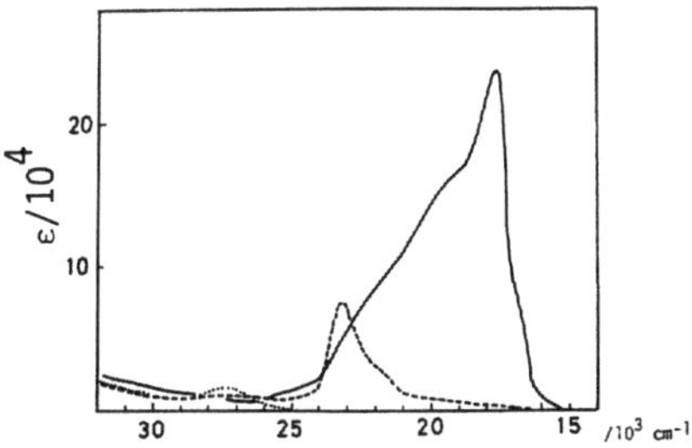

Fig.10 Crystalline spectra of DYE·2M obtained from Kramers-Krong transform of reflectivity which are obtained with the (100) face. Polarization of light along the b- (--------) and c-axes (———)

Fig.11 Crystalline absorption spectra of DYE·DC obtained from the reflectivity measurement and Kramers-Kronig transformation. The a-axis spectrum is obtained with (01$\bar{1}$) face and the c-axis spectrum is measured with the (100) plane

exciton splitting. However, no sharp J type absorption band is found along any axis. Apparently the value of B is larger for DYE·2M than DYE·DC, but still B is fairly large for DYE·DC, hence the electron overlap effect —— namely the CR interaction seems important for the appearance of the J band.

The last example of the crystalline spectra is that of DYE·1M, where a blue shifted H* band is observed instead of red shifted sharp J band. Such a sharp blue shifted band has been observed in cyclic bis-thiacarbocyanine dye of high concentration solution and it has been called as H* band [16]. The H* band in solution has been considered as due to a microcrystalline absorption.

In Fig.8(b) the projection of crystalline structure of DYE·1M is shown onto the (010) plane. The face to face dimer stack is characteristic of this crystal and the long-axes of the molecules are inclined either +21° or -21° to the ac plane. The faced molecules take bow shape and the dimer pairs are related by the screw-axis rotation along the b-axis. Although the electron exchange effect may be limited within the dimer pair and the interplanar spacings are 4.35 A for DYE·1M and 4.42 A for DYE·E, the CR interaction should occur between the stacked molecules.

The crystalline spectra of DYE·1M is shown in Fig.12 which is measured with the (100) and (01$\bar{1}$) planes. The c-axis spectrum is largely blue shifted and it is sharpened anomalously. Although the blue shifted vibronic bands may give a single absorption peak by the mechanism (b) expressed previously, but the interaction with the CR state should be important because the π electron overlap occurrs strongly along the c-axis. The spectrum along the b-axis is slightly red shifted , and that of the a-axis is blue shifted with keeping the band shape of solution spectrum. These band shifts are well explained by the calculation of surface exciton splitting.

The crystalline structure of DYE·E is isomorphous with DYE·1M, and the crystalline spectra is quite similar. However, the H* band is slightly broad and weak as compared to that of DYE·1M, presumably because the π electron overlap effect. Further details of the crystalline spectra will be presented in the forthcoming paper [17].

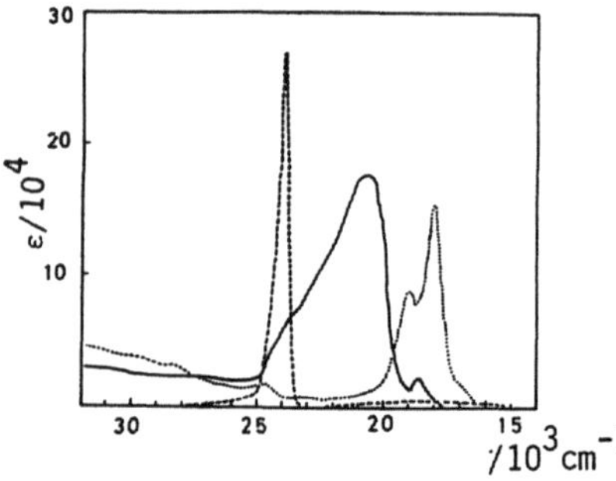

Fig.12 Crystalline spectra of DYE·1M obtained from Kramers-Kronig transform of reflectivity. The spectrum along the a-axis (————) is taken with the (01$\bar{1}$) face, along the b-(··············) and c-(-------------) axes are measured with the (100) face

Thus the fundamental theory of the band splitting of the dye aggregate is actually confirmed in the present experiment on DYE·2M, DYE·DC, DYE·A, DYE·1M and DYE·E crystals. Moreover a more detailed mechanism of the CR interaction between the stacked dyes is developed by comparing the π electron overlap integrals.

5. Other Related Phenomena of the J Band

Earlier solution studies on the J aggregate have been reviewed in articles by HERZ [16] and STURMER and HESELTINE [18]. Relatively a limited number of molecules are reported to give the J aggregate in solution. BUCHER and KUHN [19] studied on mixed monolayers of dyes such as 1,1'-distearyloxymonomethine and of arachidic acid and found the J aggregate characterized by a sharp absorption and fluorescence of zero Stokes shift. However, the band shift from the monomer absorption peak to the aggregate peak is only 959 cm-1 , which is not larger than the molecular vibrational energy. Thererefore it is not convincing that the criterion of $B > S\hbar\omega$ is actually valid.

A sharp crystalline absorption band was reported by WEISER and his collaborators on some pentamethinium cyanine dye and a conspicuous directional dispersion was observed [20].

In our laboratory we [21] have been studying on a lot of spectral and structural problems on many thiacarbocyanine dyes, but no sharp J type band is observed even though broad CR bands are overlapped with the MX bands. Presumably the π electron overlap is not sufficient enough to produce the effective CR and MX interaction.

Quite recently the crystalline structure and spectra of 5,5'dichloro-3,3'-diethyloxocarbocyanine are studied by our group [22], and we found the CR band separately at lower energy region, but its energy is far from the MX band that an effective mixing does not take place to cause the sharp J band. The molecules form a dimer of unusual short contact in the crystal, hence the CR energy level will be lowered.

In case of bis-2-(5-methyl-1,3-benzdithiylium)-trimethine cyanine perchlorate [23] the MX and CR bands interact on the (010) face and different spectra are obtained along the c- and a-axes, where the mixing of two levels produces two excited states of different polarization.

Recently an aggregate photoconductor is developed for electrophotographic applications [24], in which a J type aggregate of aryl-substituted thiapyrylium salt is used for the photosensitizer. The charge transfer transition will be involved in the aggregate which will enhance the photoconductivity of the material.

6. Theoretical Consideration on the Origin of the J Band

The concept of the J band may be taken in different ways by people belonging to different fields. One may call the J band by a red shift of the monomer band, but the spectral changes must be more precisely specified. The red shift, the disappearance of the higher vibronic bands, the sharpening of the first band and zero Stokes shift of fluorescence are significant characteristics inherent to the J band.

Theoretical considerations to explain these complex changes are discussed below. On simple exciton band Toyozawa [25] described the half band width Γ in terms of intra-band scattering mechanism, and Γ_n in (2) is given by

$$\Gamma_n = \frac{2\pi}{\hbar} \sum_k d_{k,k'}^2 / B_n \tag{4}$$

where $d_{k,k'}$ is the scattering matrix element and B_n is the exciton band width. Following this mechanism the exciton band will be sharper when the band width is wider.

In the foregoing discussion we have shown that the simple exciton picture might not be sufficient to explain the J band and the mixing of the MX band with the CR band will help the appearance of the J band. In addition the matrix element connecting between these two states is important and the transfer integral t between the LUMO or HOMO of dye molecules must be significant to have the interaction of these two states. However, the inclusion of t integral in the matrix element may not necessarily mean that the band width B is increased effectively.

In Fig.13 the exciton band structure for the MX and CR coexistent states is depicted, where the band width of the CR state is narrower than that of the MX state because the transition to the CR state has smaller transition moment. After the mixing of these two states flat energy levels are produced, and the probability of iso-energetic band to band scattering will be greatly enhanced. The half band width Γ_n' will be given by

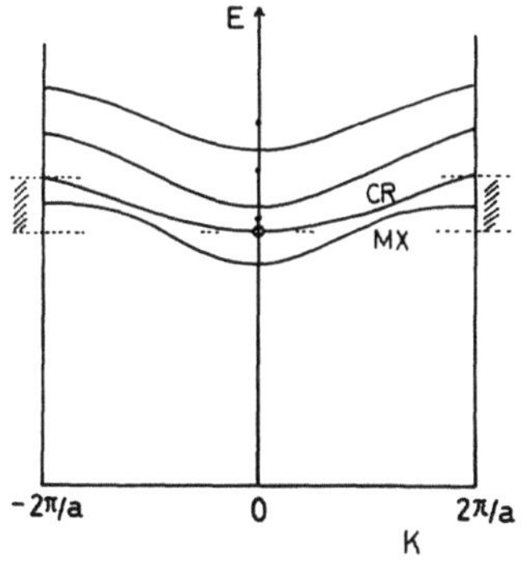

Fig.13 Exciton band structure of the mixed MX and CR bands. The 0,0, 0-1 and 0-2 vibronic bands are drawn with a CR band which is mixed with the 0-0 band

$$\Gamma_n' = \frac{2\pi}{\hbar} \sum_k t_{kk'}^2 / \delta_n \qquad (5)$$

where δ_n is the energy interval of the overlapped region or the nearly iso-energetic region. The increase of Γ_n' means that the vibronic excited state decays so fast that the transition to these state is not accessible.

The 0-0 transition has no other decaying route of iso-energetic process, therefore the absorption intensity will be concentrated in this band. In other words the Franck-Condon factor will be conserved only for the 0-0 transition.

Summarizing these arguments we may propose a mechanism of band sharpening as follows. The red shifted sharp J band may appear when the energy level of the CR state is coincident with the MX level and the molecular overlap is sufficient enough to cause the mixing of the CR and MX states. The MX 0-0 state becomes sharp because higher vibronic bands as well as the associated low frequency modes are blurred off by the mixing with the CR state.

References

1 E.E.Jelly: Nature 138, 1009 (1936)
2 G.Scheibe: Angew.Chem.49, 563 (1936)
3 W.T.Simpson, D.L.Peterson: J.Chem. Phys. 26, 588 (1957)
4 J.S.Briggs, A.Herzenberg: Mol. Phys. 21, 865 (1971)
5 H.Sumi: J.Phys. Soc. Jpn. 38, 825 (1975)
6 A.P.Marchetti, C.D.Salzberg, E.I.P.Walker: J. Chem. Phys. 64, 4693 (1976) Photogr. Sci. Eng. 20, 107 (1976)
7 M.R.Phillpott: J. Chem. Phys. 58, 588 (1973)
8 E.E.Koch, A.Otto: Chem. Phys. 3, 384 (1974)
9 E.Daltrozzo, G.Scheibe, K.Gschwind, F.Haimerl : Photogr. Sci. Eng. 18, (1974)
10 S.Makio, N.Kanamaru, J.Tanaka: to be submitted
11 D.L.Smith, H.R.Luss: Acta Cryst. B28, 2793 (1972)
13 A.A.Muenter: J .Phys. Chem. 80, 2178 (1976)
13 K.Nakatsu, H.Yoshioka, H.Morishita: Acta Cryst. B33, 2181 (1977)
14 H.Yoshioka, K.Nakatsu: Chem. Phys. Lett. 11, 255 (1971)
15 T.Kaneda, S.Yoon, J.Tanaka: Acta Cryst. B33, 2065 (1977)
16 A.H.Herz: Photogr. Sci.Eng. 18, 323 (1974)
17 J.Tanaka, M.Tanaka, S.Yoon, T.Kaneda: to be submitted
18 D.M.Sturmer, D.W.Heseltine: The Theory of Photographic Process 4th Ed., p. 194 ed. T.H.James (MacMillan, New York, 1977)
19 H.Bücher, H.Kuhn: Chem. Phys. Lett. 6, 183 (1970)
20 H.J.Hesse, W.Fuhs, G.Weiser, L.von Szentpaly: Phys. Stat. Sol. b 76, 817 (1976)
21 J.Tanaka, C.Katayama, E.Uno, N.Suzuki, T.Niwa: un-published results
22 J.Tanaka, K.Asaoka, C,Katayama: unpublished results
23 C.Tanaka, J.Tanaka, C.Katayama: unpublished results
24 W.J.Dulmage, W.A.Light, S.J.Marino, C.D.Salzberg, D.L.Smith, W.J.Staudenmayer: J. Appl. Phys. 49, 5543 (1978)
25 Y.Toyozawa, Proc.IVth Intern. Conf. Vac.UV. Radiation Phys. p 317, ed. E.E.Koch, R.Haensel, C.Kunz (Pergamon, Oxford,1974)

Part V

Molecular System

Infrared Laser Chemistry

E. Yablonovitch

Division of Applied Sciences, Harvard University
Cambridge, MA 02138, USA

1. Introduction

In the article we shall discuss a new type of photochemistry which involves infrared photons instead of visible or UV photons, and which occurs by vibrational excitation rather than electronic transitions. It has been firmly established during the past few years that polyatomic molecules can absorb sufficient energy from an infrared laser pulse that a true unimolecular reaction takes place without benefit of collisions. This may be a dissociation reaction or an isomerization in which the molecular structure is rearranged. Because a chemical activation energy is 3 eV and the infrared photon for the CO_2 laser wavelength is about 0.1 eV, the molecule is absorbing 30 or more photons.

Table 1 gives an incomplete list of some of the chemical reactions which have been studied. It is now felt that *any* polyatomic molecule will react under sufficiently intense infrared radiation. Recently there have been reports of the dissociation of molecules as small as triatomics [1].

The interest in the field was greatly enhanced by the discovery of R.V. AMBARTSUMIAN and V.S. LETOKHOV that the laser induced dissociation reaction is isotope selective [2]. This, of course, has whetted the appetite of photochemists. They are excited by possible applications in the dissociation, cleavage, isomerization and synthesis of complex molecules, let alone the question of isotope selectivity.

They are also led to hope for bond-selective chemistry. Consider, for example, the halogen substituted methane molecule:

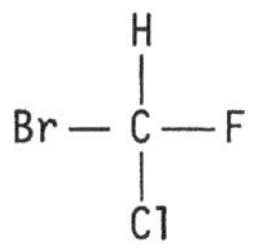

With an infrared wavelength exciting the C-H stretching vibration, they might hope to eject the H atom and obtain the CFClBr radical. With infrared wavelengths corresponding to the C-F, C-Cl and C-Br bonds, the other three respective dissociation channels might be activated. However, the realization of such dreams has proven more difficult than expected; we shall discuss the reasons for this below.

Present address: EXXON Research and Engineering Company, Linden,NJ 07036, USA

Table 1 Unimolecular reactions induced by infrared multiphoton excitation

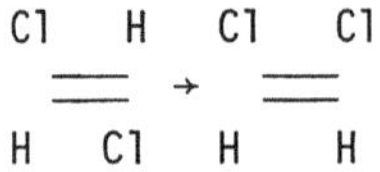	K. Nagai, M. Katayama: Chem. Phys. Lett. 51, 329 (1977)
$CClF_3 \rightarrow CF_3 + Cl$	D.F. Dever, E. Grunwald: J. Am. Chem. Soc. 98, 5055 (1976)
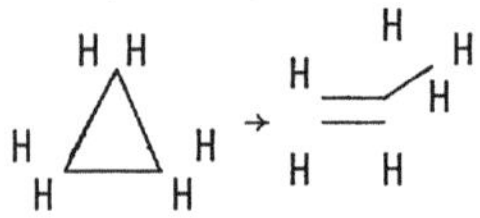	R.B. Hall, A. Kaldor: Bull. Am. Phys. Soc. 23, 73 (1978)
$CF_2Cl_2 \rightarrow$ fragments	A.T. Lin, A.M. Ronn: Chem. Phys. Lett. 49, 255 (1977)
$SeF_6 \rightarrow SeF_5 + F$	J.J. Tiee, C. Wittig: Appl. Phys. Lett. 32, 236 (1978)
$CH_3NC \rightarrow CH_3CN$ (not strictly IR-laser induced)	K.V. Ready, M.J.Berry: Chem. Phys. Lett. 52, 111 (1977)
$CH_3NC \rightarrow$ fragments	M.L. Lesiecki, W.A. Guillory: J. Chem. Phys. 66, 4317 (1977)
$BH_3PF_3 \rightarrow$ fragments	K.R. Chien, S.H. Bauer: J. Phys. Chem. 80, 1405 (1976)
$C_2H_3Cl \rightarrow C_2H_2 + HCl$	F.M. Lussier, J.I. Steinfeld: Chem. Phys. Lett. 50, 175 (1977)
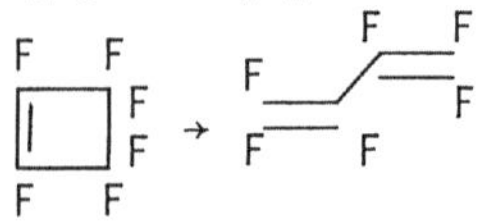	A. Yogev, R.M.J. Benmair: Chem. Phys. Lett. 46, 290 (1977)
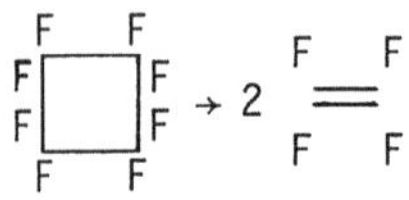	J.M. Preses, R.E. Weston Jr., G.W. Flynn: Chem. Phys. Lett. 46, 69 (1977)
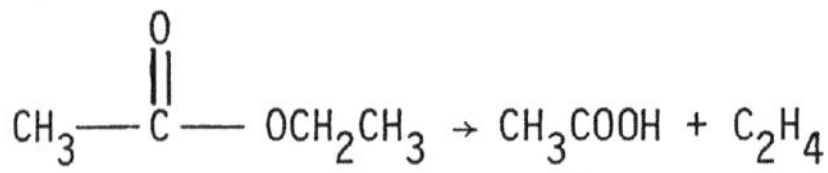	W.C. Danen, W.D. Munslow, D.W. Setser: J. Am. Chem. Soc. 99, 6961 (1977)
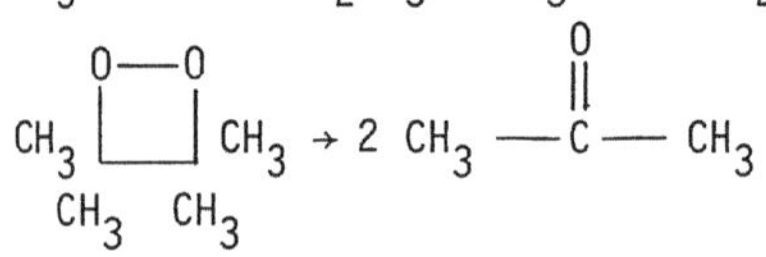	Y. Hass, G. Yahov: Chem. Phys. Lett. 48, 63 (1977)
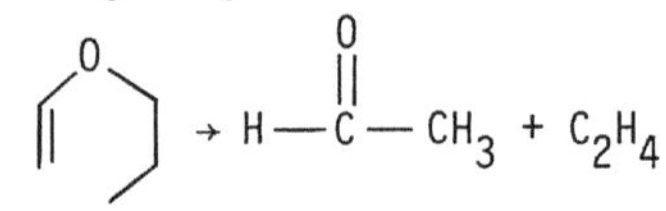	R.N. Rosenfeld, G.I. Brauman, J.R. Baker, D.M. Golden: J. Am. Chem. Soc. 99, 8063 (1977); D.M. Brenner: private communication
$NH_3 \rightarrow NH_2 + H$	J.D. Campbell, G. Hancook, J.B. Halpern, K.H. Welge: Opt. Commun. 17, 38 (1976)
$CF_3I \rightarrow CF_3 + I$	S. Bittenson, P.L. Houston: private communication

From the physicist's point of view the system of one isolated molecule and a bath of monochromatic infrared photons is an intriguing one. What type of nonlinear optical effect could be responsible for the absorption of 30 or more infrared quanta? Does the energy remain in one mode, or will it be transferred to and shared with other modes? If so, will the distribution be statistical, and could the molecular behavior be ergodic - and on what time scale? How few degrees of freedom may a molecule have and still be capable of a statistical description? Will the dividing line occur for triatomics or quadratomics? Infrared laser chemistry leads to some intriguing problems in quantum electronics, physical chemistry and statistical mechanics.

The early work of AMBARTSUMIAN and LETOKHOV on SF_6 dissociation showed that the isotope selectivity followed immediately from the 17 cm^{-1} shift in the infrared spectrum due to the mass difference between $S^{32}F_6$ and $S^{34}F_6$. Since then a large number of experiments have been performed and have led to a good physical understanding of infrared multiphoton excitation. Double or multiple laser frequencies have been employed. Intensity, pulse duration, gas pressure and delay time between pulse have been varied. The chemical products have been analyzed by IR spectroscopy, gas chromatography and mass spectroscopy. In addition, a number of specialized techniques were used:
(a) The kinetic energy distribution of the dissociation fragments has been measured in a molecular beam [3].
(b) The reaction fragments have been interrogated by resonance fluorescence excitation [4].
(c) Ultraviolet and visible luminescence from subsequent reactions of free radicals which are produced [5].
(d) Infrared luminescence from the reaction products [6].
(e) Intermolecular isotopic branching ratio [7].

In this review we shall discuss the theoretical concepts that are consistent with the dominant experimental features of laser chemistry.

2. The Quasicontinuum

A diatomic molecule may be excited from the ground state to its first vibrational level by a resonant infrared photon. In general, the vibrational anharmonicity will necessitate a lower frequency for the absorption of a second photon, a still lower one for the third photon, and so on. The reach a level, $\nu = 30$, corresponding to a typical dissociation energy, 29 different frequencies therefore appear to be required. While available evidence suggests that indeed many selected infrared frequencies would be necessary to dissociate such a *diatomic* molecule, the situation in *polyatomic* molecules is inherently different.

One way to approach the problem of the polyatomic molecule is to imagine making an exact solution of the time-dependent Schrodinger equation. This would require a knowledge of the energy, matrix elements, selection rules and exact wavefunction for each level up to the dissociation energy - an all-but-impossible task.

Fortunately, a simplification occurs. In polyatomic molecules the density of states grows rapidly with energy E, due to the rapid expansion in the volume of accessible phase space. The number of possible permutations and combinations of vibrational modes increases rapidly with available energy. The increasing density of levels is shown in Fig.1a.

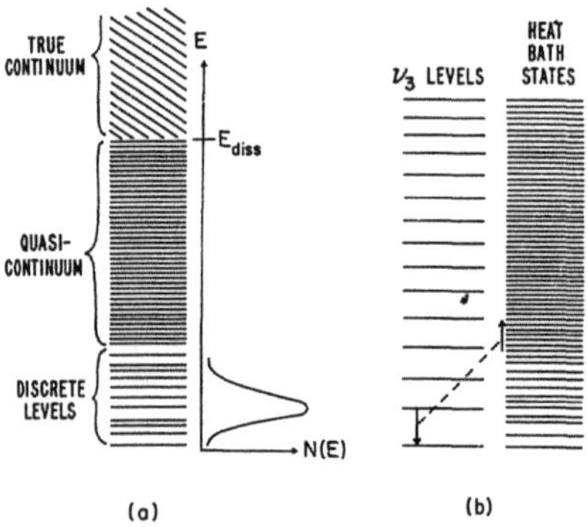

Fig.1 (a) The true quantum states of the exact vibrational Hamiltonian showing the three regimes of vibrational excitation: discrete levels, quasi-continuum and true continuum. The initial population-distribution function N(E) is the product of the vibrational density of states $\rho(E)$ and the roomtemperature Boltzmann factor. In the intramolecular heat-bath picture; (b) the energy levels of the infrared-active ν_3 mode are shown separately from the energy levels of the remaining modes, which form a quasicontinuous heat bath. Anharmonic coupling terms cause phase-fluctuation damping (with time constant T_2) and lifetime broadening (T_1) of the ν_3 mode. An energy relaxation process of the T_1 type is shown.

The key point is that at a sufficiently high density of states, Fermi's Golden Rule becomes valid. In it the transitions are described in terms of constant rates, and the full Schrodinger equation reduces to a set of rate equations [8]. This criterion determines the energy region we call the *quasicontinuum* to distinguish it from the *discrete levels* where the full machinery of the Schrodinger equation is required. The two regions are labelled in Fig.1a.

The basic condition for the validity of the Golden Rule is that the transition rate should be neither too fast nor too slow:

$$[h\rho(E)]^{-1} \ll \text{transition rate} \ll T_2^{-1} \tag{1}$$

where $\rho(E)$ is the density of available states and h/T_2 is the width of the distribution of oscillator strength. When $\rho(E)$ is large enough, a rate-equation description is possible. This conclusion is especially useful in large polyatomics, where the thermal energy at room temperature is sufficient to boost most of the molecules into the quasicontinuum. In that instance, from the initial condition onward, we may use a rate equation [9] description:

$$\frac{dW_n}{dt} = K_n^a I W_{n-1} + K_n^e I W_{n+1} - (K_{n+1}^a + K_{n-1}^e) I W_n - K_n^{diss} W_n \tag{2}$$

where W_n is the probability of being in that group of states $nh\omega$ above the starting level, K_n^a and K_n^e are absorption and stimulated emission coefficients into that group, and K_n^{diss} is the reaction rate which differs from zero only for those states above the activation energy. Because the rates, except for K_n^{diss}, are all proportional to the light intensity I, it may be divided from the right hand side of Eq.(2) to show explicitly that the temporal evolution depends only on intensity × time = energy fluence (Joules/cm^2). This remarkable conclusion is confirmed by data of P. KOLODNER and coworkers [10] on the fractional dissociation of SF_6 as a function of pulse duration. The laser energy is held fixed as the peak power increases by a factor 200. The fraction dissociated increases only 30%, showing that it is almost independent of peak power. The reaction yields in infrared multiphoton dissociation depend very strongly on energy fluence but only weakly on peak power.

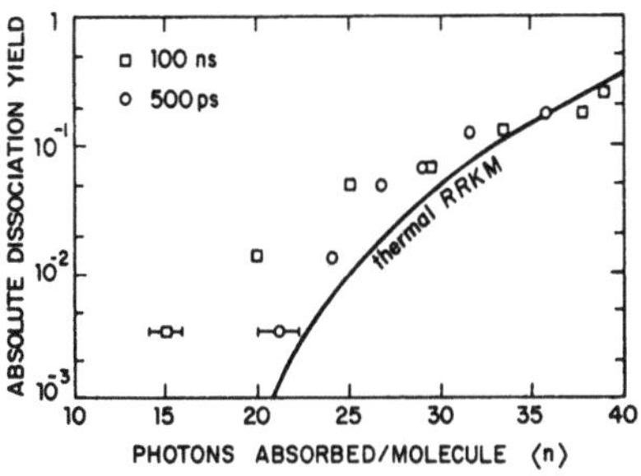

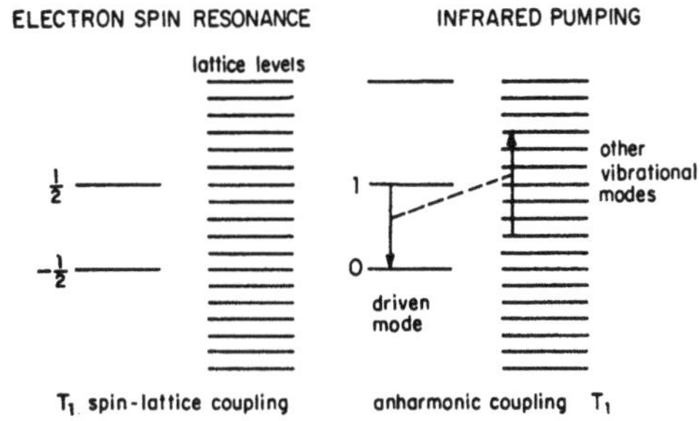

Fig.2 The probability of dissociating a molecule with a laser pulse is a function of the mean energy of excitation per pulse, $\langle n \rangle \hbar\omega$. This graph gives information relating to the shape of the energy-distribution function for multiple photon absorption. The experimental points for two pulse durations, 100 nsec and 500 psec, are compared with a theoretical curve based on the assumption of a thermal equilibrium distribution of the same energy.

Fig.3 The electron spin resonance analogy for infrared multiphoton pumping.

This conclusion is demonstrated even more forcefully in some molecular-ion cyclotron resonance experiments by BEAUCHAMP et al.[11]. The ion $[(C_2H_5)_2O]_2\ H^+$ was dissociated by 1 Watt/cm^2 acting for 1 second! Therefore, it is perhaps justifiable to consider the energy fluence scaling as confirmed over nine orders of magnitude of dynamic range.

In general the rate equation (2) may be solved numerically [9]. For the case of constant cross section, there exists an analytical solution - simply a canonical Boltzmann distribution [8]. Figure 2 plots dissociation yield versus laser energy deposited [12]. The experimental points are compared with the predictions of a thermal distribution. The agreement is rather good for 500 psec pulse excitation and less so for the 100 nsec case. Due to the effects of the discrete level bottleneck, the distribution is smeared out in the latter case.

In view of the rather simple picture which emerges, an analogy illustrated in Fig.3 can be very helpful in our understanding of infrared multiphoton pumping. The analogy is with electron spin resonance (ESR). In ESR a spin 1/2 system interacts with a lattice. As the spin system is driven by a microwave field, it transfers its energy to the lattice at the $1/T_1$ spin-lattice relaxation rate. In this way it is possible to actually heat the lattice, which could, in turn, produce a shift and a broadening of the spin resonance. In addition, the lattice causes a T_2-type phase relaxation.

The infrared pumping process is analogous. In Fig.1b the infrared active mode is separated from the other vibrational modes which are regarded as a heat bath, albeit a tiny one. As the laser drives the infrared active mode, it transfers its energy to the other modes at a $1/T_1$ relaxation rate due to anharmonic coupling. Since the heat bath modes are few, they have relatively small heat capacity and can increase their temperature significantly. This can broaden and shift the resonance due to anharmonicity. In addition, the anharmonic coupling causes a T_2-type phase relaxation.

The applicability of these ideas to SF_6 is shown in Fig.4, the absorption spectrum [13] of heated SF_6. The spectrum remains well defined even though

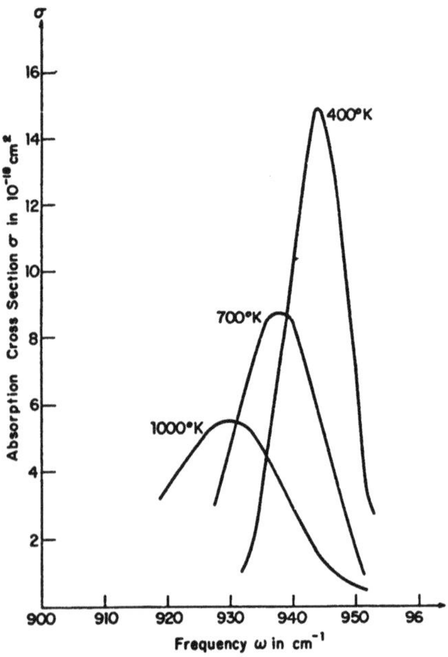

Fig.4 Absorption cross section per molecule in a thermal ensemble of SF_6. At 1000 K the molecule is boosted into the quasicontinuum, but the ν_3 absorption feature remains well defined. The surprising sharpness of the resonance is due to the nonuniform distribution of oscillator strengths. Not shown is the background absorption of about 10^{-20} cm^2 (A.V. Nowak, J.L. Lyman, ref. 13)

the molecule is thermally excited well into the quasicontinuum. It broadens and shifts with temperature as expected. The drop in absorption cross section during laser heating at 944 cm^{-1} is not a true saturation effect as would occur in a two-level system. Rather, it is due to the broadening and the shift off resonance as the molecule is excited.

It becomes important to consider the different contributions to the linewidth in the quasicontinuum. There are the inhomogeneous effects due to Doppler broadening, rotational structure and hot bands. To these must now be added the intramolecular contributions $1/T_1$ and $1/T_2$ due to the anharmonic coupling with the heat bath modes. A knowledge of T_1 is essential to determine the time scale for the randomization of energy within a molecule and for the onset of ergodicity. For this reason it is also important for mode selective chemistry. The determination of intramolecular T_1 and T_2 has hardly begun, and it will be an important challenge for nonlinear spectroscopy in the next several year.

Let us review some of the key properties of the energy region called the quasicontinuum:
(a) It is a continuum of levels, but not of oscillator strength. The infrared spectral features remain, but they consist of countlessly many dense levels, as in Fig.4
(b) Since the spectral features remain, there *is* isotope selectivity in the quasicontinuum.
(c) Exact resonance with fundamental bands is not necessary, since the background absorption from combination bands is only ∿100 times weaker. Instead of working with 1 Joule/cm^2 one can work with 100 Joule/cm^2.
(d) The coefficients in the rate equations (2) just depend on the absorption cross section at that wavelength. With a knowledge of excitation dependence of the cross section, the equations can be readily solved on a computer [9].

3. Intramolecular Relaxation

Very little is known about the time constant T_1. Numerical classical-trajectory calculations [14] appear to show that T_1 can be less than 1 psec. On the experimental side, there is also very little information. In an ingenious set of chemical-kinetics experiments, the intramolecular relaxation time was measured [15] indirectly by varying buffer-gas pressure. For the cases studied, the upper limit was estimated to be on the order of picoseconds.

In another approach the excitation of the infrared-active ν_4 mode by multiphoton pumping of SF_6 in the ν_3 line was detected [16] by means of a weak IR fluorescence in the ν_4 band of frequencies. Due to the technological constraints it was only possible to establish an upper limit of 1 μsec on the time required to transfer energy into the ν_4 mode.

In an important experiment J.P. MAIER and collaborators [17] employed picosecond infrared pulses to probe the redistribution of energy within the huge coumarin-6 molecule. They placed an upper limit of 4 psec on the T_1 lifetime of the molecule in the vapor phase at 305°C. It would be especially interesting to repeat this type of experiment in a smaller molecule with less thermal energy. Because a knowledge of T_1 under a variety of different conditions is essential for mode-selective chemistry, more quantitative answers will likely be forthcoming in the next few years.

In a time resolved infrared-infrared double resonance experiment, H.S. KWOK et al. [18] showed that the ν_3 mode of SF_6 transfers its energy to the other modes within the time resolution of the experiment, 30 psec.

Related questions are addressed by overtone spectroscopy in poly molecules, where high vibrational levels are excited by means of a single-photon absorption rather than multiphoton absorption. The spectrum of the fifth overtone band of the C-H stretch in benzene in the gas phase [19] was obtained by a laser-acoustic technique. At present, the relative contributions of T_1 lifetime broadening and T_2 phase fluctuation broadening are unknown. Future developments will be quite fascinating as we learn to distinguish T_1 lifetime effects, T_2 phase-fluctuation effects, and in nonrigid molecules the effects of slow fluctuations.

4. Opportunities in Chemical Physics

We have already alluded to the potential applicability of infrared multiphoton chemistry in producing mode-selective chemical reactions. The pumping rate must be made high enough to compete with T_1. In addition, the rise time of the pulse must be short enough to prevent a premature nonselective reaction [20]. With T_1 as short as it is, these conditions will be difficult to achieve. The 30-psec CO_2 laser pulses produces by optical free-induction decay [21] will probably play a role.

Mode-selective nonstatistical heating is only one possible application, but it may be difficult to achieve. A number of other unique attributes of multiphoton-induced chemistry present themselves, even under ergodic conditions:

Multiphoton heating can be ultrafast, leading to a higher temperature than can be achieved otherwise. This would permit a high-energy reaction channel to compete more effectively with a low-energy cahnnel. The relative yields of the different reaction products could be controlled in this way.

Infrared multiphoton heating is collisionless, so unimolecular reaction channels can be made to compete more effectively with bimolecular or collisional channels.

Most important, IR laser heating can operate on one component of a mixture, leaving the other components cold.

The latter, of course, is the essence of the isotope-separation capability, but it has exciting chemical applications as well. For example, consider two sides, A and B, of a chemical reactions:

$$A \rightleftharpoons B$$

By heating one of the components of A, it is possible to shift the equilibrium to the right, and vice versa for B. This might be rather useful in organic synthesis. Very frequently the challenge to the synthetic chemist is not in creating a molecule of a given formula, which is relatively easy, but in producing that molecule in a desired structure or isomer, which can be much more difficult. The two isomers A and B will have rather different IR spectra. Irradiation at a wavelength at which the difference in absorption cross section is large can convert the molecules to the desired isomer.

There are many other ways to exploit the remarkable capability of heating one component of a mixture.

Ultrapurification should be possible by selectively causing the undesired molecules to react [22].

sity of ground- or excited-state radicals can be generated by ochemistry and employed as reagents.

The laser can produce homogeneous catalysts, enhancing the speed of subsequent collisional reaction steps in a mixture.

One of the problems of infrared multiphoton heating is that it has been restricted mostly to the gas phase, although much interesting biological chemistry occurs in solution. The main hindering influence of the solvent is the strong relaxation, which tends to damp out the excitation and conduct away the heat. Again, ultrashort laser pulses should be able to overcome this effect and open the liquid phase to the prospect of IR photochemistry.

References

1 W.A. Guillory: Chem. Phys. Lett. to be published
2 R.V. Ambartsumian, V.S. Letokhov, E.A. Ryabov, N.V. Chekalin: JETP Lett. 20, 273 (1974)
3 M.J. Coggiola, P.A. Schulz, Y.T. Lee, Y.R. Shen: Phys. Rev. Lett. 38, 17 (1977); E.R. Grant, M.J. Coggiola, Y.T. Lee, P.A. Schulz, Y.R. Shen: Chem. Phys. Lett. 52, 595 (1977)
4 D.S. King, J.C. Stephenson: Chem. Phys. Lett. 51, 48 (1977)
5 N.R. Isenor, R.V. Merchant, R.S. Hallsworth, M.C. Richardson: Can. J. Phys. 51, 1281 (1973)
6 J.M. Press, R.E. Weston Jr., G.W. Flynn: Chem. Phys. Lett. 48, 425 (1977); C.R. Quick Jr., C. Wittig: Chem. Phys. Lett. 48, 420 (1977)
7 S.W. Benson, A.J. Colussi, R.J. Hwang, J.J. Tiee: Chem. Phys. Lett. 52, 349 (1977)

8 J.G. Black, E. Yablonovitch, N. Bloembergen, S. Mukamel: Phys. Rev. Lett. 38, 1131 (1977)
9 E.R. Grant, P.A. Schulz, Aa.S. Sudbo, Y.T. Lee: Phys. Rev. Lett. 40, 115 (1978)
10 P. Kolodner, C. Winterfeld, E. Yablonovitch: Opt. Commun. 20, 119 (1977)
11 R.L. Woodin, D.S. Bomse, J.L. Beauchamp: J. Am. Chem. Soc. 100, 3248 (1978)
12 J.G. Black, P. Kolodner, M.J. Shultz, E. Yablonovitch, N. Bloembergen: Phys. Rev. A, to be published
13 A.V. Nowak, J.L. Lyman: J. Quant. Spec. Rad. Trans. 15, 1945 (1975)
14 D.W. Noid, M.L. Koszykowski, R.A. Marcus, J.D. Macdonald: Chem. Phys. Lett. 51, 540 (1977)
15 J.D. Rynbrandt, B.S. Rabinovitch: J. Phys. Chem. 75, 2164 (1971)
16 D.S. Frankel, J.T. Manuccia: Chem. Phys. Lett. to be published
17 J.P. Maier, A. Seilmeier, A. Laubereau, W. Kaiser: Chem. Phys. Lett. 46, 527 (1977)
18 H.S. Kwok and E. Yablonovitch: Phys. Rev. Lett. 41, 745 (1978)
19 R.G. Bray, M.J. Berry: in *Proceedings* of the Conference on the Chemical Application of Lasers, University of North Carolina, Chapel Hill (1977)
20 E. Yablonovitch: Optics Lett. 1, 87 (1977)
21 E. Yablonovitch, J. Goldhar: Appl. Phys. Lett. 25, 580 (1974); H.S. Kwok, E. Yablonovitch: Appl. Phys. Lett. 30, 158 (1977)
22 J.H. Clark, R.G. Anderson: Appl. Phys. Rett. 32, 46 (1978)

Multiphoton Vibrational Pumping of Optically Prepared NO_2 Molecules[1]

I. Burak[2], J. Tsao, Y. Prior, and E. Yablonovitch[3]

Gordon McKay Laboratory, Harvard University
Cambridge, MA 02138, USA

ABSTRACT

We report the collisionless infrared multiphoton excitation of NO_2 molecules initially prepared in an electronically excited state. The infrared light induces a blue wing on the usual Stokes-shifted fluorescence spectrum. From the shape of the infrared induced fluorescence spectrum we have directly directly determined the absolute probability distribution P(n) for the net absorption of n photons.

1. Introduction

The shape of the probability distribution function P(n) for n-photon infrared excitation of molecules has been the subject of lively discussion in the literature[1-3]. In this work, we report the observation of infrared multiphoton vibrational excitation of a molecule initially prepared in an electronically excited state. By monitoring the transient visible fluorescence spectrum, we determined the energy distribution function produced by collisionless infrared multiphoton pumping.

Vibrational excitation of an electronically excited molecule has been reported earlier[4] for the ^3A_u state of biacetyl. That excitation involved a collisional energy exchange between vibrationally excited ground state molecules and optically prepared triplet molecules. The present experiment involves single photon optical excitation of NO_2 molecules from the $^2A_1$ ground state to high levels of the strongly mixed $^2B_2$ - $^2A_1$ vibronic manifold, followed by direct infrared multiphoton pumping with a very short CO_2 laser pulse. This work was prompted by an earlier report[5], in which weak double resonance signals were observed in NO_2 molecules exposed simultaneously to cw Ar^+ and CO_2 laser beams.

The experimental arrangement is shown in Fig. 1(a). A gas cell of NO_2 at a pressure $\sim$250 mtorr is subjected to a 0.5 nsec CO_2 laser pulse and an antiparallel 10 nsec dye laser pulse which overlaps the central uniform intensity region of the infrared beam. The delay between the two pulses can be adjusted from 0-30 μsec with a 30 nsec jitter. The fluorescence

[1]Supported by Office of Naval Research

[2]Permanent address: Department of Chemistry, Tel Aviv University, Israel

[3]Present address: Exxon Research & Engineering Center, Corporate Research Laboratories, Linden, New Jersey 07036

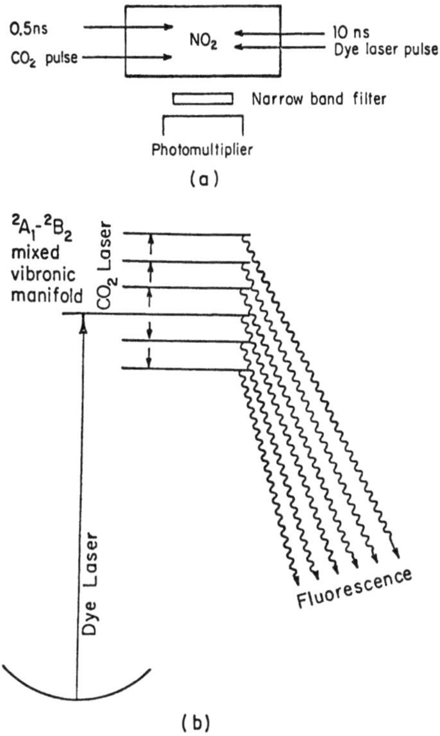

Fig.1(a) Schematic of the experimental arrangement. (b) Schematic diagram of pumped energy levels

pulse from the excited NO_2 molecules is imaged onto a photomultiplier tube through narrow bandpass filters and recorded on a dual beam oscilloscope. Fig. 1(b) illustrates the energy levels participating in the experiment.

A typical series of fluorescence signals generated by the CO_2 laser following dye laser excitation is shown in Fig. 2. A small portion of the scattered visible laser pulse leaked through the narrow band filter and was recorded at the left edge of the lower trace. The fluorescence pulse in the lower trace occurs at the instant of the delayed CO_2 pulse which is on the upper trace of Fig. 2. As can be seen the amplitude of the generated fluorescence decreases at a characteristic rate k_0' as the delay between the two pulses increases. No signals are obtained when the infrared pulse precedes the visible pulse. While ordinary visible excitation leads only to Stokes-shifted fluorescence, CO_2 laser induced signals are detected at frequencies blue-shifted with respect to the dye laser frequency ω_0. The spectral width of the induced fluorescence depends on both the CO_2 laser's energy fluence and intensity. Fluorescence signals have been recorded at frequencies blue-shifted from ω_0 by energies up to 5 times the CO_2 photon energy. The induced fluorescence spectrum is shown in Fig. 3.

The relaxation features exhibited by the multiphoton induced amplitude and fluorescence decays are comparable with the decays of ordinary fluorescence signals. Our study of the ordinary fluorescence decay can be summarized as follows:

a) The fluorescence decay times measured at pressures exceeding 100 mtorr scale inversely with the pressure.

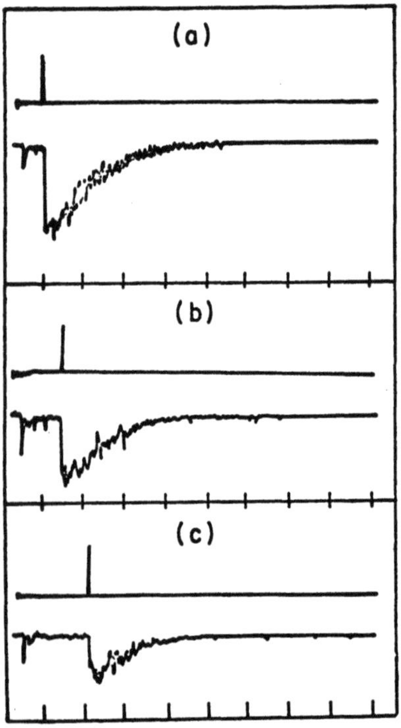

Fig.2 Upper trace: infrared laser. Lower trace: induced fluorescence signal at 470 nm following dye laser excitation at 502 nm. Delays are (a) 250 nsec, (b) 500 nsec, (c) 750 nsec. Time scale 500 nsec/division.

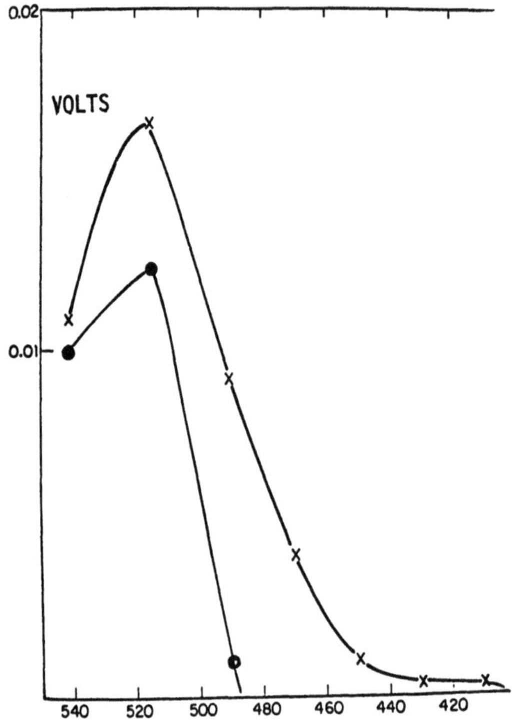

Fig.3 Dots: Ordinary fluorescence spectrum excited by dye laser at 502 nm. Crosses: Prompt spectrum induced by CO_2 laser following dye laser excitation.

b) These decay times depend strongly on the observation frequency ω. When fluorescence is studied near the excitation frequency (ω_0-$\omega \approx 500$ cm^{-1}), a quenching rate $k_0 \approx 5.5 \times 10^{-6}$ sec^{-1} $torr^{-1}$ is measured for excitation wavelengths between 400 nm and 500 nm. As the observation frequency is shifted further to the red, the decay rate decreases. For example, a quenching rate $k = 1.3 \times 10^6$ sec^{-1} $torr^{-1}$ is observed for 540 nm emission while exciting at 421 nm.

The decay of fluorescence observed at frequencies near ω_0 represents the collisional relaxation of the sharp vibronic distribution prepared by the optical pulse into lower vibronic levels. The longer decay times observed at red shifted frequencies are due to the larger number of successive vibrational decays required to deactivate the molecules. Fluorescence is observed as long as the collisionally relaxed NO_2 molecule remains in the energy region between $\hbar\omega$ and $\hbar\omega_0$. The decay rates reported in this work are similar to those reported and explained by Donnelly et al.[6]

The decay rates exhibited by the blue-shifted CO_2 induced fluorescence signals also scale with pressure. The measured decay rates vary between 0.6 k_0 and k_0, with the latter value obtained at low CO_2 laser fluences. The increase of lifetime with fluence is related to increased population in levels which require successive collisional events for complete deactivation. The amplitude versus delay relaxation measurements exhibit the same features. While the fluorescence decay monitors emission from multiphoton excited molecules with energy $>\hbar\omega$ the amplitude decay monitors the population of NO_2 molecules still capable of being excited by a multiphoton process to an energy $>\hbar\omega$. Typical values for k_0' of 0.65 k_0 and 0.95 k_0 were obtained for the amplitude decay when the observation frequency corresponded to blue shifts of one and two CO_2 photons, respectively.

A comparison of the shape of the prompt (< 30 nsec delay) collisionless infrared induced spectrum with the ordinary fluorescence spectrum permits us to extract the absolute probability P(n) for an n-photon absorption event. The distribution function P(n) may also be regarded as an energy distribution function and written as $P(E_n) = P(h\omega_0 + nh\omega_1)$, where $P(E_n)$ is the probability of occupation of that group of levels with energy near $E_n = \hbar\omega_0 + n\hbar\omega_1$ and $\hbar\omega_1$ is the photon energy of the infrared laser. In what follows, we make the important assumption that the prompt fluorescence spectrum depends only upon the populations $P(E_n)$ and not upon the method of producing those populations. Then,

$$F(\omega) \quad = \sum_{n=-\infty}^{\infty} A(\omega, E_n)P(E_n) \qquad (1)$$

where $F(\omega)$ is the prompt fluorescence signal observed at frequency ω, and $A(\omega,E_n)$ is the fluorescence spectrum produced by a population in that group of energy levels near E_n. In general, $n = 0, \pm 1, \pm 2...$corresponding both to emission and absorption of infrared photons. In practice, only a finite group of levels have non-zero population and only a finite number of fluorescence observation frequencies ω_m were monitored. Then Eq. (1) can be re-written:

$$F(\omega_m) \quad = \sum_{n} A(\omega_m,E_n)P(E_n)$$

which is in the form of a vector equation with the matrix $A_{mn} \equiv A(\omega_m,E_n)$. The matrix of coefficients A_{mn} can be measured one column at a time by performing the following auxiliary experiment using only visible light: A dye laser is tuned to photon energy E_n and its absorbed energy is measured. This determines a population $P(E_n)$. This population along with the vector of fluorescence signals observed under this condition determines a column of the matrix A_{mn}. Repeating this procedure for different E_n determines the full matrix. With a knowledge of the elements A_{mn}, the matrix can be inverted to give

$$P(E_n) \quad = \sum_{m} [A_{mn}]^{-1}F(\omega_m) \qquad (2)$$

If the fluorescence spectrum is known in sufficient detail the populations producing it can be calculated from (2). As a matter of choice, the observation frequencies were selected from the formula:

$$\omega_m \approx \omega_0 + m\omega_1 - 500\ \text{cm}^{-1}.$$

Figure 4 shows the population distribution resulting from the multiphoton infrared excitation of NO_2 molecules optically prepared by a 502 nm dye laser pulse. The CO_2 laser depletes the zeroth level and produces a falling distribution on the high energy side. Due to the small changes in

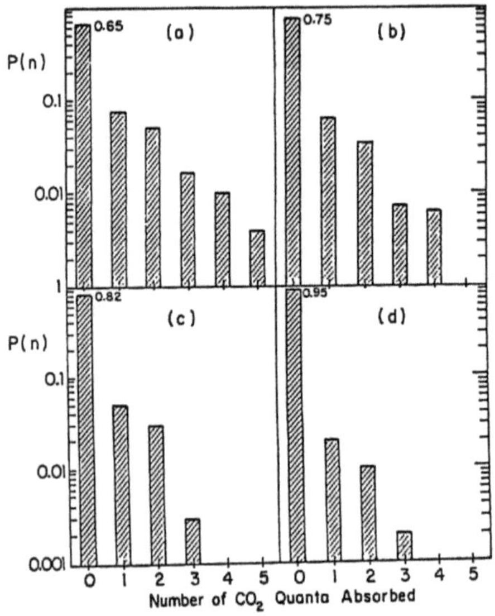

Fig.4 Histograms of probability P(n) for the net absorption of n-photons. (a) 0.5 J/cm^2, (b) 0.3 J/cm^2, (c) 0.2 J/cm^2, (d) 0.1 J/cm^2.

the ordinary Stokes-shifted fluorescence signals, the populations at energies less than $\hbar\omega_0$ could not be determined. Presumably a rather symmetrical distribution falling off at lower energies is established.

This procedure for determining P(n) is subject to the very strong assumption that the fluorescence spectrum from levels near E_n is the same whether produced by single photon or multiphoton excitation. Obviously, the selection rules and the precise levels populated may be quite different for the two processes. Nevertheless, the fluorescence spectrum is hardly affected by this difference due to averaging caused by the 1Å bandwidth of the dye laser and the 50A bandwidth of the fluorescence filters. If the differences in fluorescence spectra were important in our case, then we would have seen a fast collisional scrambling signal similar to what was observed by Donnelly et al.,[6] but contrary to our observations.

A qualitative survey of the wavelength dependence of the CO_2 laser interaction with optically prepared NO_2 molecules has also been carried out. Blue-shifted fluorescence was observed when tuning the dye laser between 450 nm and 500 nm. This insensitivity to wavelength is surprising in view of the low density of vibrational states of the triatomic NO_2. It implies that exact resonances are probably unnecessary for multiphoton excitation. Indeed infrared multiphoton dissociation has recently been reported in the triatomics OCS[7] and SO_2.[8]

In spite of the rather moderate density of states, a stochastic rate equation approach[9] may describe the temporal evolution of the population P(n). If so, the evolution can be regarded as a random walk along the energy axis with steps of one photon each. The results of Fig. 4 are consistent with the probability distribution produced by such a random walk.

We wish to thank Professor F. Kaufman for making available a preprint of his work and Jerry G. Black for assisting with the computer calculations.

References

1. E.R. Grant, P.A. Schulz, A.S. Sudbo, Y.R. Shen, and Y.T. Lee, Phys. Rev. Lett. 40, 115 (1978).
2. V.S. Letokhov and E.A. Ryabov in Laser-Induced Processes in Molecules, ed. by K.L. Kompa and S.O. Smith (Springer-Verlag, New York, 1979).
3. J.C. Stephenson, D.S. King, M.F. Goodman and J. Stone, J. Chem. Phys. 70, 4496 (1979).
4. I. Burak, T.J. Quelly and J.I. Steinfeld, J. Chem. Phys. 70, 334 (1979).
5. I.P. Herman, A. Javan and R.W. Field, J. Chem. Phys. 68, 2398 (1978).
6. V.M. Donnelly, D.G. Keil and F. Kaufman, J. Chem. Phys., in press.
7. D. Proch and H. Schroder, Chem. Phys. Lett. 61, 426 (1978).
8. S.E. Bialkowski and W.A. Guillory, Chem. Phys. Lett. 60, 429 (1979).
9. J.G. Black, E. Yablonovitch, N. Bloembergen and S. Mukamel, Phys. Rev. Lett. 38, 1131 (1977).

An Experimental Study on Unimolecular Reactions Induced by Infrared Multiple Photon Absorption

S. Tsuchiya

Department of Pure and Applied Sciences, College of General Education
University of Tokyo
Komaba, Meguro-ku, Tokyo 153, Japan

ABSTRACT

The mechanism of unimolecular reaction induced by IR multiple photon absorption (MPA) is discussed experimentally on the basis of branching reaction of cyclobutanone and vibrational state analysis of HF formed from fluoroethylenes. It is concluded that the reaction induced by MPA may be interpreted to be close to the one in thermal equilibrium.

1. Introduction

Chemical reactions induced by the multiple photon absorption (MPA) of pulsed infrared laser light are now active field of research. One of the primary interests is on the question whether MPA may offer a method to energize molecules in particular vibrational modes. If this is the case, one may dissociate molecules along a reaction pathway different from that of the thermal reaction. A number of studies have been made to elucidate this problem on the basis of experiments: (1)chemical analysis of products, (2) internal state analysis of product molecules, (3)kinetic energy and scattering angle measurements of product molecules, (4)determination of average absorbed energy per molecule, (5)monitoring of energy distribution of species excited by MPA, etc. Several review articles have been published on the reactions induced by MPA [1] ~ [7], and it becomes general consensus of the fact that the density of states of a polyatomic molecule increases significantly with increasing an energy content of the molecule to form a quasi-continuum state and that molecules are excited to this continuum through a stepwise resonant multi-photon absorption to climb up several discrete states. Once the molecules reach the quasi-continuum state, the energy fluence, not the power, is a good parameter for description of the dissociation [8].

In this brief note, we describe our recent experimental study on chemical reactions induced by MPA by use of the aforementioned methods (1) and (2) to discuss the reaction mechanism in terms of a statistical model.

2. IR Multiphoton Decomposition of Cyclobutanone

Cyclobutanone is known to decompose via two different routes at high temperatures [9]:

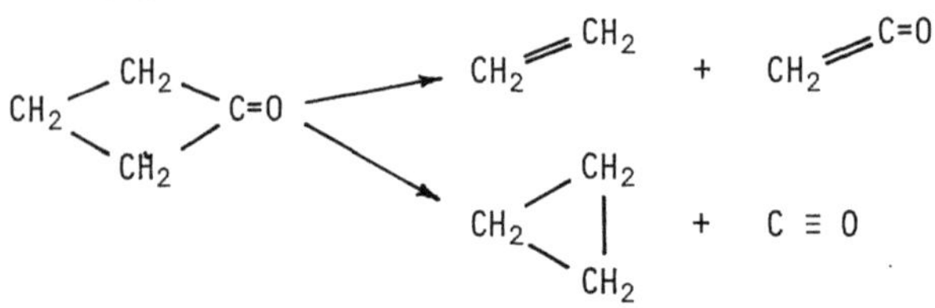

(I)

(II)

The unimolecular decomposition rate constants are determined for respective processes as

$k(I)/s^{-1} = 3.6 \times 10^{14} \exp(-52\ \text{kcal mol}^{-1}/RT)$ and

$k(II)/s^{-1} = 2.6 \times 10^{14} \exp(-58\ \text{kcal mol}^{-1}/RT)$.

If the reaction induced by MPA proceeds in the same manner as the thermal reaction, the relative significance or the ratio of two reaction channels would be a function of the energy content of cyclobutanone assuming a statistical distribution of the energy among the entire degrees-of-freedom. Thus, the observation of the branching ratio must offer some information on the energy state of cyclobutanone just before decomposition induced by MPA of a CO_2 laser.

Cyclobutanone in a reaction cell was irradiated by a line-tuned TEA CO_2 laser (Lumonics 100-2, ~ 1 J pulse^{-1}, ~ 80 ns FWHM) in two ways. The laser beam was focused by a Ge lens(f = 25 cm) at the center of a cell of 10 cm in length and 2.6 cm in diameter. The laser intensity distribution in radial direction at the focused region was almost of a Gaussian shape with 0.9 mm FWHM. In the experiment of 'parallel irradiation', a small cell with 1 cm in thickness was placed in a off-focused region of the laser beam where an effective cross section was 0.36 cm^2. This method of irradiation was preferred than the former 'focused irradiation' because of its homogeneous laser fluence throughout the irradiated area. Cyclobutanone vapor was irradiated by the 9.6 μm R(14) CO_2 laser line, which corresponds to the CH_2 rocking mode, and its reaction products were analyzed by gas chromatography and IR absorption spectrometry.

The reaction products identified were CH_2CO, C_2H_4, cyclo-C_3H_6, C_3H_6 and C_2H_2, of which the last two were obtained only in the reaction induced by 'focused irradiation'. Since the formation of C_2H_2 is significant only in 'focused irradiation' of a high laser energy fluence, and its amount relative to the other products is strongly dependent on the fluence, C_2H_2 may not be a primary product from excited cyclobutanone. Thus, the products found in this experiment support the two reaction channels (I) and (II); in the latter channel, a part of cyclo-C_3H_6 has an excess energy and isomerizes to C_3H_6.

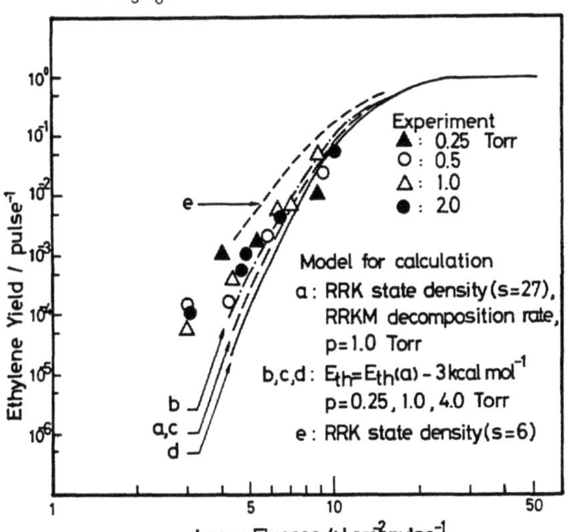

Fig. 1 Observed and calculated yields of ethylene per laser pulse as a function of laser energy fluence.

The product yield in the reaction of 'focused irradiation' showed a 3/2 power dependence on the laser fluence. This result must be interpreted as the fact that the reaction proceeds in a larger volume of the focused region with increase of the laser energy. Contrary, in 'parallel irradiation', the yield is strongly dependent on the laser energy fluence as shown in Fig. 1.

The branching ratio of the reaction channels (I) and (II) may be defined as the ratio of the produced amount of C_2H_4 to those of cyclo-C_3H_6 and C_3H_6. The determined ratio is given in Fig. 2.

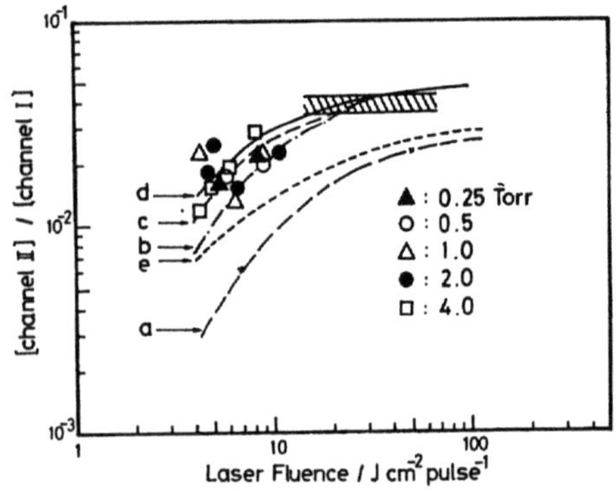

Fig. 2 Observed and calculated branching ratio of reaction channels (I) and (II). The hatched region shows the data obtained in 'focused irradiation' and the conditions of the calculated curves are the same as given in Fig. 1.

On the basis of the experimental observations, BLOEMBERGEN and YABLONOVITCH [1] justified a rate equation approach to describe MPA and induced reaction. In a quasi-continuum state, the probability W_n of molecules absorbing n IR photons obeys the following rate equation:

$$dW_n/dt = [K_n^a W_{n-1} + K_n^e W_{n+1} - (K_{n+1}^a + K_{n-1}^e)W_n]I - (K_n^I + K_n^{II})W_n$$
$$+ \text{(collision term)}, \qquad (1)$$

where the first term represents transitions due to the induced absorption and emission and the second shows the decomposition rates for two channels. Here, we assume very rapid intramolecular energy transfers, so that K_n^e/K_{n+1}^a is the same as the ratio of vibrational state densities ρ_n/ρ_{n+1}[8]. The optical absorption cross section is defined as

$$\sigma = (K_{n+1}^a - K_{n-1}^e)\hbar\omega, \qquad (2)$$

and the decomposition rate constant is the one calculated based on the RRKM theory[12]. The last term represents the collisional deactivation rate from the state with n IR photons and its reverse rate; i.e.,

$$\text{(collision term)} = -k_{col}PW_n\sum_n W_n + k_{col}P(\sum_n W_n)^2\rho_n e^{-n\hbar\omega/kT}/Q, \qquad (3)$$

where k_{col} is a deactivation rate per unit pressure and P is a pressure of cyclobutanone and $Q = \sum_n \rho_n e^{-\hbar\omega/kT}$. In the calculation, k_{col} is assumed as $10^7\ s^{-1}Torr^{-1}$. The numerical calculation to solve the coupled differential equations of (1) was performed assuming the laser pulse of a triangular shape with 80 ns FWHM and the absorption cross section of 3×10^{-20} cm^2. A change of this value may be interpreted as a change of laser energy fluence. The calculated results are compared with experimental ones in Figs. 1 and 2.

The observed product yields of reaction channel (I) are roughly in agreement with the calculation. However, the experimental yields are larger than the calculated in a low laser energy fluence which is close to a threshold laser power for the reaction. In such a low laser energy fluence, one of the prerequisites for Eq.(1) would not be satisfied since excitation of the discrete levels might be a rate determining step to reach the quasi-continuum state. In order to explain the observed branching ratio, the activation energy for channel (II) must have been reduced by 3 kcal mol^{-1}. If this is admitted, the model of Eq.(1) is satisfactory to explain the relative significance of two channels. Thus, it may be concluded that the unimolecular decomposition of cyclobutanone induced by MPA is essentially similar to the thermal reaction.

3. Energy Partitioning in HF Elimination Reaction of Fluoroethylenes

On the basis of the internal energy distribution of product molecules, it is intended to discuss the mechanism of the unimolecular reaction induced by MPA. Energized fluoroethylene decomposes to form vibrationally excited HF* photosensitized reaction:

$$CH_2CHF + Hg(6^3P_1) \longrightarrow CHCH + HF^{\dagger} + Hg(6^1S_0), \qquad \text{(III)}$$

in which CH_2CHF receives the energy of 112 kcal mol^{-1} that exceeds the activation barrier for reaction (III). The vibrational distribution of HF formed in this reaction has been determined from the measurements of IR emission spectrum[13], and the results are reproduced in Fig. 3.

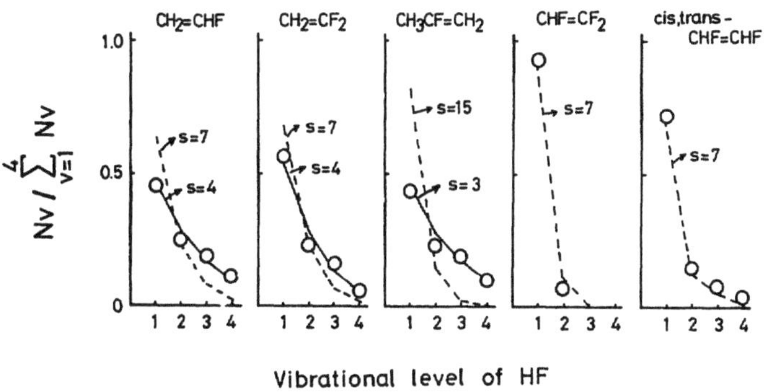

Fig. 3 The vibrational distributions of HF produced by the Hg photosensitized decomposition of fluoroethylenes. The broken lines are the statistical ones calculated assuming total number of vibrational modes in an activated complex as active modes and the solid ones are those assuming limited number of modes.

In this figure, the experimental distribution is compared with the statistical one in which the total available energy is distributed among all degrees-of-freedom of product molecules. It is found that the determined vibrational distributions of HF from three fluoroethylenes are non-statistical and those from remaining two compounds are statistical. Contrary to the Hg* photosensitized reaction, in the reaction induced by MPA, the activated fluoroethylene would cross just over the top of the exit barrier. This might change the energy partitioning in product molecules.

Fluoroethylene flowing through a reaction cell (5 cm in diameter and 40 cm in length) was irradiated by the focused laser beam by use of a Ge lens (f = 25 cm). The IR emission from the focused region was directed into a grating monochromator equipped with an InAs detector. The resolution of the monochromator was less than 100 Å which could separate the vibration-rotation lines of $R_{1-0}(1)$, $P_{2-1}(3)$, and $P_{3-2}(3)$. The CO_2 laser (Lumonics 100-2) excites the CF stretching or CH_2 rocking modes of fluoroethylenes. Figure 4 shows an example of the observed IR emission signals which is obtained by averaging signals of 256 experimental runs. Since the response time of the InAs detector is about 25 μs, an extrapolation to time

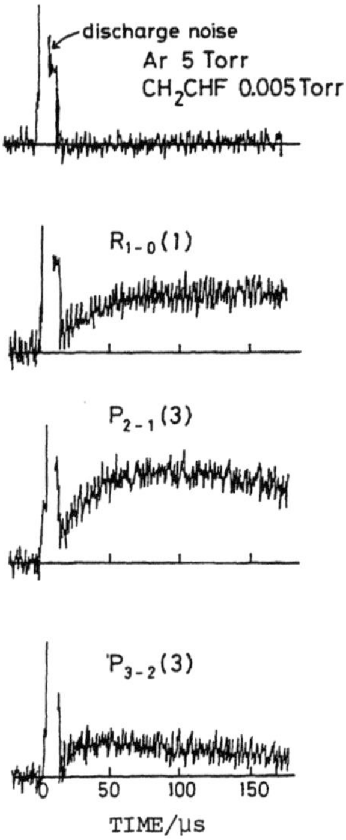

Fig. 4 An example of IR emission signal of HF produced by MPA of CH_2CHF. The signal is an averaged one of 256 IR pulses.

zero is needed to determine the initial distribution. Though a rotational non-equilibrium distribution was observed experimentally, Ar was added as a carrier gas to keep the rotational equilibrium distribution at room temperature. Addition of Ar does not change the vibrational initial distribution and enhances the HF emission [14].

The determined vibrational distributions of HF produced by MPA of four kinds of fluoroethylenes are given in Fig. 5. The results are summarized in Table 1 for comparison with those obtained in the Hg* photosensitized decomposition (PSD). It is found that relative magnitudes of energies partitioned into HF vibration resemble with each other in both reactions of MPA and PSD. This supports the conclusion that the final stages of both reactions proceeds on the same potential surface. However, the energy contents of HF vibration are much larger in PSD reactions than those obtained by MPA.

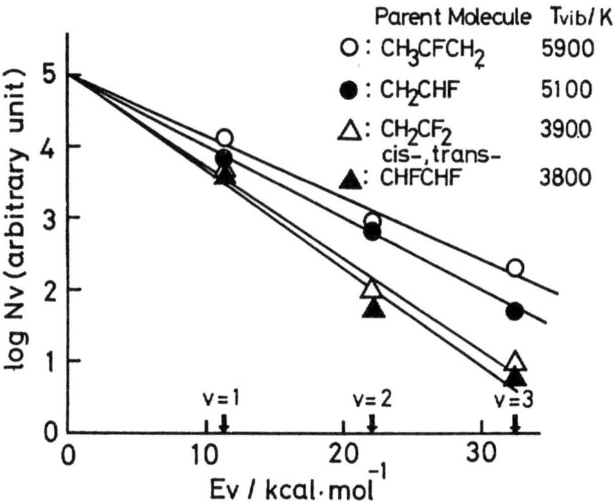

Fig. 5 Vibrational distributions of HF formed from CH_2CHF, CH_3CFCH_2, CH_2CF_2, trans- and cis-CHFCHF induced by MPA.

It was found that the vibrational distributions of HF formed from CH_2CHF, CH_3CFCH_2, and CH_2CF_2 were independent of the irradiated laser energy fluence, while in the reactions of trans- and cis-CHFCHF smaller vibrational energies were given to HF at lower laser energy fluences. In Hg* PSD of the former three compounds, the non-statistical distributions are obtained, while CHFCHF forms a distribution which is statistically expected. These findings may be attributed to the fact that there is a high activation barrier for the backward direction for reactions of CH_2CHF, CH_3CFCH_2, and CH_2CF_2, and a small barrier for reactions of trans- and cis-CHFCHF. Thus, in reactions of the former three compounds, a dynamical force which acts on HF departing from acetylene or fluoroacetylene during sliding down the barrier causes the vibrational excitation of HF. According to the recent theoretical calculation of the ab initio potential surface of CH_2CHF by KATO and MOROKUMA [15], the structure of activated CH_2CHF at the top of the barrier is close to a complex

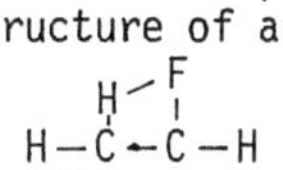

in which the bond length of HF is stretched to some extent reulting in an energy disposal into HF vibration. Contrary, due to a large endoergicity of the reaction of CHFCHF, the exit barrier height is low, so that the excess energy is distributed almost statistically.

Table 1 Vibrational temperatures and energies of HF produced by MPA and Hg* PSD of fluoroethylenes

parent molecule	$E_0-\Delta H^{a)}$/kcal mol^{-1}		T_{vib}/K	E_{vib}/kcal mol^{-1}	type of distribution
CH_2CHF	~ 50	PSD	11400	17	non-statistical
		MPA	5100	5.6	———
CH_3CFCH_2	$\gtrsim$ 50	PSD	13700	22	non-statistical
		MPA	5900	6.9	———
CH_2CF_2	~ 60	PSD	7200	9.5	non-statistical
		MPA	3900	3.6	———
cis, trans-CHFCHF	$\lesssim$ 30	PSD	5700	6.6	statistical
		MPA	3800	3.2	statistical
$CHFCF_2$	$\lesssim$ 10	PSD	2200	0.9	statistical
		MPA	2200	0.7	———

a) E_0 = activation barrier, ΔH = endoergicity in kcal mol^{-1}.

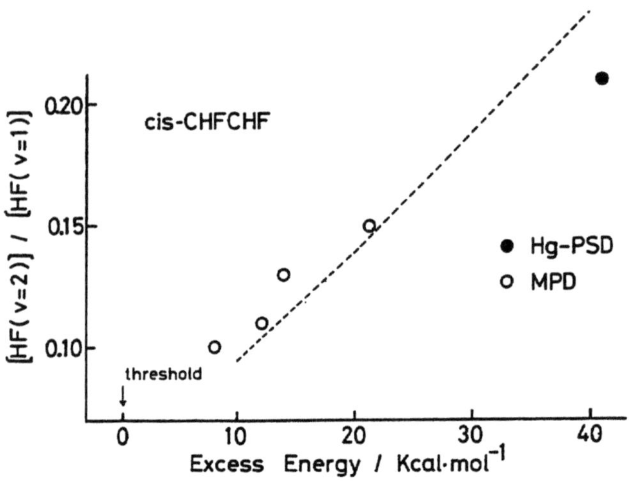

Fig. 6 The determined vibrational distribution of HF formed from cis-CHFCHF as a function of the average energy content of the activated molecules by MPA. The excess energy is defined as the average energy of activated CHFCHF minus the activation energy. The broken line represents the statistical value. The solid circle is obtained from the Hg* photosensitized reaction.

The model calculation utilizing Eq.(1) was made to determine the average energy of CHFCHF excited by MPA as a function of the laser energy fluence. In Fig. 6, the observed vibrational distributions of HF from CHFCHF are plotted against the average energy content of CHFCHF activated by MPA. The results are roughly in agreement with the statistical calculation. Thus, the stochastic statistical model may describe the energy disposal in the HF elimination reaction of fluoroethylenes if there is no high exit barrier.

4. Summary

(1) The branching ratio of reaction channels (I) and (II) of cyclobutanone excited by MPA agrees with the statistical calculation based on the RRKM theory.
(2) In the HF elimination reactions of fluoroethylenes induced by MPA, the energy disposal into HF vibration is non-statistical if a high exit energy barrier exists. However, in the reaction of CHFCHF, the energy disposal into HF vibration is statistical due to a low exit energy barrier. Thus, MPA results in a statistical quasi-equilibrium distribution of energies in activated fluoroethylene molecules.

References

1 N. Bloembergen, E. Yablonovitch: Phys. Today 31, 23 (1978).
2 V.S. Letokhov, C.B. Moore: Sov. J. Quantum Electron. 6, 129 (1976).
3 V.S. Letokhov, C.B. Moore: Chemical and Biochemical Applications of Lasers, ed. C.B. Moore, 3, 1 (1977).
4 R.V. Ambartzumian, V.S. Letokhov: Acc. Chem. Res. 10, 61 (1977).
5 A. Mooradian, T. Jaeger, P. Stokseth, ed.: Tunable Lasers and Applications (Springer, Berlin, 1976) 122, 150, 162, 177.
6 A.H. Zewail, ed.: Advances in Laser Chemistry (Springer, Berlin, 1978) 48, 71, 298, 362, 433, 455.
7 P.A. Schulz, A.S. Sudbo, D.J. Krajnovitch, H.S. Kwok, Y.R. Shen, Y.T. Lee: Ann. Rev. Phys. Chem. 30, 379 (1979).
8 J.G. Black, P. Kolodner, M.J. Schultz, E. Yablonovitch, N. Bloembergen: Phys. Rev. A19, 704 (1979).
9 A.T. Blades: Can. J. Chem. 47, 615 (1969).
10 T.H. McGee, A. Schleifer: J. Phys. Chem. 76, 963 (1972).
11 A.T. Blades, H.S. Sandhu: J. Phys. Chem. 77, 1317 (1973).
12 G.M. Breuer, R.S. Lewis, E.K.C. Lee: J. Phys. Chem. 79, 1985 (1975).
13 H. Watanabe, H. Horiguchi and S. Tsuchiya: Bull. Chem. Soc. Jpn. 53, in press.
14 C.R. Quick and C. Wittig: J. Chem. Phys. 69, 4201 (1978).
15 K. Kato, K. Morokuma: 14th Symposium on High Speed reactions, Tokyo (1979); see also S. Kato, K. Morokuma: J. Chem. Phys. 72, 206 (1980).

Theory of Collisionless Molecular Multiphoton Processes

S. Mukamel

Department of Chemical Physics, Weizmann Institute of Science
Rehovot, Israel, and

Department of Chemistry, William Marsh Rice University
Houston, TX 77001, USA

ABSTRACT

We derive reduced equations of motion (REM) which describe the dynamics of polyatomic molecules in the presence of strong infrared lasers. The derivation is made starting from the complete molecular Liouville equation and making use of the Zwanzig-Mori projection operator formalism. The complete molecular information relevant for the dynamics of molecular multiphoton processes (MMP) is expressed in terms of a hierarchy of intramolecular dipole correlation functions. We show how by invoking simple statistical assumptions (the random phase approximation and separation of time scales) this information is considerably reduced to essentially four quantities per transition: An integrated Rabi frequency $\bar{\Omega}_{nm}$, a dephasing rate Γ_{nm}, a detuning $\bar{\omega}_{nm}$ and ratios of statistical weights of the levels d_n/d_m. Depending on our choice of reduction scheme we may obtain either generalized Bloch equations or simple rate equations. The interrelation between the two and their limits of validity are precisely specified. The present formulation enables us to describe the dynamics of MMP all the way from "region I" via the quasicontinuum and up to the dissociation. Finally we present novel spectroscopic results done in supersonic beams which provide us directly with the intramolecular dephasing rates which are the key dynamical quantities in MMP.

I. Introduction

Our understanding of the dynamics of highly vibrationally excited polyatomic molecules has improved dramatically in recent years due to the development of novel spectroscopic techniques such as overtone spectroscopy [1,2], spectroscopy of supercooled molecules in beams [3,4], picosecond spectroscopy [5] etc. Of special interest are studies of molecular multiphoton processes (MMP) which are based on pumping few e.V. (30-40 photons) of energy into isolated (collision-free) polyatomic molecules via the interaction with strong infrared laser pulses in the power range of 10M watt/cm^2[6,7]. Some of the main reasons for the enormous theoretical and experimental activity in this field are:(i) This is a relatively clean and convenient way of pumping energies of chemical interest to large molecules in short times (0.5nsec-100 nsec). The excitation conditions are easily controllable by varying the laser frequency, power, fluence, using several lasers (multicolour experiments) etc. (ii) The excitation may in principle be made bond selective since the oscillator strength is not evenly distributed among the various molecular vibrational degrees of freedom. This opens the fascinating possibility of achieving laser-induced and laser controlled chemical reactions by putting the right amount of energy in the region of interest. This depends of course on the relative rates of pumping the energy to the intramolecular redistribution of energy which tends to destroy any selectivity. (iii) Unimolecular dissociation following MMP turned out to be highly isotopically selective in several cases. (iv) Understanding the mechanism of the multiphoton pumping processes provides us with a clue for monitoring the dynamics of highly vibrationally excited polyatomic molecules. This is connected to the new field of intramolecular line broadening (dephasing) and may help us clarify the validity of the existing statistical theories of unimolecular reactions [8].

The following qualitative picture [6] has emerged out of the numerous experimental and theoretical studies: The molecular energy levels are separated into three regions. In the lowest energy range (region I) the density of molecular states is very low and the laser field is interacting with isolated molecular states (coherent driving). In this region the laser power is required to overcome the molecular anharmonicities and phenomena such as threshold power, saturation behavior, isotopic selectivity and multiphoton resonances are accounted for in terms of the molecular level-scheme of region I. After the molecule has absorbed few quanta, the density of molecular states becomes very large and we can no longer describe the time evolution in terms of few isolated molecular states. This region is denoted region II or the quasi-continuum and a proper description of the molecular time evolution in this range requires a quantitative understanding of the mechanism of intramolecular energy transfer and line broadening (dephasing) of highly vibrationally excited polyatomic molecules, of which very little is known at present. Finally, when the molecule acquires enough energy for dissociation, it enters region III, where, in addition to all the complications of region II, we have to incorporate also the dynamics of unimolecular decomposition. Since numerous reviews were written recently on this subject [6,7], we shall not give here an extensive survey of the current experimental status. We shall rather try to develop a general unified framework for the theoretical description of these processes. It is clear that a complete microscopic treatment of MMP (i.e.the calculation of the entire molecular density matrix ρ) is neither feasible nor desirable. Due to the huge number of molecular states involved (10^{10} or more) such a treatment will require an enormous amount of (unavailable) information for the calculation of a density matrix whose most parts are irrelevant and redundant. The present approach [9] is based on the projection operator formalism of Zwanzig [10] and Mori [11] combined with the representation of the true molecular states. The latter enables us to formulate the problem in a form free of perturbative arguments in any intramolecular interactions. The main steps in this "hydrodynamic-like" approach [9] are: (1) The choice of a few molecular operators whose expectation values are the important variables for the dynamics of MMP; (2) the definition of an appropriate Mori projection operator onto the space spanned by these operators; and (3) the derivation of reduced equations of motion (REM) for the time evolution of these variables.

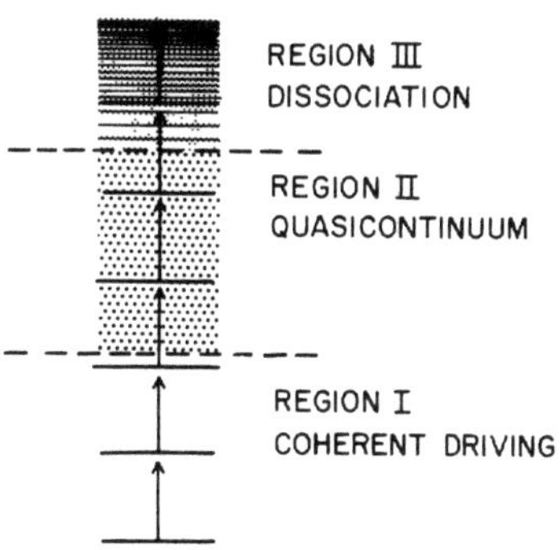

Fig.1 Energy regions for molecular multiphoton processes.

This procedure is completely general, and formally the choice of the number and type of variables is arbitrary. However, the complexity and usefulness of the resulting REM depend crucially upon a successful choice of variables which should be adjusted to the desired level of description. In Section II we present the basic reduction formalism which is based on the projection operator techniques of Mori and Zwanzig. In Section III we present the Hamiltonian for MMP and formally derive the REM for a specific choice of variables corresponding to level populations. These REM are then expanded in Section IV and the conditions under which they further reduce to simple rate equations in the quasi-continuum are analysed. In Section V we add more variables corresponding to coherences and derive another REM (generalized Bloch equations) which are valid also in region I of coherent driving. An alternative derivation of the simple rate equations is also obtained as a limiting case of these new REM. Finally in Section VI we discuss and summarize our results.

II. The basic reduction formalism

We consider a complicated system with many degrees of freedom characterized by a Hamiltonian H and a density matrix ρ whose time evolution is given by the Liouville equation:

$$\frac{d\rho}{dt} = -i[H,\rho] \equiv -iL\rho \ , \tag{1}$$

where L is the Liouville (tetradic) operator corresponding to H. Due to the complexity of the system, the information contained in the complete density matrix ρ is too detailed and its evaluation is impractical. Consequently we shall be interested only in a few quantities $\sigma_\mu(t)$ which are the expectation values of a small set of dynamical operators A_μ, i.e.

$$\sigma_\mu(t) \equiv (A_\mu,\rho(t)) \tag{2}$$

where we have defined the scalar product $S_{\mu\nu}$ of two operators as

$$S_{\mu\nu} \equiv (A_\mu,A_\nu) \equiv Tr(A_\mu^+ A_\nu) \ . \tag{3}$$

Without loss of generality we may assume that our relevant operators are orthonormal, i.e. $S_{\mu\nu}=\delta_{\mu\nu}$. The Mori-Zwanzig [10,11] projection operator technique enables us to derive reduced equations of motion (REM) which yield $\sigma_\mu(t)$ directly without having to calculate ρ. To that end we define a Mori projection P which projects onto the subspace spanned by our relevant operators A_μ:

$$PB = \sum_\mu (B,A_\mu)A_\mu \ , \tag{4}$$

and the complementary projection $Q\equiv 1-P$. We shall further assume that $\rho(0)=P\rho(0)$ so that $Q\rho(0)=0$. Making use of these definitions the REM may be written in the form [9]

$$\frac{d\sigma_\nu}{dt} = -i \sum_\mu \langle L\rangle_{\nu\mu}\,\sigma_\mu(t) - \int_0^t d\tau \sum_\mu \langle R(t-\tau)\rangle_{\nu\mu}\,\sigma_\mu(\tau) \ , \tag{5}$$

where $R(t-\tau)$ is the tetradic operator

$$R(t-\tau) \equiv L\,\exp(-iQL(t-\tau))\,Q\,L \ , \tag{6}$$

and the tetradic matrix elements $\langle L\rangle_{\nu\mu}$ and $\langle R\rangle_{\nu\mu}$ are defined as

$$\langle Y\rangle_{\nu\mu} \equiv \mathrm{Tr}(A_\nu^\dagger\, Y\, A_\mu) \quad , \quad Y=R,L \ . \tag{7}$$

Equations (5) are exact and are valid for an arbitrary choice of dynamical operators A_μ. In practice, however, the memory kernel $\langle R\rangle$ is usually evaluated in some approximate manner using an expansion in a properly chosen parameter. To that end it is sometimes advantageous to use a different form of the REM [9], i.e.,

$$\frac{d\sigma_\nu}{dt} = -i \sum_\nu \langle L\rangle_{\nu\mu}\, \sigma_\nu(t) - \sum_\nu \langle\tilde{R}(t)\rangle_{\nu\mu}\, \sigma_\mu(t) \tag{8}$$

where

$$\langle\tilde{R}(t)\rangle_{\nu\mu} = \sum_{\nu'} W(t)_{\nu\nu'}\, V^{-1}_{\nu'\mu}(t) \quad , \tag{9}$$

$$W_{\nu\nu'}(t) = \mathrm{Tr}(A_\nu^\dagger\, LQ\, \exp(-iLt) A_{\nu'}) \ , \tag{9a}$$

$$V_{\nu'\mu} = \mathrm{Tr}(A_{\nu'}^\dagger\, \exp(-iLt) A_\mu) \quad . \tag{9b}$$

Equations (5) or (8) constitute the basic reduction scheme to be used throughout the present lecture. The form (5) arises naturally when keeping the complete time ordering of the various operators and will be referred to as the COP (chronological ordering prescription). The form (8) uses only partial time ordering and will be referred to as the POP (partial ordering prescription). Equation (8), like (5), is also exact. However, once an expansion is made both equations may have very different predictions [12,9]. Furthermore, it may turn out that depending on the choice of variables one form could become advantageous. This is indeed the case for the present problem of MMP. When we try to derive simple rate equations (Sections III and IV) then the COP form (Eq. (5)) is most convenient. When we add more variables corresponding to coherences (Section V), the expansion of the COP equations becomes extremely tedious and the POP approach (Eq. (8)) is the natural way to proceed. A comparison of the two forms for general relaxation and line shape problems was made recently [12].

III. Reduced equations of motion involving populations only

We consider a polyatomic molecule interacting with a monochromatic infrared laser beam whose frequency is ω_L, under collision-free conditions. We assume that the Schrödinger equation for the isolated molecule (in the absence of the field) has been solved and that we have the complete set of molecular eigenvalues as well as the dipole matrix elements between them. Assuming that the molecule is initially cold ($kT \ll h\omega_L$), then only states with energies around $n\omega_L$, $n=0,1,2,\ldots$ are important for the multiphoton excitation process and need to be considered. We shall therefore group these relevant molecular states into levels and denote them as $\{|n\alpha\rangle\}$ with eigenvalues $E^\circ_{n\alpha}$ where n stands for the level and α runs over the states within the n'th level. We further invoke the rotating wave approximation (RWA) [13] which is very reasonable for MMP with infrared photons and which amounts to neglecting high frequency terms in the Hamiltonian which are not expected to contribute significantly to the molecular time evolution. We can thus write the combined Hamiltonian for the molecule and the field in the time independent form[9]:

$$H = H_0 + H' \quad , \tag{10}$$

where

$$H_0 = \sum_{n,\alpha} |n\alpha\rangle E_{n\alpha} \langle n\alpha| \quad , \tag{10a}$$

and

$$H' = \varepsilon \sum_{\substack{n\alpha \\ m\beta \\ m=n\pm1}} |n\alpha\rangle \mu_{nm}^{\alpha\beta} \langle m\beta| . \tag{10b}$$

Here the molecular states within the n'th level have absorbed n infrared quanta from the field, and $E_{n\alpha}=E_{n\alpha}^{\circ}-n\omega_L$ is the energy of the $|n\alpha\rangle$ state dressed (to zero order) by the field. $\mu_{nm}^{\alpha\beta}=\langle m\beta|\mu|n\alpha\rangle$ is the transition dipole between the $|n\alpha\rangle$ and $|m\beta\rangle$ states and ε is the laser field amplitude. The molecular level and coupling scheme is presented in Fig. 2.

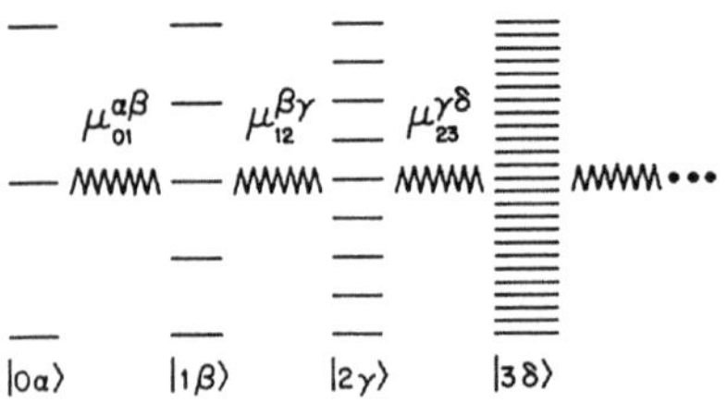

Fig.2 The coupling scheme for molecular multiphoton processes.

We shall now turn to the construction of the set of relevant operators. In a molecular multiphoton excitation experiment the quantities that are of primary interest to us are the populations of the various levels (the probability $P_n(t)$ for the molecule to absorb n photons at time t). It is thus clear that a minimal set of relevant variables should include these populations. In order to derive REM for $P_n(t)$ we shall now introduce the following set of molecular operators [9]:

$$A_{nn} = \frac{1}{\sqrt{d_n}} \sum_{\alpha} |n\alpha\rangle \langle n\alpha| \ , \qquad n=0,1,\ldots,N-1 \tag{11}$$

d_n being the number of states within the n'th level (the statistical weight of that level) and N is the total number of levels considered. The choice of the states included in A_{nn} (and consequently of d_n) is important since the ratios d_n/d_m enter explicitly in the resulting REM (Eqs. (33) or (47)). It should thus be made with physical insight and only states that are expected to participate in the dynamics of the MMP in the experimentally relevant time scale should be included in the summation [11]. The A_{nn} operators are orthonormal (with respect to their scalar product) and the populations of the various levels are

$$P_n(t) = \sqrt{d_n}\,(A_{nn},\rho(t)) = \sqrt{d_n}\,\sigma_n(t) \quad . \tag{12}$$

At time t=0 we assume that all the molecules are in the zeroth level so that $\rho(0)=P\rho(0)=1/\sqrt{d_0}\,A_{00}$. For the present set of operators we have the convenient relation $PL_0=0$ (L_0 and L' are the Liouville operators corresponding to H_0 and H' respectively). This implies that

$$QL = (1-P)(L_0+L') = L_0 + QL' \ , \tag{13}$$

which enables us to expand $\exp(-QL\tau)$ in a power series in QL' and makes the formal evaluation of the COP operator $\langle R(T-\tau)\rangle$ straightforward resulting in [9]:

$$\frac{d\sigma_\nu}{dt} = -\int_0^t d\tau \sum_\mu \langle R(t-\tau)\rangle_{\nu\mu}\, \sigma_\mu(\tau) \quad , \tag{14}$$

where

$$\langle R(t-\tau)\rangle = \langle R^{(2)}(t-\tau)\rangle + \langle R^{(4)}(t-\tau)\rangle + \langle R^{(6)}(t-\tau)\rangle + \ldots , \tag{14a}$$

$$\langle R^{(2)}(t-\tau)\rangle = K^{(2)}(t-\tau,0) \quad , \tag{14b}$$

and

$$\langle R^{(2n)}(t-\tau)\rangle = \int_0^{t-\tau} d\tau_1 \int_0^{\tau_1} d\tau_2 \cdots \int_0^{\tau_{2n-3}} d\tau_{2n-2}\, K^{(2n)}(t-\tau,\tau_1,\tau_2,\ldots,\tau_{2n-2},0) \tag{15}$$

Here

$$K^{(2n)}(\tau_1,\tau_2,\ldots,\tau_{2n}) = (-1)^{n+1}\langle L'(\tau_1)L'(\tau_2)(1-P)L'(\tau_3)L'(\tau_4)$$
$$(1-P)\ldots(1-P)L'(\tau_{2n-1})L'(\tau_{2n})\rangle \, , \tag{16}$$

where

$$L'(\tau) = \exp(iL_o\tau)L' \exp(-iL_o\tau) \quad . \tag{17}$$

Let us introduce further the n'th moment of L' as the n-time correlation function

$$M^{(n)}(\tau_1,\tau_2\ldots\tau_n) = \langle L'(\tau_1)L'(\tau_2)\ldots L'(\tau_n)\rangle \, , \tag{18}$$

in terms of which we may rewrite Eq. (16) in the form:

$$K^{(2)}(\tau_1,\tau_2) = M^{(2)}(\tau_1,\tau_2) \, , \tag{19a}$$

$$K^{(4)}(\tau_1,\tau_2,\tau_3,\tau_4) = -[M^{(4)}(\tau_1,\tau_2,\tau_3,\tau_4) - M^{(2)}(\tau_1,\tau_2)M^{(2)}(\tau_3,\tau_4)] \tag{19b}$$

etc. The $\langle\cdots\rangle$ in Eqs. (14)-(19) denotes a tetradic matrix element as defined by Eq. (7).

Equations (14) together with (15)-(19) enable us to derive closed REM for the N populations of the various levels (P_n) in terms of the tetradic NxN R matrix. Evaluation of the latter requires the calculation of the intramolecular dipole correlation functions $M^{(n)}$ (Eq. (18)) (or $K^{(n)}$ (Eq. (19)), n=2,4,... which provide us with the complete molecular information needed for the description of all MMP whenever the experimental observables are connected with populations only.

IV. Expansion of the REM - the derivation of simple rate equations

We shall now utilize the formalism of Section III to derive explicit expression for the relaxation operator $\langle R\rangle$. To that purpose we make use of the property [10]

$$\sum_n \sqrt{d_n/d_m}\langle R(\tau)\rangle_{nn,mm} = 0 \, , \tag{20}$$

which is a direct consequence of the conservation of probability, to rewrite Eq. (14) (to fourth order in ε) in the form

$$\frac{dP_n}{dt} = \varepsilon^2 \sum_{m=n\pm1} \int_0^t d\tau \langle R^{(2)}(t-\tau)\rangle_{nn,mm}$$

$$[P_m(\tau)\sqrt{d_n/d_m} - P_n(\tau)\sqrt{d_m/d_n}]$$

$$+ \varepsilon^4 \sum_{\substack{m=n\pm2\\ n\pm1}} \int_0^t d\tau \langle R^{(4)}(t-\tau)\rangle_{nn,mm} \tag{21}$$

$$[P_m(\tau)\sqrt{d_n/d_m} - P_n(\tau)\sqrt{d_m/d_n}] \ ,$$

where

$$\langle R^{(2)}(t-\tau)\rangle_{bb,aa} = (d_a d_b)^{-1/2} \langle \mu_{ab}(t-\tau)\mu_{ba}(0)\rangle + \text{c.c.} \tag{22a}$$

$$\langle R^{(4)}(t-\tau)\rangle_{cc,aa} = (d_a d_c)^{-1/2} \int_0^{t-\tau} d\tau_1 \int_0^{\tau_1} d\tau_2 \, K^{(4)}_{cba}(t-\tau,\tau_1,\tau_2,0) + \text{c.c.} \tag{22b}$$

$$K^{(4)}_{cba} = \text{I} + \text{II} + \text{III} \tag{23}$$

$$\text{I} = \langle \mu_{ab}(0)\mu_{bc}(t-\tau)\mu_{cb}(\tau_1)\mu_{ba}(\tau_2)\rangle$$

$$- \frac{1}{d_b} \langle \mu_{bc}(t-\tau)\mu_{cb}(\tau_1)\rangle\langle\mu_{ab}(0)\mu_{ba}(\tau_2) \ ,$$

$$\text{II} = \langle \mu_{ab}(0)\mu_{bc}(\tau_1)\mu_{cb}(t-\tau)\mu_{ba}(\tau_2)\rangle$$

$$- \frac{1}{d_b} \langle \mu_{bc}(\tau_1)\mu_{cb}(t-\tau)\rangle\langle\mu_{ab}(0)\mu_{ba}(\tau_2)\rangle,$$

$$\text{III} = \langle \mu_{ab}(0)\mu_{bc}(\tau_2)\mu_{cb}(t-\tau)\mu_{ba}(\tau_1)\rangle \ ,$$

and where we consider only $c=a\pm2$ (the $a\pm1$ terms are similar), i.e.

$$a=n, \quad b=n\pm1, \quad c=n\pm2 \ . \tag{24}$$

The correlation functions $\langle \cdots \rangle$ are defined as the trace of the product of operators and

$$\mu_{ab}(\tau) = \exp(iH_o\tau)\mu_{ab} \exp(-iH_o\tau)$$
$$= \sum_{\alpha\beta} |a\alpha\rangle \, \mu_{ab}^{\alpha\beta} \langle b\beta|\exp(i\omega_{\alpha\beta}\tau) \tag{25}$$

In Fig.3 we have a diagrammatic representation of $\langle R^{(4)}\rangle$. We note that there are six pathways leading from $|aa\rangle\rangle$ to $|cc\rangle\rangle$ in fourth order. Four of them (I , II and their c.c.) are passing via $|bb\rangle\rangle$ and the other two (III and its c.c.) are passing via $|ac\rangle\rangle$ or $|ca\rangle\rangle$. These two types of pathways are often called incoherent and coherent pathways, respectively. For the sub-

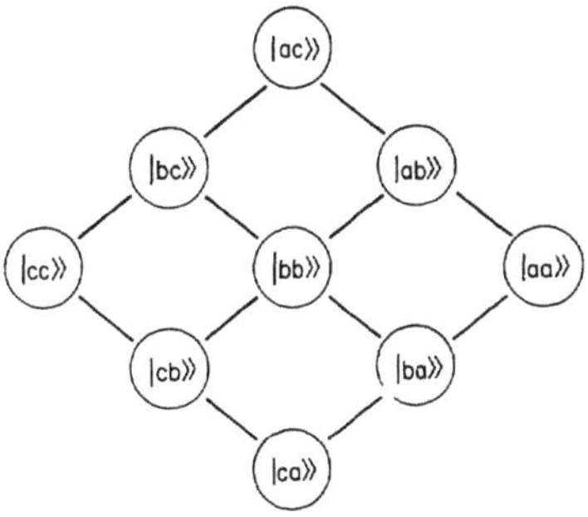

Fig.3 Diagrammatic representation of the Liouville space terms contributing to $R^{(4)}$ showing the six pathways to go from $|aa\rangle\rangle$ to $|cc\rangle\rangle$ in fourth order.

sequent manipulations we shall now introduce the integrated Rabi frequency for the ab transition, i.e.

$$\bar{\Omega}^2_{ab} = \langle R^{(2)}_{aa,bb}(0)\rangle = 2\varepsilon^2 \sum_{\alpha\beta} |\mu^{\alpha\beta}_{ab}|^2/\sqrt{d_a d_b} \quad , \tag{26}$$

and the dimensionless correlation function

$$\tilde{I}_{ab}(\tau) = \langle\mu_{ab}(\tau)\mu_{ba}(0)\rangle/\langle\mu_{ab}(0)\mu_{ba}(0)\rangle \quad , \tag{27}$$

so that we have

$$R^{(2)}_{bb,aa}(t-\tau) = \bar{\Omega}^2_{ab} \operatorname{Re} \tilde{I}_{ab}(t-\tau) \equiv \bar{\Omega}^2_{ab} I(t-\tau) \tag{28}$$

and $I(0)=1$. Equations (21)-(24) express P_n in terms of the hierarchy of intramolecular correlation functions $\langle\mu\mu\rangle$, $\langle\mu\mu\mu\mu\rangle$ etc. We shall now make use of the complexity of our system to further simplify the REM and reduce the amount of relevant molecular information. This will be done in two stages.

(i) The random-phase approximation

When each level contains many states and the various μ's have arbitrary phase we expect that

$$\begin{aligned}&\sum_{\alpha\beta\beta'\gamma} \mu^{\alpha\beta}_{ab}(\tau_1)\mu^{\beta\gamma}_{bc}(\tau_2)\mu^{\gamma\beta'}_{cb}(\tau_3)\mu^{\beta'\alpha}_{ba}(\tau_4)\\ &\quad\cong \sum_{\alpha\beta\gamma} \mu^{\alpha\beta}_{ab}(\tau_1)\mu^{\beta\gamma}_{bc}(\tau_2)\mu^{\gamma\beta}_{cb}(\tau_3)\mu^{\beta\alpha}_{ba}(\tau_4) \quad ,\end{aligned} \tag{29}$$

i.e. only the terms with $\beta=\beta'$ will contribute whereas the sum over $\beta'\neq\beta$ will average to zero. This is a form of the statistical random phase approximation. We further assume that the ab and bc transitions are uncorrelated so that Eq. (29) may be further factorized resulting in

$$\begin{aligned}&\langle\mu_{ab}(\tau_1)\mu_{bc}(\tau_2)\mu_{cb}(\tau_3)\mu_{ba}(\tau_4)\rangle\\ &\quad\cong \frac{1}{d_b}\langle\mu_{ab}(\tau_1)\mu_{ba}(\tau_4)\rangle\langle\mu_{bc}(\tau_2)\mu_{cb}(\tau_3)\rangle \quad .\end{aligned} \tag{30}$$

This implies that in the evaluation of $K^{(4)}_{cba}$ we have

$$I = II = 0 \quad , \tag{31}$$

so that

$$\langle R^{(4)}_{cc,aa}(t-\tau)\rangle = \bar{\Omega}^2_{ab}\,\bar{\Omega}^2_{bc}\int_0^t d\tau_1\int_0^{\tau_1} d\tau_2$$

$$\tilde{I}_{cb}(t-\tau-\tau_2)\tilde{I}_{ba}(\tau_1)\ . \tag{32}$$

(ii) Separation of time scales - the Markovian limit

Suppose we have the desirable situation whereby $P_n(\tau)$ is changing on a time scale much slower than that of $\langle R(t-\tau)\rangle$. (This means that $P_n(\tau)$ consists of a complete set of slow variables) [10,11]. In this case the Markovian limit holds whereby on a coarse grained time scale ($\tau>\tau_c$) we may safely assume that $\langle R(t-\tau)\rangle$ acts as a δ-function inside the integrations (Eq. (21)) so that it may be factored out and we get the simple rate equations

$$\frac{dP_n}{dt} = \sum_{m=n\pm1} W^{(2)}_{nm}(P_m\sqrt{d_n/d_m} - P_n\sqrt{d_m/d_n}) + \sum_{\substack{m=n\pm2\\ n\pm1}} W^{(4)}_{nm}(P_m\sqrt{d_n/d_m} - P_n\sqrt{d_m/d_n})\ , \tag{33}$$

where

$$W^{(k)} = \int_0^\infty d\tau \langle R^{(k)}(\tau)\rangle\ . \tag{34}$$

Using Eqs. (22) and (34) we have

$$W^{(2)}_{nm} = \int_0^\infty d\tau\langle R^{(2)}(\tau)\rangle_{nn,mm} \equiv \bar{\Omega}^2_{nm}\,\tau_c\ , \tag{35}$$

where τ_c is the dipole correlation time (typical time scale of $I_{nm}(\tau)$), i.e.

$$\tau_c \equiv \int_0^\infty d\tau\ I_{nm}(\tau)\ . \tag{36}$$

Regarding $W^{(4)}$ we have utilizing Eqs. (32) and (34)

$$W^{(4)}_{ca} = \bar{\Omega}^2_{ab}\,\bar{\Omega}^2_{bc}\int_0^\infty d\tau\int_0^{\tau} d\tau_1\int_0^{\tau_1} d\tau_2\ \tilde{I}_{cb}(\tau-\tau_2)\tilde{I}_{ba}(\tau_1) + \text{c.c.} \tag{37}$$

It is clear that the only contributions to the integral (37) come from the region $0<\tau,\tau_1,\tau_2 \lesssim \tau_c$, since otherwise the integrand vanishes. We thus have

$$W^{(4)}_{ca} \cong \bar{\Omega}^2_{ab}\,\bar{\Omega}^2_{bc}\,\tau_c^3\ . \tag{38}$$

Similar arguments will hold also for the higher order correlation functions (34) so that we can write

$$W^{(2k)} \cong W^{(2)}(\bar{\Omega}\tau_c)^{2k-2} \qquad k=2,3,\ldots \tag{39}$$

Making use of the estimates (35) and (39) we may now state the condition for the applicability of the Markovian limit as follows

$$\eta \equiv \bar{\Omega}\tau_c \ll 1\ . \tag{40a}$$

When Eq. (40) holds the following things happen (i) the time scale of variation of $P(\tau)$ is $(\bar{\Omega}^2\tau_c)^{-1}$ which will be $\gg\tau_c$. This justifies replacing $\langle R\rangle$ by

W (separation of time scales). (ii) All the higher order rate constants will become much smaller than $W^{(2)}$, i.e.

$$W^{(2k)}/W^{(2)} = \eta^{2k-2} \ll 1 , \tag{40b}$$

so that we may ignore them and retain $W^{(2)}$ only in Eq. (33). Finally we note that if we assume

$$\tilde{I}_{nm}(\tau) = \exp(-i\bar{\omega}_{nm}\tau - \Gamma_{nm}\tau) , \tag{41}$$

then

$$\tau_c = \mathrm{Re}\int_0^\infty d\tau\, \tilde{I}_{nm}(\tau) = \frac{\Gamma_{nm}}{\bar{\omega}^2_{nm}+\Gamma^2_{nm}} \tag{42}$$

and

$$W^{(2)}_{nm} = \frac{\bar{\Omega}^2_{nm}\,\Gamma_{nm}}{\bar{\omega}^2_{nm}+\Gamma^2_{nm}} \quad . \tag{43}$$

In concluding this section we note that we have demonstrated here how, when rate equations apply, the relevant molecular information necessary for the description of MMP reduces essentially to four numbers per transition: an integrated Rabi frequency $\bar{\Omega}$, a dephasing rate Γ, a detuning $\bar{\omega}$ and the ratio of statistical weights d_n/d_m.

V. REM for populations and coherences - derivation of generalized Bloch equations

In Sections III and IV we have derived REM for the populations in MMP and showed how in the Markovian limit they reduce to simple rate equations. The basic reduction procedure of Section II is, however, more general and enables us to derive a closed set of REM for any arbitrary set of chosen variables. As we have already pointed out, the choice of the right number and kind of variables is a crucial step in the derivation of the REM since their simplicity and applicability depend on a successful choice.

We shall now construct and analyze a different set of REM for MMP by the addition of more variables corresponding to coherences (i.e. time derivatives of the population variables). This is done due to the following reasons:

(1) It is clear that at the early stages of the molecular driving (Region I) the random phase and separation of time scales assumptions made in the derivation of the rate equations do not hold and we should in fact solve the exact Schrödinger equation with few states and coherent driving. Although we can in principle retain the populations as our only variables and expand the evolution operator <R> in higher powers in the field, it is much easier to add few variables corresponding to coherences and get a simple Markovian equation which will be in the form of a general (multilevel) Bloch equation. Thus the explicit inclusion of coherences as variables in Region I results in a considerable simplification of the description in this region.

(2) In spectroscopic studies other than multiphoton excitations (e.g. ordinary line shapes [14], coherent transients [15] and resonance fluorescence [16,4]), the experimentalist usually probes directly the time evolution of coherences and their damping (dephasing) rates (e.g. an ordinary line shape is the Fourier transform of the correlation function of the molecular coherence) [14]. In the present REM for the populations, the dephasing rates are "buried" inside the kernel <R>. By using a less-reduced description including coherences, we are

able to see clearly the role of coherences in the dynamics of MMP and we can use the results of other spectroscopic experiments to evaluate the parameters appearing in our REM.

(3) Conceptually, the addition of coherences enables us to look at MMP from a different viewpoint and to gain a better insight into the meaning of the Markovian limit and the "reduction of information" that occurs there. We shall be able to provide an alternative derivation to the rate equations (33) which will demonstrate how the explicit inclusion of coherence variables becomes redundant in this case.

The construction of the coherence variables is done by considering the cummutators of the population variables A_{nn} (Eq. (11)) with H' [9]. The k quantum coherence A_{nm} where $|n-m|=k$ is defined as

$$A_{nm} = \frac{1}{\gamma_{nm}} \sum_{\alpha\beta} |n\alpha\rangle \, \nu_{nm}^{\alpha\beta} \langle m\beta| \qquad |n-m|=k \tag{44}$$

where (taking m>n)

$$\nu_{nm}^{\alpha\beta} = \sum_{\substack{\gamma\delta\ldots \\ \beta'}} \mu_{n,n+1}^{\alpha\gamma} \, \mu_{n+1,n+2}^{\gamma\delta} \cdots \mu_{m-1,m}^{\beta'\beta} \quad , \tag{45}$$

and

$$\gamma_{nm}^2 = \sum_{\alpha,\beta} |\nu_{nm}^{\alpha\beta}|^2 \quad , \tag{46}$$

(for k=1 we take $\nu=\mu$). The number of coherence operators to be included in our REM will be determined for our convenience and we may stop at any k_{max} ($k=1,2,\ldots,k_{max}$). Using the set of relevant operators (11) and (44) we may now evaluate the REM making use of the formulation of Section II. This was done in detail in Ref. 9 making use of the POP formulation. (We note that for this choice of operators the Mori projection P does not commute with L_0 and $PL_0 \neq 0$. As a result the expansion of exp(-iLt) in powers of L' becomes more convenient than the expansion of exp(-iQLt) and this is the reason for the adoption of the POP formulation in this case). The resulting equations, to first order in H' and after invoking some simplifying assumptions, are [9]

$$\frac{dP_n}{dt} = \frac{-i\bar{\Omega}_{n,n+1}}{\sqrt{2}} (\sigma'_{n+1,n} - \sigma'_{n,n+1}) - \frac{i\bar{\Omega}_{n,n-1}}{\sqrt{2}} (\sigma'_{n-1,n} - \sigma'_{n,n-1}), \tag{47a}$$

$$\frac{d\sigma'_{n,n+1}}{dt} = [-i\bar{\omega}_{n,n+1} - \Gamma_{n,n+1}]\sigma'_{n,n+1} - i\bar{\Omega}_{n,n+1} (P_{n+1}\sqrt{d_n/2d_{n+1}} - P_n\sqrt{d_{n+1}/2d_n}) + \frac{i\bar{\Omega}_{n+1,n+2}}{\sqrt{2}} \sigma'_{n,n+2} - \frac{i\bar{\Omega}_{n,n-1}}{\sqrt{2}} \sigma'_{n-1,n+1} \quad , \tag{47b}$$

$$\begin{aligned} \frac{d\sigma'_{nm}}{dt} = {} & [-i\bar{\omega}_{nm} - \Gamma_{nm}]\sigma'_{nm} + \frac{i\bar{\Omega}_{m,m+1}}{\sqrt{2}} \sigma'_{n,m+1} \\ & + i\bar{\Omega}_{m,m-1}\sqrt{d_m/2d_{m-1}} \, \sigma'_{n,m-1} \\ & - i\bar{\Omega}_{n,n+1}\sqrt{d_n/2d_{n+1}} \, \sigma'_{n+1,m} - i\bar{\Omega}_{n,n-1} \, \sigma'_{n-1,m}/\sqrt{2} \quad (m>n+1) \quad . \end{aligned} \tag{47c}$$

(In (47c) we have taken $m>n+1$. The REM for $\sigma'_{nm}(m<n-1)$ and for $\sigma'_{n+1,n}$ are simply the complex conjugates of (47c) and (47b)). Here

$$\sigma'_{nm}=(d_n\, d^2_{n+1}\, d^2_{n+2} \cdots d^2_{m-1}\, d_m)^{1/4}(A_{nm},\rho(t)) \tag{48}$$

(The first factor in Eq. (48) is the statistical weight of the nm coherence (the number of possible pathways to go from n to m). The following main assumptions were made [9] in the derivation of Eqs. (47).

(1) The random phase approximation

Four time dipole correlation functions which appear in $\langle\tilde{R}\rangle$ (Eq. (9)) were factorized as was done in Eq. (30).

(2) Separation of time scales - the Markovian assumption

The correlation functions $\tilde{I}_{nm}(\tau)$ are expected to exhibit a complicated behaviour at short times τ_c ($\tau_c \sim$ inverse of typical molecular frequencies) but to reduce to the simple exponential form (Eq. (41)) asymptotically for long times. This behaviour is typical for dipole correlation functions as calculated for other solvable model systems (impurities in solids [17], pressure broadening [18] and stochastic models of line shapes [14]). If our variables in the REM change on a time scale much slower than τ_c we may (on a coarse-grained time scale $t>\tau_c$) replace $\tilde{I}_{nm}$ by its asymptotic form (Eq. (41)). $\bar{\omega}_{nm}$ and Γ_{nm} are defined in terms of the asymptotic behaviour of the logarithmic derivative of $\tilde{I}_{nm}(t)$, i.e.

$$-i\bar{\omega}_{nm} - \Gamma_{nm} \equiv \lim_{t\to\infty} \frac{1}{\tilde{I}_{nm}} \frac{d\tilde{I}_{nm}}{dt} . \tag{49}$$

When the dephasing rates Γ_{nm} are fast compared to the driving $\bar{\Omega}_{nm}$, we may invoke a steady state assumption for the coherences (i.e. set $d\sigma'_{nm}/dt=0$ in (47)), solve for σ'_{nm} and substitute back into the equations for the populations. As a result our REM assume the form of simple rate equations corresponding to incoherent driving (Eqs. (33)). If we consider only single quantum coherences (i.e. set $\sigma'_{nm}=0$ for $|n-m|>1$) we have

$$\sigma'_{n,n+1} = \frac{-i\bar{\Omega}_{n,n+1}}{\sqrt{2}(i\bar{\omega}_{n,n+1}+\Gamma_{n,n+1})} \cdot \left(P_{n+1}\sqrt{d_n/d_{n+1}}-P_n\sqrt{d_{n+1}/d_n}\right) , \tag{50}$$

which when substituted back into (47a) results in Eq. (33) where $W^{(2)}$ is given by Eq. (43) and $W^{(4)}=0$. Thus we have here an alternative derivation of the simple rate equations (33).

The present derivation of (33) provides us with a new insight regarding the significance of the simple rate equations. As is clearly seen from (47), the dynamics of molecular multiphoton processes is governed by the competition between the driving terms $\bar{\Omega}_{nm}$ which tend to build higher coherences and the dephasing rates Γ_{nm} which tend to destroy them. When

$$\eta \equiv \bar{\Omega}_{nm}\tau_c = \frac{\bar{\Omega}_{nm}\,\Gamma_{nm}}{\bar{\omega}^2_{nm}+\Gamma^2_{nm}} <<1 , \tag{51}$$

(weak driving) we may solve for the steady state of the k quantum coherences perturbatively resulting in $\sigma'_{n,n+k}=O(\eta^k)$ and the contribution of the k quantum coherences to the rate equation will be $O(\eta^{2k})$. We recall that in the previous derivation of the simple rate equations (Sec. IV) we had $W^{(2k)}=O(\eta^{2k})$. It is thus obvious that the k quantum coherences play here the same role of $W^{(2k)}$ in the previous derivation. The fact that $W^{(2k)}$ $k>1$ become irrelevant in the Markovian limit (Eq. (40)) is thus a manifestation of the fact that the steady state values of the higher order coherences $\sigma'_{n,n+k}(k>1)$ are negligibly small in this case so that the MMP are insensitive to the detailed dynamics of the multiquantum coherences.

VI. Discussion

We shall now summarize our main results and discuss their physical significance.

(i) The systematic reduction scheme

We have constructed an appropriate set of molecular operators (Eqs. (11) and (44)) relevant for the description of MMP and making use of the Zwanzig-Mori projection operator formalism we were able to derive REM for their time evolution (Eqs. (33) or (47)). The relevant molecular information necessary as an input to our REM is expressed in terms of an infinite hierarchy of n-time dipole correlation functions $M^{(n)}$ (Eqs. (18),(19)).

(ii) The Bloch versus rate equations

We have examined two alternative choices of relevant molecular operators. The first (Secs. III and IV) includes population variables only and leads in the Markovian limit to simple rate equations. The second choice (Sec. V) includes also coherence variables and leads in the Markovian limit to Bloch equations. We should bear in mind that the Markovian limit actually means separation of time scales between the relevant (P) and the other (Q) variables, and it has a different meaning when we change the definition of P. Thus invoking the Markov assumption in both equations is not equivalent and therefore the resulting REM (Eqs. (33) and (47)) are not the same. Only when we further assume $\eta<<1$ (fast dephasing), then the Bloch equations (47) reduce to the simple rate equation (33). We have thus established two routes for the derivation of simple rate equations. The first is direct using the COP equations and the second is via the Bloch equations, making use of the POP formalism. The second method gives us a deeper insight on the dynamical meaning of the rate equations. The first derivation is however less restrictive. As we are not considering explicitly the multiquantum coherences we do not have to invoke unnecessary assumptions regarding their time evolution. This is demonstrated in the fact that in order to get the Bloch equations we had to assume that $\Gamma_{nm}(t)$ exhibits an exponential behaviour (Eq. (49)) whereas in the first derivation all we had to assume was that $\Gamma_{nm}(t)$ decays fast (short τ_c) regardless of its exact functional form.

(iii) The reduction of molecular information

When the simple rate equations (33) hold, there is an enormous reduction in the amount of molecular information necessary for an adequate description of MMP. Not only that all higher order correlation functions become insignificant (40b), but even the details of the two time correlation functions $\Gamma_{nm}(t)$ become irrelevant and all we need is their time integral (Eq. (36)). The molecular quantities that enter the rate equations are the following four numbers per consecutive transition: (i) $\overline{\Omega}_{nm}$, the integrated dipole for the nm transition times the field amplitude; (ii) Γ_{nm}, inverse correlation time (i.e. a dephasing rate); (iii) a detuning frequency $\bar{\omega}_{nm}$, and (iv) ratios of statistical weights of the levels d_n/d_m.

The significance of $\bar{\Omega}_{nm}$ as defined in (26) lies in the reasonable assumption [9] that for large molecules where only few degrees of freedom are coupled with the radiation field, $\bar{\Omega}_{nm}$ will be approximately independent of the number of states involved (i.e. d_n and d_m). This arises from an approximate oscillator strength sum rule [9]. Regarding the dephasing rates Γ_{nm} it is clear from their definition (49) that they are associated with the energetic spread of the states within the n and m molecular levels. (This may be verified using simple models) [9]. The ratio d_n/d_m enters as a detailed balance relation: (the rate of $m \to n$)/(the rate of $n \to m$) equals d_n/d_m. The Bloch equations contain the same types of parameters but there are more of them (one dephasing rate for each multiquantum coherence).

It is clear that during the lower stages of the molecular excitation process (region I) we have actually to solve the entire Liouville equation (coherent driving). The Bloch equations then provide the adequate description. In the absence of reduction (each level $|n\alpha>$ contains only one state) then $\tilde{I}_{nm}(t)=\exp(-i\bar{\omega}_{nm}t)$ and there is no dephasing $\Gamma_{nm}=0$ (Eq. (49)). The Bloch equations are then equivalent to the complete Liouville equation. As the excitation builds up, the dephasing rates increase and in the quasicontinuum we expect simple rate equations to hold. The condition $\eta=\bar{\Omega}\tau_c<<1$ is expected to be met typically after the absorption of a few infrared quanta. Since even for the strongest laser fields used in MMP [6,7] we have $\bar{\Omega}<10cm^{-1}$ whereas the spread of molecular states τ_c^{-1} is expected to be around $100cm^{-1}$ quite early [1].

(iv) The REM as a starting point for simple models

In the construction of the REM we have assumed for the sake of formal convenience that we know the exact molecular eigenstates $|n\alpha>$. We may enjoy this assumption only while remaining in the formal stage since the true molecular eigenstates are not known even for a single triatomic molecule. This, however, should not prevent us from using the REM, on the contrary. The REM are a convenient starting point for molecular calculations which should be focused on evaluating the necessary dipole correlation functions in some approximate manner. These correlation functions are a property of the isolated molecule (without the laser field) and when plugged into the REM will result in the entire multiphoton dynamics. A reasonable way to proceed is to use some of the classical or semiclassical methods which are well established in molecular dynamics calculations [19], for the evaluation of the necessary correlation functions. This could be much easier and more appropriate than the evaluation of the exact molecular eigenstates.

(v) Intramolecular vibrational redistribution and laser-chemistry

The way we chose our variables (Eqs. (11) and (44)) is convenient for studying the total energy absorbed by the molecule via MMP. The intramolecular vibrational redistribution (IVR) processes are hidden inside the dephasing rates Γ_{nm}. As a demonstration we may think of a situation where we have a zero order molecular basis set (say the harmonic basis) in which in each level there is only one state (a "doorway state") which carries oscillator strength and is coupled radiatively to the adjacent levels. This state is further coupled by intramolecular coupling to the rest of the molecular states within the same level so that it undergoes irreversible decay with rate Γ_n corresponding to intramolecular vibrational redistribution (IVR). In this idealized case, $\tilde{I}_{nm}(t)$ will exhibit an exponential behaviour(Eq. (41)), where

$$\Gamma_{nm} = \frac{1}{2}(\Gamma_n + \Gamma_m) \tag{52}$$

is the mean IVR rate of the $|n\rangle$ and $|m\rangle$ doorway states. Thus the dephasing rate in this case is connected directly to the T_1 relaxation rates corresponding to IVR of the zero order molecular states. If we wish to consider more realistic models (it is clear that there is more than one doorway state in each level) and to follow explicitly the intramolecular vibrational redistribution processes, we have to use a more fine grained set of variables. This may be done by partitioning each level variable (Eq. (11)) into several variables corresponding to different distribution of the available energy among the various molecular degrees of freedom. This requires, of course, a more detailed knowledge of the intramolecular interactions. The possibilities of achieving laser-induced and laser-controlled chemical reactions are crucially dependent on the competition between the energy pumping and IVR and the present formalism is suitable for studying these processes in detail.

(vi) Direct measurements of intramolecular line broadening in polyatomic molecules

An important result of the present formalism is the incorporation of the relevant molecular information in the form of the intramolecular dephasing rates Γ_{nm} (Eq. (49)). It is clear that studies of MMP are not ideal for careful examination of Γ_{nm} since in each experiment we have many levels involved and the experimental observables are extremely averaged and far from being microscopic. In recent years novel spectroscopic techniques were developed, which enable us to measure Γ_{nm} directly [3,4]. This information may be used in our REM to predict the outcome of multiphoton experiments. Ordinary molecular

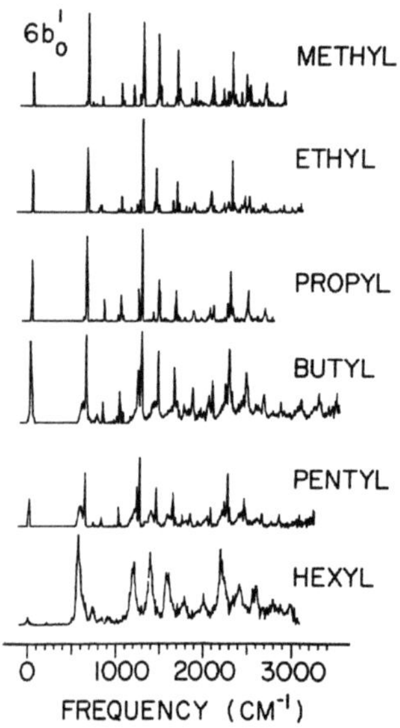

Fig.4

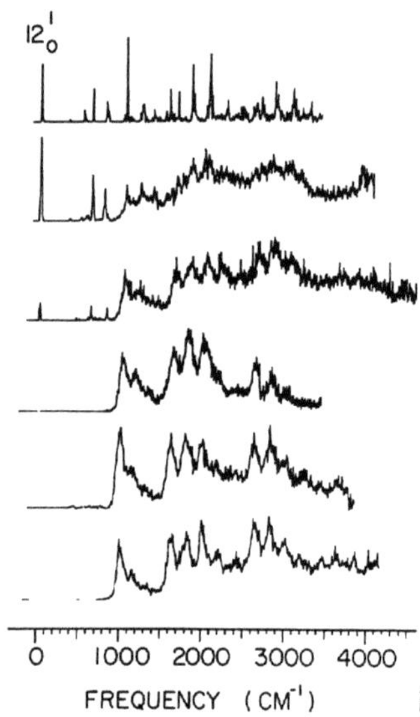

Fig.5

Figure 4. Fluorescence spectra from $6b^1_0$ excitation of n alkyl benzenes (R-C_6H_5 where $R=CH_3,\ldots C_6H_{13}$) in a supersonic beam (Ref. 4b). The available vibrational energy is 530cm^{-1}.

Figure 5. Fluorescence spectra from 12^1_0 excitation of the same molecules of Fig. 4 (Ref. 4b). The available vibrational energy is 930cm^{-1}.

line shapes as measured in cells and at room temperature are usually inhomogeneously broadened due to rotational envelopes, sequence congestion (overlapping transitions), Doppler broadening etc. and do not yield any dynamical information regarding intramolecular processes. It is possible to overcome these difficulties by performing high resolution spectroscopy of cold molecules (1-2°K) in supersonic beams. This technique eliminates the inhomogeneous broadening and results in homogeneously broadened line shapes. A beautiful example of this type of spectroscopy is shown in Figs. 4 and 5 where we show the fluorescence spectra of a series of molecules consisting of a benzene ring with a hydrocarbon side chain (R-$C_6H_5$where R=CH_3 (methyl), C_2H_5 (ethyl) etc. up to C_6H_{13} (hexyl)), obtained in a supersonic beam [4]. The vibrations which show up in these spectra belong essentially to the benzene ring (since the electronic transition is of the π electrons of the ring) and are changing very little within this series of molecules. The side chain provides the low frequency vibrations necessary for IVR. This series is a very successful choice since we have a clear separation of the molecular modes into a "system" and "bath" where the former show in the spectra and the later induce relaxation. By moving along the series we are able to change the bath without affecting the system. The two figures differ by the excitation wavelength. In Fig. 4 the excitation is to the 6b mode and in Fig. 5 to the 12 mode and the amount of available vibrational energy is $\sim$530cm^{-1} and $\sim$930cm^{-1} respectively. We note how the spectra become more complicated with the increase in the size of the side chains (R) and with the available energy. A careful analysis of such spectra [4] gives us a direct measure of IVR and the resulting dephasing rates may be used as an input to our REM for MMP.

Acknowledgements

The support of The Robert A. Welch Foundation, the National Science Foundation and the Westinghouse Educational Foundation Grant of the Research Corporation is gratefully acknowledged.

References

1 R. Bray and M.J. Berry, J. Chem. Phys. 71, 4909 (1979).
2 D.F. Heller and S. Mukamel, J. Chem. Phys. 70, 463 (1979).
3 D.H. Levy, L. Wharton and R.E. Smalley, in Chemical and Biochemical Applications of Lasers, Vol. II, ed. C.B. Moore (Academic Press,1977).
4 (a) J.B. Hopkins, D.E. Powers and R.E. Smalley, J. Chem. Phys. (in press).
(b) J.B. Hopkins, D.E. Powers, S. Mukamel and R.E. Smalley, J. Chem. Phys. (in press).
(c) S. Mukamel and R.E. Smalley, J. Chem. Phys. (in press).
5 A. Laubereau and W. Kaiser,in Chemical and Biochemical Applications of Lasers, Vol. II, ed. C.B. Moore (Academic Press, 1977).
6 See papers in: (a) Tunable Lasers and Applications, ed. A. Mooradian, T. Jaeger and P. Stokseth (Springer, New York, 1976).
(b) Multiphoton Processes, ed. J.H. Eberly and P. Lambropolous (Wiley, New York, 1978).
(c) Adv. Chem. Phys. (1980). (Proceedings of the Laser Chemistry Conference, Israel 1978).
7 For recent reviews see: R.L. Woodin, D.S. Bomse and J.L. Beauchamp,in Chemical and Biochemical Applications of Lasers, Vol. IV, ed. C.B. Moore (Academic Press, 1979);
N. Bloembergen and E. Yablonovitch, Phys. Today, May, p. 23 (1978).
8 S.A. Rice,in Excited States, Vol. 2, ed. E.C. Lim (Academic Press, 1975).
9 S. Mukamel, Phys. Rev. Lett. 42, 168 (1979);
J. Chem. Phys. 70, 5834 (1979);
71, 2012 (1979).

10 R. Zwanzig, Physica 30, 1109 (1964).
11 (a) H. Mori, Prog. Theoret. Phys. 33, 423 (1965).
(b) M. Tokuyama and H. Mori, ibid 55, 411 (1976).
12 S. Mukamel, Chem. Phys. 37, 33 (1979).
13 W.H. Louisell, Quantum Statistical Properties of Radiation (Wiley, New York, 1973).
14 (a) R. Kubo, in Fluctuation, Relaxation and Resonance in Magnetic Systems, ed. D. Ter Haar (Oliver & Boyd, Edinburgh, 1962).
(b) R. Kubo, in Adv. Chem. Phys. XV (1969).
15 (a) R.G. Brewer and R.L. Shoemaker, Phys. Rev. Lett. 27, 631 (1971).
(b) T.J. Aartsma, J. Morsink and D.A. Wiersma, Chem. Phys. Lett. 42, 520 (1976); 47, 425 (1977); 49, 34 (1977).
16 J. Fiutak and J. Van Kranendonk, Can. J. Phys. 40, 1085 (1962).
17 V. Hizhnyakov and I. Tehver, Phys. Stat. Sol. 21, 755 (1967).
18 P.W. Anderson, Phys. Rev. 80, 511 (1950).
19 W.H. Miller, Adv. Chem. Phys. Vol. 30, p. 77, ed. H.P. Lawley (Wiley, 1975).

Unified Theory of Collisional Redistribution in Near-Resonance Light Scattering

S. Mukamel

Department of Chemical Physics, Weizmann Institute of Science
Rehovot, Israel, and

Department of Chemistry, William Marsh Rice University
Houston, TX 77001, USA

ABSTRACT

The tetradic scattering formalism is used to solve for the spectral distribution of light scattered from collisionally perturbed molecules with binary interactions. Using a systematic density expansion it is shown that the necessary four-time N-particle dipole correlation functions may be rigorously expressed in terms of three two-time single-particle correlation functions, two of which are associated with the absorption and emission line broadening, and the third is a cross correlation function. The present results provide a unified treatment valid all the way from the impact to the statistical limits.

I. Tetradic scattering formulation of resonance light scattering

In the preceding lecture (TCMP) we have developed a general formalism suitable for the description of multiphoton processes of relaxing systems. The theoretical analysis is performed in two steps: The first is to express the quantities of interest in terms of an appropriate set of correlation functions. This step is formal and does not depend on the details of the system. The second step is to actually calculate the correlation functions using microscopic models. It is at this stage where the fine details of the specific system need to be considered. The problem considered in TCMP namely multiphoton processes of polyatomic molecules, is very complicated and as a result the second step could be treated only approximately using statistical arguments such as the random phase approximation. In this lecture we shall consider the simplest multiphoton problem (i.e. two-photon) where the solution may be carried out rigorously to the very end including the exact evaluation of the necessary correlation functions. As we shall see, the first (formal) step is strikingly similar to what was done in the preceding lecture although the problem to be considered is very different and the difference appears only in the second step of evaluating the correlation functions.

We consider a molecule having three levels (the "system") which undergoes near-resonance light scattering (RLS) and is collisionally perturbed by a low pressure bath of inert molecules which do not interact with the radiation field and with each other. The total Hamiltonian for the combined system + bath + the radiation field within the rotating-wave approximation, which describes the process $|a\omega_1\rangle \rightarrow |b\,vac\rangle \rightarrow |c\omega_2\rangle$ (see Fig. 1) is [1]

$$H = H_s + H_1 + V \quad , \tag{1}$$

$$H_s = |a\rangle(\varepsilon_a + \omega_1 - \tfrac{i}{2}\gamma_a)\langle a| + |b\rangle(\varepsilon_b - \tfrac{i}{2}\gamma_b)\langle b| + |c\rangle(\varepsilon_c + \omega_2 - \tfrac{i}{2}\gamma_c)\langle c| \tag{1a}$$

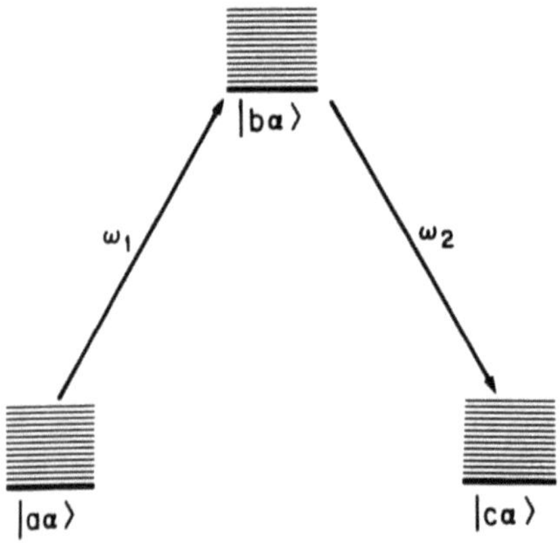

Fig.1 The molecular levelscheme for resonance light scattering

$$H_1 = |a>F_a(Q_B)<a|+|b>F_b(Q_B)<b|+|c>F_c(Q_B)<c| \quad , \tag{1b}$$

$$V = |a>\mu_{ab}(Q_B)<b|+|b>\mu_{bc}(Q_B)<c|+|b>\mu_{ba}(Q_B)<a|+|c>\mu_{cb}(Q_B)<b| . \tag{1c}$$

Here H_s is the zero-order system Hamiltonian, H_1 describes the bath interacting with the system and V is the coupling of the system to the radiation field. ε_i, γ_i and $F_i(Q_B)$ denote the energy, inverse lifetime, and the interaction with the bath of our system molecule in the i'th state i=a,b,c; $\omega_1(\omega_2)$ are the frequencies of the incoming (outgoing) photons. $\mu_{ij}(Q_B)$ is the dipole operator which depends on the coordinates (Q_B) of the N perturber bath molecules. For simplifying the notation we denote here the direct product states of the system + the radiation field $|a\omega_1>$, $|b\,vac>$ and $|c\omega_2>$ simply by $|a>$, $|b>$ and $|c>$ respectively. The system interacting with the bath will be described following the approach of JABLONSKI and BARANGER [2-4]. We place our "system" molecule in the center of a macroscopic sphere with volume Ω and assume that the bath consists of N perturber molecules whose interaction with the system is a sum of binary contributions (the binary interaction approximation, BIA), i.e.

$$F_i(Q_B) = \sum_\nu F_{i\nu}(Q_{B\nu}) \quad , \qquad i=a,b,c \tag{2}$$

where ν stands for the ν'th bath particle. Denoting the eigenstates of $F_i(Q_B)$ by $|i\alpha>$ (i=a,b,c) where α is the collection of all bath quantum numbers, with eigenvalues ε_α we immediately have within the BIA

$$|i\alpha> = |i> \prod_\nu |\alpha_\nu> \quad , \qquad \varepsilon_\alpha = \sum_\nu \varepsilon_{\alpha\nu} \quad . \tag{3}$$

The RLS cross section for this model system was solved in the past [5] assuming separation of time scales between the system and the bath, and its behavior in the two extremes of fast collision (impact) limit and slow collision (static) limit, relative to the line broadening, is well understood. In addition, a stochastic formulation of the problem was recently developed by TAKAGAHARA, HANAMURA and KUBO [6] which interpolates between the two limits. It is our purpose here to present the exact solution to the RLS problem defined by the Hamiltonian (Eq. 1) without invoking any further dynamical approximations.

The photon scattering cross section will be evaluated using the tetradic scattering formalism [7-9]. This approach is based on formulating scattering theory in terms of density matrices in Liouville space (rather than wavefunctions in Hilbert space) and is particularly convenient for performing averages over irrelevant degrees of freedom. The tetradic T matrix $\mathcal{T}$ is defined [7-9] as

$$\mathcal{T}(\omega) = \mathcal{V} + \mathcal{V}\, \mathcal{G}(\omega)\, \mathcal{V} \tag{4}$$

where $\mathcal{V}$ is the tetradic operator corresponding to V, i.e. $\mathcal{V}=[V,\ ]$, and $\mathcal{G}(\omega)$ is the tetradic Green's operator for the entire system and bath

$$\mathcal{G}(\omega) \equiv (\omega-L+i\eta)^{-1} = -i\int_0^\infty d\tau\, \tilde{\mathcal{G}}(\tau)\exp(i\omega\tau) \, . \tag{5}$$

Here $\tilde{\mathcal{G}}(\tau)=\exp(-iL\tau)$ is the propagator in Liouville space and $L(L_1,L_0)$ being the Liouville operators corresponding to $H\ (H_1,H_0)$ i.e. $L=[H,\ ]$ etc. For the subsequent manipulations we define also $\mathcal{G}_0(\omega)$ and $\tilde{\mathcal{G}}_0(\tau)$ by replacing L by L_0 in Eq. (5). The RLS cross section is given (up to a proportionality constant) by [1,6]

$$I(\omega_1,\omega_2) = -i\langle \mathcal{T}(0)_{cc,aa}\rangle \equiv -i\sum_{\alpha\gamma} \langle\langle {}^{\gamma\gamma}_{cc}|\mathcal{T}(0)|{}^{\alpha\alpha}_{aa}\rangle\rangle P(\alpha) \, . \tag{6}$$

Here $\langle\cdots\rangle$ denotes a thermal average of a tetradic operator over the bath degrees of freedom (averaging over all possible initial bath states and summing over all final bath states), i.e.

$$\langle A_{ij,kl}\rangle \equiv \sum_{\alpha\beta} \langle\langle {}^{\beta\beta}_{ij}|A|{}^{\alpha\alpha}_{kl}\rangle\rangle P(\alpha) \, . \tag{7}$$

$|{}^{\alpha\beta}_{ij}\rangle\rangle$ is the tetradic state corresponding to $|i\alpha\rangle\langle j\beta|$ and $P(\alpha)$ is the equilibrium distribution of bath states,

$$P(\alpha) = \exp(-\varepsilon_\alpha/kT)/\mathrm{Tr}(\exp(-F_a/kT)) \, . \tag{8}$$

Upon expansion of $\mathcal{T}$ to fourth order in $\mathcal{V}$ (which is the lowest order contributing to Eq. (6), see Fig. 3 of TCMP) we get

$$\mathcal{T}(\omega) = \mathcal{V}\, \mathcal{G}_0(\omega)\mathcal{V}\, \mathcal{G}_0(\omega)\mathcal{V}\, \mathcal{G}_0(\omega)\mathcal{V} \, , \tag{9}$$

which may be recast in the fully time-ordered form

$$\mathcal{T}(\omega) = (-i)^3\int_0^\infty d\tau \int_0^\tau d\tau_1 \int_0^{\tau_1} d\tau_2\, \exp(i\omega\tau)\mathcal{V}\, \tilde{\mathcal{G}}_0(\tau-\tau_1)\mathcal{V}\, \tilde{\mathcal{G}}_0(\tau_1-\tau_2)\mathcal{V}\, \tilde{\mathcal{G}}_0(\tau_2)\mathcal{V} \, , \tag{10}$$

or alternatively

$$\mathcal{T}(\omega) = (-i)^3\int_0^\infty d\tau_1 \int_0^\infty d\tau_2 \int_0^\infty d\tau_3\, \exp[i\omega(\tau_1+\tau_2+\tau_3)]\cdot\mathcal{V}\, \tilde{\mathcal{G}}_0(\tau_1)\mathcal{V}\, \tilde{\mathcal{G}}_0(\tau_2)\mathcal{V}\, \tilde{\mathcal{G}}_0(\tau_3)\mathcal{V}. \tag{11}$$

Utilizing the form of the radiative coupling V (Eq. (1c)) and making use of Fig. 2 of TCMP, we notice that there are six pathways in Liouville space to go from $|aa\rangle\rangle$ to $|cc\rangle\rangle$ in fourth order. However, utilizing the general Liouville conjugation symmetry [8,9]

$$\mathcal{G}_{ab,cd}(\omega) = -\mathcal{G}^*_{ba,dc}(-\omega) \, , \tag{12}$$

we see that there are only three independent paths. We thus have, using Eq.(11),

$$I(\omega_1,\omega_2) = \int_0^\infty d\tau_1 \int_0^\infty d\tau_2 \int_0^\infty d\tau_3 [\mathrm{I} + \mathrm{II} + \mathrm{III}] + \mathrm{C.C.} \equiv I_1+I_2+I_3+\mathrm{C.C.} \tag{13}$$

where

$$I = \exp[-i\ \omega_{cb}\ \tau_1 - i\ \omega_{ab}\ \tau_3 - \frac{\gamma_a}{2}\tau_3 - \frac{\gamma_b}{2}(\tau_1+2\tau_2+\tau_3) - \frac{\gamma_c}{2}\tau_1]K_I(\tau_1,\tau_2,\tau_3), \tag{13a}$$

$$II = \exp[i\ \omega_{cb}\ \tau_1 - i\omega_{ab}\ \tau_3 - \frac{\gamma_a}{2}\tau_3 - \frac{\gamma_b}{2}(\tau_1+2\tau_2+\tau_3) - \frac{\gamma_c}{2}\tau_1]K_{II}(\tau_1,\tau_2,\tau_3), \tag{13b}$$

and

$$III = \exp[i\ \omega_{cb}\ \tau_1 - i\omega_{ac}\ \tau_2 - i\omega_{ab}\ \tau_3 - \frac{\gamma_a}{2}(\tau_2+\tau_3) - \frac{\gamma_b}{2}(\tau_1+\tau_3) - \frac{\gamma_c}{2}(\tau_1+\tau_2)]K_{III}(\tau_1,\tau_2,\tau_3)\ . \tag{13c}$$

Here

$$K_I(\tau_1,\tau_2,\tau_3) = \sum_{\substack{\alpha\beta\\ \beta'\gamma}} \langle\langle{}^{\gamma\gamma}_{cc}|V(\tau_1+\tau_2+\tau_3)|^{\gamma\beta}_{cb}\rangle\rangle\ \langle\langle{}^{\gamma\beta}_{cb}|V(\tau_2+\tau_3)|^{\beta'\beta}_{bb}\rangle\rangle\ \langle\langle{}^{\beta'\beta}_{bb}|V(\tau_3)|^{\alpha\beta}_{ab}\rangle\rangle\ \langle\langle{}^{\alpha\beta}_{ab}|V(0)|^{\alpha\alpha}_{aa}\rangle\rangle P(\alpha) \tag{14a}$$

$$= \langle V_{ab}(0)V_{bc}(\tau_1+\tau_2+\tau_3)V_{cb}(\tau_2+\tau_3)V_{ba}(\tau_3)\rangle,$$

$$K_{II}(\tau_1,\tau_2,\tau_3) = \sum_{\substack{\alpha\beta\\ \beta'\gamma}} \langle\langle{}^{\gamma\gamma}_{cc}|V(\tau_1+\tau_2+\tau_3)|^{\beta'\gamma}_{bc}\rangle\rangle\ \langle\langle{}^{\beta'\gamma}_{bc}|V(\tau_2+\tau_3)|^{\beta'\beta}_{bb}\rangle\rangle\ \langle\langle{}^{\beta'\beta}_{bb}|V(\tau_3)|^{\alpha\beta}_{ab}\rangle\rangle\ \langle\langle{}^{\alpha\beta}_{ab}|V(0)|^{\alpha\alpha}_{aa}\rangle\rangle P(\alpha) \tag{14b}$$

$$= \langle V_{ab}(0)V_{bc}(\tau_2+\tau_3)V_{cb}(\tau_1+\tau_2+\tau_3)V_{ba}(\tau_3)\rangle\ ,$$

and

$$K_{III}(\tau_1,\tau_2,\tau_3) = \sum_{\substack{\alpha\beta\\ \beta'\gamma}} \langle\langle{}^{\gamma\gamma}_{cc}|V(\tau_1+\tau_2+\tau_3)|^{\beta'\gamma}_{bc}\rangle\rangle\ \langle\langle{}^{\beta'\gamma}_{bc}|V(\tau_2+\tau_3)|^{\alpha\gamma}_{ac}\rangle\rangle\ \langle\langle{}^{\alpha\gamma}_{ac}|V(\tau_3)|^{\alpha\beta}_{ab}\rangle\rangle\ \langle\langle{}^{\alpha\beta}_{ab}|V(0)|^{\alpha\alpha}_{aa}\rangle\rangle P(\alpha)\ , \tag{14c}$$

$$= \langle V_{ab}(0)V_{bc}(\tau_3)V_{cb}(\tau_1+\tau_2+\tau_3)V_{ba}(\tau_2+\tau_3)\rangle\ .$$

In Eqs. (14) we have introduced the definitions

$$V(\tau) = \exp(iL_1\tau)V\ \exp(-iL_1\tau)\ , \tag{15a}$$

$$V(t) = \exp(iH_1t)V\ \exp(-iH_1t)\ , \tag{15b}$$

and

$$\langle V_{ab}(t_1)V_{bc}(t_2)V_{cb}(t_3)V_{ba}(t_4)\rangle \equiv \sum_{\substack{\alpha\beta\beta'\\ \gamma}} P(\alpha)V^{\alpha\beta}_{ab}(t_1)V^{\beta\gamma}_{bc}(t_2)V^{\gamma\beta'}_{cb}(t_3)V^{\beta'\alpha}_{ba}(t_4). \tag{16}$$

We have also defined $\omega_{ab}\equiv\varepsilon_a+\omega_1-\varepsilon_b$, $\omega_{cb}\equiv\varepsilon_c+\varepsilon_2-\varepsilon_b$ and $\omega_{ac}\equiv\omega_{ab}-\omega_{cb}$.

II. Microscopic evaluation of the correlation functions - the density expansion

In Section I we have expressed the RLS cross section in terms of the four-time N particle correlation functions K_i (Eqs. (14)). We are now in a position to evaluate K_i rigorously in terms of single particle correlation functions. Within the BIA (Eqs. (2) and (3)), $K_i(\tau_1,\tau_2,\tau_3)$ are factorized to a product of N identical factors i.e.

$$K_i(\tau_1,\tau_2,\tau_3) = |\mu_{ab}|^2|\mu_{bc}|^2\, k_i^{N}(\tau_1,\tau_2,\tau_3) \qquad i=I,II,III \tag{17}$$

where k_i are four-time single-particle correlation functions defined in an analogous manner to K_i (Eqs. (14)), where α,β,β' and γ denote a single particle state (rather than a macroscopic N particle state) and V_{ab}, V_{bc} are replaced by $V_{ab}/\mu_{ab}=|a\rangle\langle b|$ and $V_{bc}/\mu_{bc}=|b\rangle\langle c|$ respectively. We have also defined $\mu_{ab}\equiv\mu_{ab}(\infty)$ and $\mu_{bc}\equiv\mu_{bc}(\infty)$.

The evaluation of k_i will be made making use of the fact that Ω is a large parameter so that k_i may be expanded in powers of $1/\Omega$. This expansion (as may be seen from Eq. (17)) corresponds to a density expansion of K_i (in powers of of N/Ω). To that end we first note that $F_{i\nu}(Q_{B\nu})$ i=a,b,c are finite only in a small $Q_{B\nu}$ region of microscopic dimensions. As a result, the bath states corresponding to $|a\alpha\rangle$, $|b\beta\rangle$ and $|c\gamma\rangle$ will be almost identical. Consequently the diagonal matrix elements $\langle a\alpha|V|b\alpha\rangle$ will be close to unity. In fact it is clear that [1-4]

$$|\langle a\alpha|V|b\alpha\rangle|^2 = 1 - O(\tfrac{1}{\Omega}) \quad , \tag{18a}$$

so that

$$\sum_{\beta\neq\alpha} |\langle a\alpha|V|b\beta\rangle|^2 = O(\tfrac{1}{\Omega}) \ , \tag{18b}$$

and for large Ω, the dominant contribution to the four-time correlation functions $k_i(\tau_1,\tau_2,\tau_3)$ will come from the completely diagonal sequence $\alpha=\beta=\beta'=\gamma$, which will be O(1). This contribution, however, has almost no time dependence since $E_\alpha\simeq E_\beta\simeq E_{\beta'}\simeq E_\gamma$ and thus it appears only in $k_i(0,0,0)$. The next largest terms will correspond to all possible sequences $\alpha\beta\beta'\gamma$ where the bath state changes twice (single changes do not contribute to (14)). Since each k_i is a product of four operators there will be six such terms (corresponding to all possibilities of choosing two terms out of four). All these terms will be $O(1/\Omega)$. We can thus write

$$k_i(\tau_1,\tau_2,\tau_3) = 1 - \frac{1}{\Omega} W_i(\tau_1,\tau_2,\tau_3) + O(\frac{1}{\Omega^2}) \ , \tag{19}$$

where W_i is O(1). As a result then in the thermodynamic limit ($\Omega\to\infty$, $N\to\infty$, N/Ω finite) we get

$$\begin{aligned} K_i(\tau_1,\tau_2,\tau_3) &= |\mu_{ab}|^2|\mu_{bc}|^2[1 - \frac{1}{\Omega} W_i(\tau_1,\tau_2,\tau_3)]^N \\ &\to|\mu_{ab}|^2|\mu_{bc}|^2 \exp(- \frac{N}{\Omega} W_i(\tau_1,\tau_2,\tau_3)) \quad , \quad i=I,II,III \end{aligned} \tag{20}$$

where

$$\begin{aligned} W_I(\tau_1,\tau_2,\tau_3) &= g_{aba}(\tau_3) + g^*_{cbc}(\tau_1) + g_{abc}(\tau_1+\tau_2+\tau_3) \\ &+ g^*_{abc}(\tau_2) - g_{abc}(\tau_2+\tau_3) - g^*_{abc}(\tau_1+\tau_2) \quad , \end{aligned} \tag{21a}$$

$$W_{II}(\tau_1,\tau_2,\tau_3) = g_{aba}(\tau_3) + g_{cbc}(\tau_1) + g_{abc}(\tau_2+\tau_3) + g^*_{abc}(\tau_1+\tau_2) - g_{abc}(\tau_1+\tau_2+\tau_3) - g^*_{abc}(\tau_2) \,, \tag{21b}$$

and

$$W_{III}(\tau_1,\tau_2,\tau_3) = g_{abc}(\tau_3) + g^*_{abc}(\tau_1) + g_{aba}(\tau_2+\tau_3) + g_{cbc}(\tau_1+\tau_2) - g_{abc}(\tau_1+\tau_2+\tau_3) - g^*_{abc}(-\tau_2) \;. \tag{21c}$$

In Eqs. (20) and (21) we have expresssed the four-time correlation functions (14) in terms of the three two-time correlation functions g_{aba}, g_{cbc} and g_{abc}, defined as follows:

$$g_{aba}(\tau) = \Omega[1-\langle V_{ab}(0)V_{ba}(\tau)\rangle/\langle V_{ab}(0)V_{ba}(0)\rangle] = \frac{\Omega}{|\mu_{ab}|^2}\sum_{\alpha\beta} P(\alpha)|V^{\alpha\beta}_{ab}|^2\,(1-\exp(i\omega_{\beta\alpha}\tau)) \,, \tag{22a}$$

$$g_{cbc}(\tau) = \Omega[1-\langle V_{cb}(0)V_{bc}(\tau)\rangle/\langle V_{cb}(0)V_{bc}(0)\rangle] = \frac{\Omega}{|\mu_{bc}|^2}\sum_{\alpha\beta} P(\alpha)|V^{\alpha\beta}_{cb}|^2\,(1-\exp(i\omega_{\beta\alpha}\tau)) \,, \tag{22b}$$

and

$$g_{abc}(\tau) = \Omega[1-\langle V_{ab}(0)V_{bc}(\tau)\rangle/\langle V_{ab}(0)V_{bc}(0)\rangle] = \frac{\Omega}{\mu_{ab}\,\mu_{bc}}\sum_{\alpha\beta} P(\alpha)V^{\alpha\beta}_{ab}\,V^{\beta\alpha}_{bc}(1-\exp(i\omega_{\beta\alpha}\tau)) \;. \tag{22c}$$

We should bear in mind that the Ω factor in Eqs. (22) simply cancels the $1/\Omega$ dependence of the sum so that $g(\tau)=O(1)$. Equations (13) together with Eqs. (20)-(22) constitute our final result for the RLS cross section. In conclusion we note the following:

1) Formal analogy with molecular multiphoton processes

In Section I we have expressed the RLS cross section $I(\omega_1,\omega_2)$ in terms of the four-time correlation functions K_i of the dipole operator (Eqs. (14)). It is interesting to note that K_I, K_{II} and K_{III} of the present paper correspond exactly to the same pathways that contributed to the I , II and III terms which appeared in the evaluation of $M^{(4)}$ in TCMP (see Eqs. 23 and Fig. 3 of TCMP). This is not surprising in view of Eqs.(10) and (11) which relate the tetradic T matrix (in fourth order) to $M^{(4)}$ of TCMP. (In this lecture we have used Eq. (11) whereas in TCMP we have used the alternative form 10). We should also note that the same relations hold also for higher order terms where $M^{(2n)}$ is associated with $T^{(2n)}$.

2) The relevant molecular information

The main result of the present formulation of RLS is that the four-time N-particle dipole correlation functions appearing in the general expressions for the RLS cross section (Eqs. (14)) may be rigorously expressed (within the BIA) in terms of the three two-time single particle correlation functions $g_{aba}(\tau)$, $g_{cbc}(\tau)$ and $g_{abc}(\tau)$ (Eqs. (22)). The former two are associated with the spectral broadening of the line shapes corresponding to the ab and bc transitions $S_{ab}(\omega_1)$ and $S_{bc}(\omega_2)$ respectively [4,10,11] i.e.

$$S_{ab}(\omega_1) \propto \mathrm{Re}\int_0 d\tau \, \exp(i\omega_{ab}\tau)\exp[-\frac{N}{\Omega} g_{aba}(\tau)] \, , \tag{23a}$$

$$S_{bc}(\omega_2) \propto \mathrm{Re}\int_0^\infty d\tau \, \exp(i\omega_{bc}\tau)\exp[-\frac{N}{\Omega} g_{cbc}(\tau)] \, . \tag{23b}$$

However, $g_{abc}(\tau)$ is a cross-correlation function characteristic of the two photon RLS process and does not appear in ordinary line shape formulations [4,10,11]. We have thus reduced the complete molecular information relevant for RLS and expressed it in terms of these three simple correlation functions. The evaluation of $g_{aba}(\tau)$ and $g_{cbc}(\tau)$ may be achieved by one of the many standard methods developed for ordinary line shapes. The quantum mechanical evaluation of $g_{abc}(\tau)$ (22c) is a straightforward generalization of these methods.

3) Fast collisions - the impact limit

In the impact (Markovian) limit of fast collisions (relative to the line broadening) [1-6] we take $g(\tau)$ to be linear in τ. In this case the triple integrals [13] are factorized into three single integrals in τ_1, τ_2 and τ_3 since

$$K_i = K_{i1}(\tau_1)K_{i2}(\tau_2)K_{i3}(\tau_3) \qquad i=I,II,III \tag{24}$$

and we get [1]

$$I(\omega_1,\omega_2) = \frac{|\mu_{ab}\,\mu_{bc}|^2}{(\varepsilon_a+\omega_1-\varepsilon_b)^2+(\hat{\Gamma}+\frac{1}{2}\gamma_b)^2} \tag{25}$$
$$[2\pi\delta(\varepsilon_a+\omega_1-\varepsilon_c-\omega_2) + \frac{4\hat{\Gamma}}{\gamma_b}\,\frac{\hat{\Gamma}+\frac{1}{2}\gamma_b}{(\varepsilon_b-\varepsilon_c-\omega_2)^2+(\hat{\Gamma}+\frac{1}{2}\gamma_b)^2}] \, .$$

This result was previously derived by several authors [5,6,1]. $I(\omega_1,\omega_2)$ consists, in this case, of a narrow coherent Raman component and a broad incoherent redistribution term. Equation (25) holds in the impact limit when levels $|a\rangle$ and $|c\rangle$ are assumed similar, i.e. $F_a=F_c$ so that $g(\tau)=g_{aba}(\tau)=g_{cbc}(\tau)=g_{abc}(\tau)=\hat{\Gamma}\tau$ and we have also assumed $\gamma_a=\gamma_c=0$ (long lifetime of $|a\rangle$ and $|c\rangle$). The Raman component is not broadened since the bath cannot dephase the a-c transition as we have assumed $F_a=F_c$ and we have ignored the lifetime broadening of these levels.

4) Slow collisions - the quasistatic limit

In the inverse (static) limit of slow collisions relative to the broadening, we need consider only the short-time behavior of $g(\tau)$ which is $g(\tau)=\Delta^2\tau^2/2$

In this case (unlike the Markovian limit) the integrals (13) no longer factorize in τ_1 and τ_3. Instead the natural variables for the integrations will now be $\tau_1 \pm \tau_3$ and τ_2. Upon changing variables and performing two of the integrations we get [1]

$$I(\omega_1,\omega_2) = \frac{8\pi}{\gamma_b} [\int_0^\infty dy \exp(-\gamma_b y - \frac{1}{2} \Delta^2 y^2) \cos \omega_{ab} y] \delta(\varepsilon_a + \omega_1 - \varepsilon_c - \omega_2). \quad (26)$$

No redistribution terms appears in this case and the RLS spectrum consists of the Raman δ-function component whose absolute intensity is affected by the inhomogeneous line broadening.

5) The stochastic Gaussian model

The results of the stochastic Gaussian model of frequency modulation, developed by TAKAGAHARA, HANAMURA and KUBO [6] may be obtained from our general expressions by taking $F_a=F_c \neq F_b$, $\gamma_a=\gamma_c=0$ and using the Brownian oscillator correlation function for $g(\tau)$ [10], i.e.

$$g_{aba}(\tau) = g_{cbc}(\tau) = g_{abc}(\tau) = \frac{\Delta^2}{\Gamma^2} (\exp(-\Gamma\tau) - 1 + \Gamma\tau) \ . \quad (27)$$

6) On the validity of the factorization assumption

SZÖKE, CARLSTEN and RAYMER [12] (SCR) have recently studied the RLS from strontium vapor in argon. They were able to detune the excitation wavelength up to 1000 cm^{-1} off resonance both above and below the atomic transition. The relative magnitude of the coherent and incoherent components was found to be strongly asymmetric with respect to the sign of the detuning (i.e. it depends on the sign of $(\varepsilon_a+\omega_1-\varepsilon_b)$). This is in contrast to the Markovian expression (25). SCR then made a phenomenological assumption that Eq. (25) should be modified in the non-Markovian limit by replacing the Lorentzian corresponding to incoherent scattering by the emission line shape $S_{bc}(\omega_2)$ and taking $\hat{\Gamma}(\omega)$ to be a frequency dependent. The assumption of SCR is equivalent to solving for the tetradic T matrix [9] by invoking the factorization approximation [13] which amounts to setting

$$<T(0)> = <V\, G_0(0) V\, G_0(0) V\, G_0(0) V> \simeq V <G_0(0)> V\, <G_0(0)> V <\, G_0(0)> V \quad (28)$$

where each $<G_0(0)>$ is an ensemble-averaged reduced molecular propagator defined using an appropriate projection operator. Equation (28) is formally equivalent to the factorization (24). Using our general expressions (20,21) we note that the factorization assumption ((24) or (28)) is exact in the Markovian limit ($g(\tau)$ linear). However, once this is not the case, the integrand in (13) is no longer factorized in terms of τ_1,τ_2 and τ_3. In fact, we have seen that in the static limit $\tau_1+\tau_3$, $\tau_1-\tau_3$ should be used as the natural times within the integrations [13]. It is thus formally inconsistent [13] to invoke the separation approximation far from the Markovian limit.

7) Extension to other problems

The present formulation applies as it stands also to collisional redistribution in two-photon absorption which is formally equivalent to RLS. It is also possible, using the same density expansion to solve rigorously for the n-time correlation functions, $n>4$. This enables us to extend the unified theories to strong field phenomena, e.g. saturation in absorption and RLS. This extension was made recently [14].

Acknowledgements

The support of the Robert A. Welch Foundation, the National Science Foundation and the Westinghouse Educational Grant of Research Corporation is gratefully acknowledged.

References

1 S. Mukamel, J. Chem. Phys. 71, 2884 (1979).
2 A. Jablonski, Phys. Rev. 68, 78 (1945).
3 M. Baranger, Phys. Rev. 11, 481 (1958).
4 J. Szudy and W.E. Baylis, J. Quant. Spect. Rad. Trans. 15, 641 (1975).
5 D.L. Huber, Phys. Rev. 158, 843 (1967); 170, 418 (1968); 178, 93 (1969); 187, 392 (1969); B1, 3409 (1970).
6 T. Takagahara, E. Hanamura and R. Kubo, J. Phys. Soc. Japan 43, 802 (1977); 811 (1977); 1522 (1977); ibid 44, 742 (1978); 728 (1978).
7 U. Fano, Phys. Rev. 131, 259 (1963).
8 A. Ben-Reuven, Adv. Chem. Phys. 33, 235 (1975).
9 A. Ben-Reuven and S. Mukamel, J. Phys. A8, 1313 (1975).
10 R. Kubo in Fluctuation, Relaxation and Resonance in Magnetic Systems, ed. D. Ter Haar (Oliver and Boyd, Edinburgh, 1962).
11 R. Kubo, Adv. Chem. Phys. XV (1969).
12 J.L. Carlsten, A. Szöke and M.G. Raymer, Phys. Rev. A15, 1029 (1977).
13 G. Nienhuis and F. Schuller, Physica C92, 397 (1977).
14 I. Rabin and S. Mukamel, in press.

Energetics of the Photodissociation of Polyatomic Molecules

K.F. Freed

The Department of Chemistry and The James Franck Institute
The University of Chicago
Chicago, IL 60637, USA

ABSTRACT

We present a quantum mechanical theory of the decomposition processes of polyatomic molecules, an important relaxation channel for electronically or thermally excited molecular systems. Both direct photodissociation and weak predissociation are described from a unified standpoint. An essential feature of the polyatomic molecule dissociation is the fact that the excess energy may be partitioned among the vibrational and rotational degrees of freedom in addition to the electronic and translational ones which are the only degrees of freedom accessible in diatomic molecule dissociations. The central focus of the theory is, therefore, on the description of the vibrational, translational, rotational, etc. energy distributions of the decomposition fragments and on the factors, such as the bound and dissociative potential energy surfaces, the excitation wavelength, etc., influencing these distributions. The theory describes both the sudden Franck-Condon process, leading to the electronic transition from a bound to a dissociative state, and the half-collision wherein the fragments mutually interact while they recede from each other on the dissociative potential energy surface.

1. Introduction

One of the important decay channels of excited molecular states involves decomposition. Molecular dissociation processes play a vital role in many chemical systems that are fundamental to the description of chemical reactions, of atmospheric processes, and astrophysical and biological problems. Despite the central nature of dissociation phenomena, they have rarely been studied separately as isolated primary events in polyatomic molecules.

The physics of the dissociation of diatomic molecules is fairly straightforward when only a single repulsive potential curve enters [1-5]. The receding atoms depart with a kinetic energy that is determined by energy conservation, while the angular distribution of the atoms depends on the lifetime of the excited molecules and on their initial rotational energy. When more repulsive curves are involved, transitions may occur between these curves leading to various curve crossing problems which are of considerable interest in the theory of collision dynamics [6-8].

In the case of polyatomic molecules, on the other hand, the molecular fragments may contain internal vibrational and rotational excitation (as well as electronic excitation). The excitation energy of the fragments could, in principle, be employed to perform useful work by providing population inversions to run photochemical lasers [9] or generate vibrationally excited species that can serve as the reactants for subsequent chemical reactions. It is important to recognize that vibrationally excited dissociation fragments can overcome activation barriers to permit reactions that otherwise might not be accessible to the unexcited molecule. Here we consider the theoretical descriptions of the energy disposal in the dissociation of polyatomic molecules, the dependence of the product energy distributions upon the initial quantum states of the system,and the state selected angular distributions of the photoproducts.

Photodissociation represents the initial primary process in many photochemical reaction networks. However, because of technical limitations, the study of this initial step has, in the past, been limited to an analysis of the secondary, tertiary, etc., products arising from reactions of the photofragments with scavenger molecules. This type of work provides valuable information concerning the nature of the photodissociation species and of the kinetics of radical reactions. The bulk nature of the experiments leads to a loss of information concerning the vibrational and rotational energy content of the photofragments. Knowledge of these energy distributions may become crucial under conditions where vibrational relaxation is slow compared to the rates of chemical reactions, since vibrational energy in the photofragments can considerably alter the chemical reaction rates. The presence of such excited species may be of importance in the understanding of the processes associated with photochemical air pollution and with the combustion of gasoline, etc. It is thus of considerable interest to isolate and analyze the energy distribution in this primary dissociation process since researchers in the field of chemical reactions are actively studying the influence of reactant internal energy on chemical reaction rates.

If we widen our horizons and study the dependence of the photodissociation fragment's internal energy distribution on the initial state and/or the wavelength of the exciting light, we can pose the question as to whether it is possible to control the products of photochemical reactions by proper state and wavelength selection. Likewise, it is of interest to determine conditions to maximize isotope effects in photodissociation as possible sources for isotope separation and as a useful means for generating additional data and checking theories. It is also of importance to consider the optimal conditions for the production of excess energy in particular vibrations for use in the development of photochemical lasers and for the production of hot reagents for chemical purposes.

Thermal unimolecular decompositions provide another important case in which the decomposition process is central. The theory of unimolecular decompositions, the RRKM theory, is, perhaps, the most successful available theory of chemical reaction phenomena [10]. This theory is presented as allegedly being predicated on the assumption that vibrational energy is rapidly randomized within a molecule on time scales that are rapid compared to inverse of decomposition rates. It should be emphasized that the theories could equally well begin by assuming that the vibrational energy is not randomized [11].In this alternative viewpoint existence of a statistical distribution within the energized molecules is taken to be a consequence of the random nature of the collisional excitation process, and a random lifetime assumption is invoked. Indeed, chemical activation experiments [12]display results where there is some memory of the initial mode of preparation of the energized species, whereas computer calculations [13]imply that thermal collisions need not necessarily produce statistical distributions of energized molecules. Predissociation offers a unique possibility for the study of internal energy randomization or lack thereof. A molecule can be prepared in different nearly isoenergetic initial predissociating levels with considerably different initial vibrational energy distributions. The extent to which this difference is translated into varying fragment internal energy distributions is a measure of the absence or incompleteness of intramolecular energy randomization.

Despite the great importance and scientific interest in the internal energy distributions of dissociation fragments, previous technological limitations are responsible for there being relatively few experiments of this nature. Improvements in laser technology now provide the possibility for new advances in this area, and the near future should bring substantial gains in the experimental area. It is,perhaps, relevant here to note briefly some of the experimental methods that have heretofore been applied to the problem [14].

In photofragment spectroscopy [15-17] a molecular beam is crossed with a photon beam. The dissociation fragments are then analyzed with time-of-flight mass spectroscopy. This analysis provides information on the kinetic energy distribution of the recoiling fragments. Then energy conservation considerations can be utilized to deduce information concerning internal energy distributions. Fragment angular distributions can provide information concerning the direction of the molecular transition moment involved in the photon absorption and the lifetime of the dissociating state. Photofragmentation spectroscopy has the virtue of operating in the collision-free region, but it has the disadvantage of not directly determining internal state distributions.

Hot dissociation fragments often emit radiation from which the internal energy distribution can be directly determined. Infrared chemiluminescence has long been used as a tool for probing the vibrational energy distribution of the products of chemical reactions [18] and it can likewise be employed in cases of photodissociation. Infrared emission rates are sufficiently slow that it generally is not possible to use it to study molecules under collision-free conditions. However, by judicial design of the experiment it should often be possible to determine the fragment vibrational distribution even though the initial rotational distribution is obliterated by collisions. Laser-induced fluorescence [19-21] (or possibly stimulated Raman scattering) can be used to probe internal state distributions on a much more rapid time scale to consider dissociation fragments in the collision-free regime. Likewise, far ultraviolet excitation techniques [22-29] can be used to provide photofragments in electronically excited states whose optical emission yields a sensitive probe of the internal energy distributions. The observed fragment populations can be used to check the accuracy of computed or semiempirical repulsive potential surfaces and to assess approximations in calculations of dynamics on these surfaces. Also, the fragment populations can aid in semiempirical determinations of these surfaces.

Here we describe various polyatomic decomposition processes in an attempt to understand the basic features governing the internal energy and angular distribution of the fragments. The specific details of photodissociation dynamics are, of course, intimately tied to the shapes of the relevant excited repulsive potential energy surfaces, but accurate information concerning these surfaces is generally unavailable. Highly accurate ab initio electronic energy calculations are rather difficult and expensive, and they are only beginning to be applied to excited unbound surfaces [30-32]. Although semiemperical methods are widely used to generate ground potential energy surfaces [33] the methods are not believed to be reliable for the excited repulsive surfaces. Thus, we begin our analysis with a lack of the relevant data concerning the potential energy surfaces. Hence, it is natural to introduce simple models for these surfaces for a number of reasons. First, the use of simple, but physically general, models enables the elucidation of general trends. Second, the experimental data can be used to provide a semiempirical determination of the parameters of the model. In most instances the currently available experimental data are only sufficient to fit model potentials accurately with at most one or two adjustable parameters, so the use of more general models merely obscures the essential physical trends. When more detailed data becomes available, however, the predictions of these simple models will lose their quantitative accuracy. Then it will be necessary to employ more general models and/or ab initio surfaces. Because so little is currently known about the dynamics of molecular decomposition processes, we feel that the use of these simple models is quite adequate to establish a basic zeroth order description.

There are a whole family of dissociation processes and these include photodissociation, collision-induced dissociation, electron bombardment dissociation and unimolecular dissociation. All of these processes may involve either a direct dissociation or predissociation. The predissociation process is one in which the molecule begins or is excited to a metastable predissociating state that is not a true eigenstate of the molecular Hamiltonian. A term in this full Hamiltonian acts as a perturbation coupling the metastable level to a repulsive electronic state whereon the fragments separate. Direct dissociation occurs when the molecule begins or is excited to a purely repulsive electronic state or to one in which the energy, initially present in the reaction coordinate, exceeds that necessary for dissociation. It is also possible to observe an intermediate case where there is coherent excitation to both a metastable predissociating level and a repulsive electronic state and where there are terms in the molecular Hamiltonian coupling the zeroth-order metastable and purely repulsive states. It is desirable to provide a unified theory of all of these different dissociation processes. This can be accomplished by separating those factors associated with the preparation of the "initial" state of the system and by demonstrating the identity of the description of the subsequent decomposition. Because of the greater initial state selectivity afforded by the photoinduced dissociations, most of the specific applications of the theory and comparisons with experiment are presented for this case. This particular emphasis should not be interpreted as arising from any inherent deficiency of the theory in dealing with the other types of dissociation phenomena.

2. Introduction to the Physics of Photodissociation Processes

The quantum mechanical description of the physics of photodissociation processes is basically quite straightforward. Consider the case where the molecule begins in some bound rovibrational state $|i\rangle$ of the ground electronic manifold. Radiation is incident on the system, and transitions occur due to the radiation matter interaction V. In the limit of weak light intensity the transition rate from the bound state $|i\rangle$ to a dissociative state $|f\rangle$ can be evaluated by use of the golden rule of time dependent perturbation theory. Thus the transition rate from $|i\rangle$ to $|f\rangle$ is given by the familiar expression,

$$\Gamma_{f \leftarrow i} = \hbar^{-1} |\langle f | V | i \rangle|^2, \quad (1)$$

where the continuum wavefunction is normalized to 2π times a delta function in energy.

In the Born-Oppenheimer approximation the wavefunctions for the initial and final state separate into a product of an electronic factor and a vibrational-rotational factor. The integrations in (1) are carried out first over the electronic and rotational degrees of freedom. Often the coupling V depends only weakly on the coordinates of the nuclei, so this dependence may be ignored or, if necessary, it may be treated directly [4]. Invoking the Condon approximation where this dependence is ignored, the transition rate reduces to the integral over nuclear degrees of freedom,

$$\Gamma_{f\leftarrow i} = \hbar^{-1}\left|\overline{V}\int \chi_f^*(\underset{\sim}{Q})\chi_i(\underset{\sim}{Q})d\underset{\sim}{Q}\right|^2 \qquad (2)$$

where χ_f and χ_i are now the wavefunctions for the nuclear motion alone in the two electronic states, and $\overline{V}$ gives the average electronic coupling strength for the transition. Hence, the description of molecular photodissociation is reduced to the evaluation of the Franck-Condon overlap integral (2), involving a bound initial state wavefunction $\chi_i(\underset{\sim}{Q})$ for the nuclear motion and an unbound final state wavefunction $\chi_f(\underset{\sim}{Q})$ for the nuclear motion. The program for the computation of (2) is the determination of the two nuclear wavefunctions and then the subsequent evaluation of the overlap integral.

The quantum mechanical description of weak predissociation, i.e., predissociation through sharp non-overlapping states, proceeds almost identically. Because the individual metastable predissociating levels are non-overlapping, sufficiently monochromatic radiation can produce the system initially in one vibronic metastable level $|i\rangle$. An operator V' couples this zeroth order level $|i\rangle$ to continuum states $|f\rangle$, thereby inducing the predissociation. The sharpness of the predissociating levels suggests that the coupling operator V' is weak enough to enable the golden rule to be applied to the calculation of the predissociation rate. Hence the predissociation rate from $|i\rangle$ to $|f\rangle$ is identical to (1) with the intramolecular coupling operator V' replacing the radiation matter interaction V. Invoking the Condon approximation again reduces the predissociation rate to the expression (2) with $\overline{V}'$ replacing V. The only change in the description of weak predissociation is the difference in the coupling operator and the fact that $|i\rangle$ now refers to the initially prepared predissociating level. For somewhat stronger couplings V' the predissociating levels $\{|i\rangle\}$ begin to overlap. Then purely monochromatic radiation must produce a linear superposition of these initial predissociating levels. It is quite straightforward to introduce this linear superposition and to evaluate the overall photodissociation rate when the operator V' is still sufficiently small that it may be treated by perturbation theory. The only basic difference is that the transition rate involves cross terms between the $i\to f$ and $i'\to f$ nuclear overlap integrals where $|i\rangle$ and $|i'\rangle$ are overlapping predissociating levels. The theory becomes somewhat more involved when the operator V' becomes strong enough that a simple perturbation development is no longer acceptable or when there is direct oscillator strength from the ground state levels to both predissociating levels and the adjacent dissociative continuum [35]. For simplicity we center our discussion on the simpler situations governed by (1) and (2) involving direct photodissociation or predissociation through long lived intermediate states.

2.1 The Diatomic Molecule

The simplest possible example of molecular photodissociation occurs in the diatomic molecule. There is only one vibrational degree of freedom in the initial state, and this correlates with one nuclear scattering degree of freedom on the dissociative potential energy curve. Thus, the integral in (2) is a simple one-dimensional bound-continuum integral. Given the photon energy $\hbar\omega$ and the minimum energy required for dissociation E_{diss}, the asymptotic recoil kinetic energy of the nuclear fragments in the center of mass system is fixed by energy conservation as

$$E_{kin} = \hbar\omega - E_{diss}. \tag{3}$$

Hence the final state nuclear wavefunction to be placed in (2) is the continuum wavefunction on the final potential curve with relative kinetic energy E_{kin}. For sharp predissociation the energy of the initial predissociating level E_i replaces $\hbar\omega$ in the energy conservation equation (3). Hence, for the description of direct photodissociation or weak predissociation in a diatomic molecule, it is only necessary to evaluate one-dimensional bound-continuum overlap integrals. The availability of computers makes this a rather trivial task to perform numerically and, as discussed below, accurate semiclassical approximations are available to give an analytic approximation to these integrals [35,36].

2.2 Photodissociation in Polyatomic Molecules

Polyatomic molecules have more than one vibrational degree of freedom. For the moment ignore the rotational degrees of freedom in both the initial bound and the final dissociative states in order to greatly simplify the mathematical development. This implies that we neglect those bending vibrations in the initial state which correlate with relative interfragment orbital rotations on the dissociative potential energy surface. This simple model is termed the collinear model of photodissociation as it assumes that the dissociation proceeds parallel to the bond which is being broken between two atoms in the molecule. The introduction of a proper description of all of the rotational degrees of freedom severely complicates the mathematical description of the photodissociation (as well as all collisional processes) and precludes the treatment of molecules containing more than three or perhaps four atoms. On the other hand, as shown below, the collinear model does not suffer from this difficulty; it can be used to treat larger polyatomic molecules.

We now proceed to describe the initial and final state nuclear wavefunctions to enable the evaluation of the transition rate (2). The initial bound or predissociating state nuclear motions can be described from a knowledge from the vibrational spectroscopy of these states in terms of normal or perhaps local modes of vibration. We introduce a set of basis harmonic oscillator wavefunctions $\{\psi_{n_j}(Q_j)\}$ for the normal or local modes of vibration, where Q_j is the jth vibrational coordinate in the bound initial state $|i\rangle$, and n_j is the vibrational quantum number for this mode. Residual anharmonicities in the electronic state $|i\rangle$ can be incorporated by expressing the initial nuclear wavefunction $\chi_i(\underset{\sim}{Q})$ as the linear superposition of basis functions,

$$\chi_i(\underset{\sim}{Q}) = \sum_{\{n_j\}} C_{\{n_j\}} \prod_{j=1}^{N} \psi_{n_j}(Q_j). \tag{4}$$

The coefficients $C_{\{n_j\}}$ may be determined by perturbation theory or by matrix diagonalization as is necessary in the particular situation.

The final state potential energy surface is characterized by a "reaction coordinate" along which the motion is unbounded. Let this coordinate be designated Q_1'. Then there are a remaining set of vibrational coordinates $\{Q_j' | j \neq 1\}$ which are orthogonal to the reaction coordinate. Motions along these orthogonal directions are described by the use of a set of harmonic oscillator basis functions $\{\psi_{n_j}'(Q_j')\}$. The nuclear wavefunction on this unbound potential energy surface is then represented as a superposition of the basis functions,

$$\chi_f(\underset{\sim}{Q}) = \sum_{\{n_j'\}} S_{E\{n_j'\}}(Q_1') \prod_{j=2}^{N} \psi_{n_j'}(Q_1'), \tag{5}$$

where $S_{E\{n_j'\}}(Q_1')$ are continuum wavefunctions to be determined by a scattering calculation on the dissociative potential energy surface.

Introducing the basis function expansions (4) and (5) for the initial and final state nuclear wavefunctions, the nuclear overlap integral reduces to a sum of overlap integrals involving the basis nuclear wavefunctions,

$$\int d\underset{\sim}{Q}\chi_f(\underset{\sim}{Q})\chi_i(\underset{\sim}{Q}) = \sum_{\{n_j\},\{n_j\}} C_{\{n_j\}} \int S^*_{E\{n_j'\}}(Q_1')\,\psi_{n_1}(Q_1)$$

$$\times \prod_{j=2}^{n} \psi_{n_j'}(Q_j')\,\psi_{n_j}(Q_j)\,d\underset{\sim}{Q} \equiv \sum_{\{n_j\},\{n_j\}} C_{\{n_j\}} \langle E\,\{n_j'\}|\{n_j\}\rangle\ . \tag{6}$$

Hence, it suffices for us to be able to evaluate the individual integrals $\langle E\,\{n_j'\}\ |\ \{n_j\}\rangle$ in the summation in (6). The use of harmonic oscillator basis functions has obviously simplified matters to the utmost. However, there is a severe complication in evaluating (6) which stems from the fact that the reaction coordinate Q_1' is generally not the same as any of the natural vibrational coordinates $\{Q_j\}$ appropriate to the description of the nuclear motion on the initial state potential energy surface. Likewise the orthogonal final state nuclear coordinates $\{Q_j'\ |\ j \neq 1\}$ are again quite different from the initial state natural vibrational coordinates $\{Q_j\}$. Hence, the primitive overlap integrals on the right hand side of (6) are <u>multidimensional nonseparable integrals</u>. A first attempt at numerical evaluation of these integrals for the simplest case of n = 2 found that the integrals were slowly convergent even when the integrals were taken out to large distances, so the attempt was aborted because of the large amount of computer time required [37].

2.3 Evaluation of the Multidimensional Franck-Condon Integrals

It can be shown, however, that it is possible to exactly reduce the primitive overlap integrals exactly to simple one dimensional bound-continuum integrals of a form very similar in nature to that obtained in the description of photodissociation in diatomic molecules [35, 36, 38].As this subject has been reviewed by Freed and Band [35] only the final results are sketched.

Since the two sets of coordinate systems $\{Q_j\}$ and $\{Q_j'\}$ are designed to represent the small amplitude vibrations in the initial bound state and the small amplitude vibrations in directions orthogonal to the reaction coordinate, respectively, there is a simple linear transformation between the two coordinate systems,

$$\underset{\sim}{Q} = \underset{\sim}{C}\,\underset{\sim}{Q}' + \underset{\sim}{b} \tag{7}$$

where $\mathcal{C}$ is an N x N matrix of constants and $\underline{b}$ is an N dimensional column matrix of constants. Utilizing the transformation (7) enables us to convert the primitive overlap integrals exactly to the one dimensional form [35,36, 38],

$$< E\{n_j'\} \mid \{n_j\} > = \int P_{\{n_j'\},\{n_j\}}(Q_1')g_{fi}(Q_1')S^*_{E\ \{n_j'\}}(Q_1')dQ_1', \qquad (8)$$

involving only the reaction coordinate Q_1'. In (8) $P_{\{n_j'\},\{n_j\}}(Q_1')$ is a polynomial in Q_1' of order $n_{tot} = n_1 + \sum_{j=2} (n_j + n'_j)$, and $g_{fi}(Q_1')$ is a normalized Gaussian in Q_1' which can be obtained from performing the vibrationless integral over all degrees of freedom except Q_1',

$$< E\ \{0\}|\{0\}> \equiv \eta \int g_{fi}(Q_1')\ S^*_{E\ \{n_j'\}}(Q_1')dQ_1'. \qquad (9)$$

2.4 The Effective Oscillator Concept

Since $g_{fi}(Q_1')$ arises through the transformation (7), it is clear that $g_{fi}(Q_1')$ depends on all of the vibrations for both the initial and final potential energy surface. The Gaussian is centered about the position r_0 which likewise is dependent on all the initial and final state vibrations. Utilizing the reduced mass μ_1' for the reaction coordinate Q_1' enables us to convert the width of the Gaussian $g_{fi}(Q_1')$ into an effective force constant k_0 if we take the normalized Gaussian $g_{fi}(Q_1')$ to be the ground state wavefunction in a mathematically convenient effective oscillator. The derivation of formulas for r_0 and k_0 can be accomplished quite straightforwardly once the transformation (7) and the bound and dissociative frequencies, force constants, etc., are provided. Recursion relations have been given for the evaluation of the polynomials $P_{\{n_j'\},\{n_j\}}(Q_1')$ where the coefficients depend on both initial and final state vibrational quantum numbers $\{n_j'\}$, $\{n_j\}$.

Since we have identified the normalized Gaussian $g_{fi}(Q_1')$ as a ground state wavefunction in an effective oscillator, the overlap integrals (8) can be written in a form which more closely resembles the simple diatomic molecule example. This proceeds by expressing the polynomial $P_{\{nj\},\{nj\}}(Q_1')$ as a linear combination of Hermite polynomials H_p which are associated with the eigenfunctions of this effective oscillator,

$$P_{\{n_j'\},\{n_j\}}(Q_1') = \sum_{p=0}^{n_{tot}} a_p(\{n_j'\},\{n_j\})\ H_p(\xi_1'Q') \qquad (10)$$

where $(\xi_1')^2 = \mu_1'\omega_1'/\hbar$ with ω_1' the effective oscillator frequency, and the coefficients $a_p(\{n_j'\},\{n_j\})$ can be determined from recursion relations [35, 36, 38]. Now, (8) is in the form of a sum of diatomic molecule-like one-dimensional bound continuum overlap integrals which are straightforward to evaluate numerically if desired.

It should be noted that earlier theoretical descriptions of photodissociation processes of polyatomic molecules despaired at the nonseparability of the primitive overlap integrals $\langle E\{n_j'\}|\{n_j\}\rangle$ of (6). Therefore, they introduced the drastic approximation of assuming that the reaction coordinate Q_1' is identical to one of the normal modes of vibration in the initial bound electronic state [39,40]. This then separates the one-dimensional diatomic-like bound-continuum overlap integral from the remaining oscillator integrals. Consequently these theories have been termed "quasidiatomic models" because they have reduced the complicated polyatomic problem to one that is essentially equivalent to a diatomic problem [15,37,34-42]. Unfortunately, potential energy surfaces are generally not so cooperative to the assumptions of the quasidiatomic model because the reaction coordinate Q_1' is generally considerably different from any of the initial state normal vibrations. The transformation (7) shows that Q_1' is generally a linear combination of a number of the vibrational degrees of freedom in the initial bound state. Nevertheless, the general theory above shows that even given the essential nonseparability introduced by the transformation (7) it is still possible to reduce the multidimensional overlaps to simple sums of diatomic molecule-like bound-continuum overlap integrals.

This reduction of the multidimensional Franck-Condon integrals to one-dimensional form (8) enables us to provide a simple pictorial representation of the photodissociation overlap amplitudes [35,38]. We can simply plot the repulsive interaction between the two atoms whose bond is being broken on the final potential energy surface. Then, the continuum wavefunction $S_E(Q_1')$, appropriate to this potential energy curve, can be represented on the figure. On the same figure we may represent the effective oscillator which has $g_{fi}(Q_1')$ as the ground state wavefunction. The relative energies of the minimum of the effective oscillator and the repulsive potentials are unimportant since (8) involves a linear combination of the effective oscillator wavefunctions, thereby indicating that a simple diatomic-like application of energy conservation constraints (3) is meaningless for the effective oscillator that describes the full multidimensional world of the polyatomic molecule potential energy surfaces. The primitive overlap integrals now have been reduced to a calculation of the overlap of effective oscillator wavefunctions with the continuum wavefunctions $S_E(Q_1')$ for the potential curve along the reaction coordinate.

Some particular examples are illustrated in Fig. 1 for varying relative positions along the reaction coordinate of the repulsive curve and the effective oscillator. The top curve on the left in Fig. 1 displays a situation in which the turning point for motion on the repulsive curve lies at larger Q_1' than the dominant portion of the ground state wavefunction in the effective oscillator well. As the continuum wavefunction is taken to lower and lower kinetic energies, the turning point on the repulsive curve proceeds further and further away from the effective oscillator, so the bound-continuum overlap integral from the ground state of the effective oscillator becomes exponentially smaller and smaller. This translates into a preference for producing the fragments with all the available energy in the relative translational motion. Equivalently this corresponds to a vibrational energy distribution which is peaked about vanishing vibrational energy in the fragments. The top right hand vibrational distribution in Fig. 1 illustrates this for the simple case of collinear triatomic dissociation. The bottom curves on the left in Fig. 1 show an example where the turning point for motion on the repulsive curve lies to the left of the effective oscillator minimum, and a similar analysis indicates that this situation corresponds to the vibrational population inversion depicted on the right hand bottom side of Fig. 1. An intermediate case is described by the middle of Fig. 1 where the fragment vibrational energy distribution is peaked about a nonzero vibrational energy.

The calculation of the position and width of the effective oscillator enables us to obtain a qualitative understanding of the vibrational energy distributions on the dissociative surface that are produced in the sudden Franck-Condon rearrangement process by the simple inspection of diagrams like those in Fig. 1. Accurate analytical approximations to the one dimensional bound-continuum integrals (8) have been provided [35,36]. For more detailed investigations of the values of the transition amplitudes, as mentioned above, the integrals (8) are readily evaluated numerically if desired.

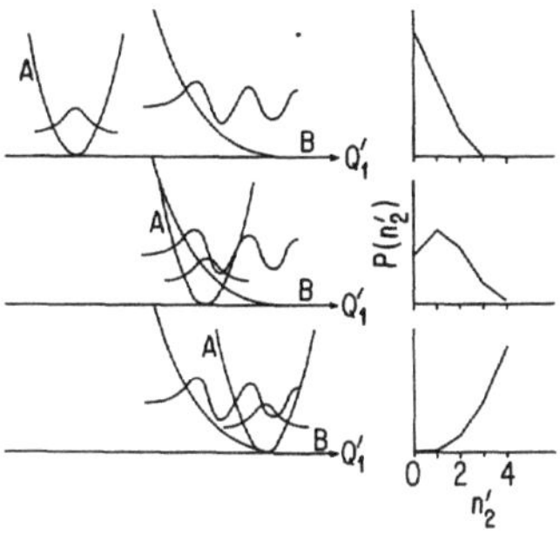

Fig.1 Schematic respresentation of various possible relative positions of the effective oscillator potential (A) with respect to the interfragment zeroth order repulsive potential (B). The ground state wavefunction in the effective oscillator and a continuum wavefunction for the repulsive curve are also displayed. The overlap of these functions yields the relative vibrational energy distributions given to the right for each case.

2.5 The Half-Collision Process

It should be emphasized that the evaluation of the full transition amplitude involves the calculation of the integrals (6) which necessitates the determination of the initial nuclear wavefunctions (4) and the continuum wavefunctions (5). While many scattering programs are available, they generally do not retain the information concerning the continuum wavefunction (5) although this information is implicitly contained in the calculation. Nevertheless, a minor modification of existing scattering routines should enable the determination of the information necessary for the calculation of the photodissociation rates. It can be shown that it is really not necessary to numerically tabulate the full final surface continuum wavefunction (5) to evaluate the photodissociation rates (2). Provided that the appropriate modifications in scattering programs are made, only the asymptotic properties of $S_E(Q_1')$ are required as $Q_1' \to \infty$ in order to calculate the photodissociation rates. (See discussion in Sect. 4.) Thus, it is clear that Fig. 1 and the subsequent analysis of the partitioning of the vibrational energy among the vibrational degrees of freedom provides a view of how the initial state vibrational energy is converted to nascent vibrational energy on the dissociative potential energy surface by the sudden Franck-Condon process (direct photodissociation or predissociation) which transports the molecule from the bound to the repulsive surface. The scattering then determines how this nascent vibrational energy distribution on the repulsive surface is converted into the energy distributions of the fragments due to the interfragment interactions that occur on the dissociative surface during the "half-collision" that the fragments are receding from each other. Figure 1 shows how the change in natural coordinate systems in going from the bound to the dissociate surface, from $|i\rangle$ to $|f\rangle$, can be an important source for pumping energy into the vibrational degrees of freedom of the fragments. The half-collision can lead to further vibrational relaxation or vibrational excitation of the fragments.

The dissociative potential energy surfaces of polyatomic molecules are very poorly known. Ab initio calculations are beginning to appear for some small triatomic examples [30-32] but very little is known in the general case even in a semiempirical vein. The use of a fixed set of vibrational coordinates on the dissociative surface is a simple approximation which is not necessarily optimal due to the fact that, as a bond is being broken in the dissociation process, the other bonds in the fragments often are becoming strengthened. Therefore, as the half-collision proceeds along the reaction coordinate Q_1', the vibrations $\{Q_1'|j \neq 1\}$ orthogonal to the reaction coordinate are changing their character. A more satisfactory description of the vibrations on the dissociative surface involves the use of a set of natural collision coordinates $\{Q_j'(Q_1')|j \neq 1\}$ for this dissociative potential energy surface which are varying along the reaction coordinate Q_1'. This implies that the matrices $\tilde{C}$ and $\tilde{b}$ in (7) are now functions of the reaction coordinate. This situation then makes the primitive overlap integrals $\langle E\{n_j'\}|\{n_j\}\rangle$ even more highly nonseparable. If the standard collisional natural collision coordinates [43] are utilized, the integrations over $Q_j', j \neq 1$, cannot be performed analytically, and the gain in a more physical description of the vibrations on the repulsive surface is completely offset by the resultant essentially nonseparable integrals. However, it is possible to introduce a new set of natural collision coordinates which still enables the primitive integrals to be reduced to one-dimensional form [44]. Unfortunately these natural collision coordinates are not an orthogonal set of coordinates. This slight disadvantage is probably small given the ability to reduce the multidimensional integrals to simple one-dimensional forms, but this point remains to be studied numerically in more detail.

3. Comparison with Experiments Using the Collinear Approximation

There are a number of impediments to a complete comparison between theoretical predictions and experimental results for the product energy distribution in photodissociation. It is only fairly recently that measurements have been made of the fragment energy distributions in photodissociation of triatomic molecules. The scattered data are really insufficient to offer a critical test of the theories of molecular dissociation. The majority of the experimental data has been obtained by excitation in the vacuum ultraviolet at one or two specific wavelengths whose values are determined by the availability of atomic resonance lamps. These data are difficult to analyze because the nature of the electronically excited states in these energy regions are not well understood and also the haphazard choice of atomic resonance lines does not necessarily produce excitation at the most desirable wavelength, nor do different atomic lines produce excitation in the same electronic band system. Recent work by SIMONS and coworkers [29] and others [21] is attempting to make a systematic sweep of the absorption wavelength, and it is hoped that experiments of this type will produce sufficient information to confront the theoretical predictions. The majority of the remaining experimental data has been obtained via photofragment spectroscopy which only provides information about the kinetic energy distribution of the fragments. The technique of laser induced fluorescence of the photofragments offers the possibility of providing complete internal state energy distributions of the products, and recent experiments by BARONAVSKI and MCDONALD [21] attest to the potential powers of this method.

3.1 The Dissociative Potential Surface

Hence we begin our attempt at a comparison between theory and experiment with a lack of sufficient experimental information. On the other hand, molecular dynamics calculations rest on a knowledge of the relevant potential energy surfaces for the process. As noted previously these are only beginning to be computed by ab initio methods for the lowest excited electronic states of a few triatomic molecules. Those states probed by the vacuum ultraviolet excitation and the interesting case of ICN have not been attacked by full scale theoretical calculations. For these reasons and because of our desire to understand the qualitative features of the dependence of the internal energy distributions in the products upon the nature of potential energy surfaces, the nature of the initially excited state, and the wavelength of the exciting radiation, we have adopted a semiempirical procedure whereby simple model potentials are introduced and the parameters in the potentials are fit to the available experimental data. These fitted potential curves are then used by the theory to make additional predictions concerning the internal energy distributions which would be obtained at wavelengths which had not been probed by the experiment and/or energy distributions in degrees of freedom which have not been probed experimentally.

Since the majority of the experiments have been applied to triatomic molecules we restrict attention to this interesting yet simplistic case. It should, however, be emphasized that the general theory (4)-(10) applies equally well to larger molecules. Within the collinear approximation there are only two degrees of freedom for the triatomic potential energy surface. For linear molecules these two degrees of freedom involve the distance from the end atoms to the central atom or any linear combination thereof. For the bound initial state the relevant coordinates are the two stretching normal modes of vibration (or even local modes), while on the dissociative surface the proper asymptotic coordinates are the vibration of the diatomic fragment and the motion of the departing atom with respect to the center of mass of the diatomic fragment. These two coordinate systems do indeed satisfy the transformation properties (7).

A potential energy surface for the dissociative state having the smallest number of adjustable parameters is readily constructed as follows: The potential energy associated with the diatomic fragment is taken to be identical to the potential energy for the isolated fragment. In addition there is a repulsive interaction between the two atoms whose bond is being broken in the dissociation process. (It is also possible to introduce an interaction between the two end atoms, but this introduces additional adjustable parameters.) Adopting the simplest harmonic oscillator approximation for the diatomic fragment, this potential energy surface becomes

$$V(Q_1' , Q_2') = \tfrac{1}{2} k(Q_2')^2 + V_1(R_{23}), \tag{11}$$

where Q_2' is the vibrational coordinate for the diatomic fragment, k is the force constant for the isolated diatomic fragment, and R_{23} is the separation between the two atoms whose bond is being broken in the dissociation. The force constant k is known from spectroscopic data on the fragments, and a more realistic Morse type potential could be utilized instead of the harmonic potential (11) provided the relevant experimental data on the spectroscopy of the fragments is available. The interaction $V_1(R_{23})$ must be repulsive at short range and must tend asymptotically at large R_{23} as an inverse power van der Waals attraction. As the repulsive part of the potential is expected to be exponential, V is chosen to have the form

$$V_1(R_{23}) = C_0\exp(-DR_{23}) - C \text{ for } R_{23} \leq R_{23}^0 \tag{12}$$

with the van der Waals tail for $R_{23} > R_{23}^0$.

The scattering states with the potential energy surface (11) and (12) have been studied extensively in treatment of collinear atom-diatom collision processes. Since the effects of the sudden Franck-Condon rearrangement process, where the molecule is suddenly transported from the bound to the dissociative surface, had not previously been treated, we have adopted a simple semiclassical description of the scattering on the potential energy surface (11). This treatment involves a simple forced oscillator model of the half-collision process [45]. The forced oscillator models are known to be reasonably accurate for atom-diatom collinear collision processes when the vibrationally inelastic transition probabilities are sufficiently low. Other workers have utilized first order K-matrix methods [42,46,47] which are generally of comparable accuracy for the full collision process. Some full coupled channel scattering treatments have also been carried out [48]. Our interest here is in the gross factors governing product energy distributions so the simple semiclassical model is adequate especially in view of the lack of sufficient experimental data to make a full scattering calculation worthwhile at this juncture.

3.2 Experimental Data and Comparison with Theory

MELE and OKABE have performed experiments on the HCN photodissociation,

$$HCN + h\nu \rightarrow H + CN\ (B^2\Sigma^+), \tag{13}$$

using Xe and Kr resonance flash lamp photon sources. At the time of the theoretical analysis, the absorption spectra were unavailable, so it was not known whether both absorption wavelengths produce excitation in the same electronic manifold nor was the nature of the electronically excited state known. Thus, as an illustrative example it was assumed that the dissociation proceeds as a direct dissociation with both wavelengths exciting the same electronic state. Subsequent work has shown that this is probably not the case [49]. Nevertheless, one can utilize the MELE and OKABE experimental data as an illustrative model of the manner in which the excited repulsive potential energy curve V_1 of (11) can be fit to experimental data. As shown below, the comparison is extremely instructive.

Figure 2 displays the relative probability of producing CN $(B^2\Sigma^+)$ state with the vibrational quantum number n_2' for excitation with the xenon flash lamp with photon energy of 76820 cm^{-1} and for the krypton flash lamp with excitation energy of 81220 cm^{-1}. The solid lines in Fig. 2 give the calculated relative vibrational distribution and the dots provide the experimental values. The effective oscillator potential is presented in curve A of

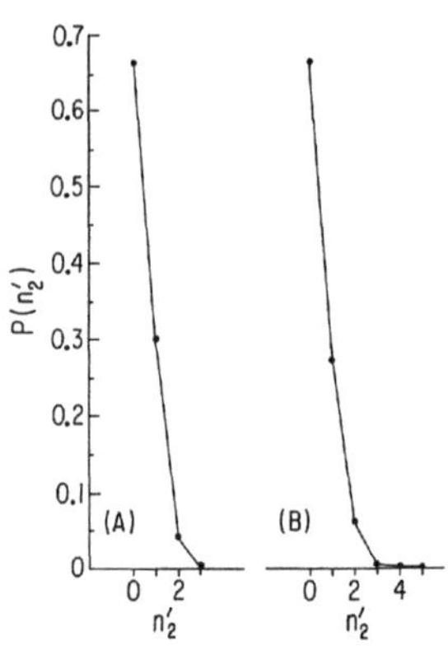

Fig.2 Relative vibrational distributions for CN $(B^2\Sigma^+)$ for the direct photodissociation of HCN. Points are data from MELE and OKABE [23] for (A) Xe line excitation and (B) Kr line excitation, while lines are calculated values for potentials displayed in Fig. 3.

Fig. 3 as a function of Q_1'. The parameters of the effective oscillator potential are uniquely determined within the model (11) from the spectroscopic constants for the stretching vibrations in ground state of HCN and for the vibrations of the $B^2\Sigma^+$ state of the CN fragment. We have found that given the experimental values for the vibrational energy distribution at two different wavelengths it is possible to obtain an infinite set of potential energy curves (12) which all provide the same relative vibrational populations of Fig. 2 [35,38]. Four such potential energy curves are given in Fig. 3 as curves B1 through B4. Since the majority of the CN molecules are produced in the ground and first excited vibrational levels, they emerge from the krypton excitation with the asymptotic kinetic energies indicated on the right in Fig. 3 by zero and one, respectively. The classical turning points on the different repulsive curves B1 through B4 vary at these energies by over one Å. Hence it is quite clear that experimental data on the relative vibrational energy distributions produced at two excitation wavelengths is wholly insufficient, in general, to uniquely determine even the simple model potential (12).

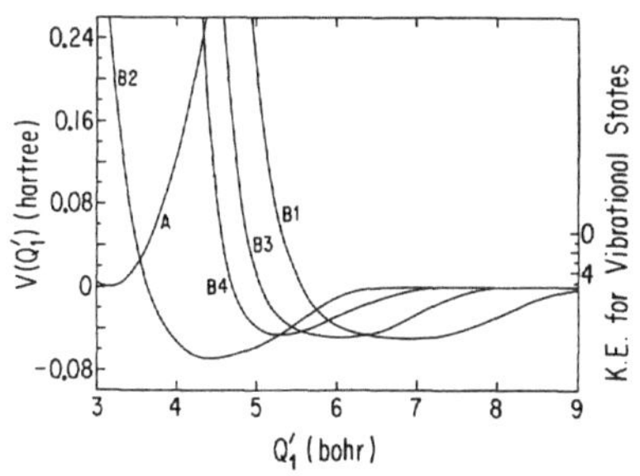

Fig.3 Potential energy curves for HCN direct photodissociation. A is the effective harmonic oscillator potential determined from HCN ground state and CN ($B^2\Sigma^+$) spectroscopic constants. Curves B are various repulsive potentials between H and CN ($B^2\Sigma^+$) which all reproduce the data in Fig. 2. The maximum asymptotic interfragment kinetic energies are depicted on the left for the various fragment vibrational states with Kr lamp excitation.

The predicted photodissociation hydrogen-deuterium isotope effect and decay rate are markedly different as we range from curve B1 to curve B2 with classical turning point at smaller separations Q_1' between hydrogen and the center of mass of CN. In going from B1 to B2 the photodissociation rate changes by 33 orders of magnitude, the predicted hydrogen-deuterium isotope effect changes by 13 orders of magnitude, and the relative importance of the final state interactions increases from being of negligible importance for curve B1 to providing a contribution to the production of the final vibrational excitation in the case of B2. It would be possible to narrow down the range of possible potential curves if data had been available concerning the absorption spectra, as this would locate the Franck-Condon region of the repulsive curve, the energy at which the repulsive curve lies above the minimum of the effective oscillator potential. (This, of course, presupposes that both excitation wavelengths excite a given band system, and this is now known to probably be incorrect.)

The hydrogen-deuterium isotope effect on the photodissociation rate is on the order of 10^{13} for curve B1. While potential curve B1 is not a physically realistic possibility in this case, it does motivate the question as to what are the conditions which would produce large isotope effects in photodissociation. The potential energy curves in Fig. 3 indicate one possible mechanism for large isotope effects in photodissociation. The relevant Franck-Condon factors of (8) involve the overlap integrals between the eigenfunctions of the effective oscillator and the continuum functions associated with the potential energy curves. It is clear that at the energy corresponding to fragments in the ground vibrational state of $CN(B^2\Sigma^+)$ the overlap between the effective oscillator levels and the continuum functions is extremely poor; the transition from curve A to curves B1, B3 or B4 proceeds via a tunneling process. Hence, the Franck-Condon factor depends exponentially on the square root of the reduced mass, μ, of the leaving atom with respect to the diatomic fragment. If μ_1 and μ_2 are the reduced masses for the two isotopes, the ratio of the photodissociation rates for the two isotopes is of the form $\exp[K(\sqrt{\mu_1}-\sqrt{\mu_2})]$ where K is a collection of molecular constants. The isotope effect is maximized towards the threshold for the photodissociation, but of course here the intensity of the photodissociation is minimal. Hence, in order to maximize the isotope effect a balance must be maintained between the magnitude of the isotope effect and the magnitude of the photodissociation rate. Perhaps this mechanism for large isotope effects in photodissociation is related to the well known fact in the spectroscopy of small organic molecules that hydrogen atom predissociation is often suppressed by deuteration, the deuterated molecule displaying a sharp rotational spectrum whereas the electronic spectrum for the protonated compound being diffuse by virtue of hydrogen atom predissociation.

Another illustrative example that has been studied, again using the simple assumption of direct photodissociation from a linear initial state, is the case of CO_2 photodissociation. In that case the data of LEE and JUDGE [25] display substantial vibrational excitation in the carbon monoxide fragments. Because the fragments in different vibrational states exit with different asymptotic relative kinetic energies between the atom and the diatom, the experiments with broad vibrational distributions sample the repulsive potential over wider energy ranges. Thus, given information about the product energy distribution at only two wavelengths it has only been possible to date to find a single potential energy curve which reproduces the experimental results [35].(There is no real point in a more detailed analysis as the dissociation is probably not a simple direct dissociation from a linear initial state. In addition, a proper treatment of CO_2 dissociation requires incorporation of symmetry constraints [50].) It is interesting to note that in the CO_2 example the vibrational excitation of the CO fragment derives roughly one half of its energy from the sudden Franck-Condon rearrangement process due to the change in the nature of the natural coordinate system between the bound and dissociative surface, with the other half of the vibrational excitation of the fragment arising from the half-collision on the upper surface in which the departing oxygen atom exerts forces on the CO to produce additional excitation in the CO fragment. Hence, this illustrative example emphasizes the fact that a full theoretical description of the photodissociation requires a proper treatment of both the sudden Franck-Condon process and the vibrational relaxation that can occur during the half-collision on the repulsive potential energy surface.

The use of the simple collinear model and the comparison between it and inadequate experimental data are useful in a number of respects: Firstly, it shows how the change in the natural coordinate system on the two potential energy surfaces, the transformation (7), can provide an important source for pumping energy into the vibrational degrees of freedom of the fragments of the decomposition process. Likewise, the repulsive interactions between fragments on the upper potential energy surface can also produce excitation or even deexcitation of the vibrational degrees of freedom of the fragments. The latter had previously been invoked as the only method of producing vibrational excitation of the fragments in the simple quasidiatomic-like models, whereas it has now been amply demonstrated that the Franck-Condon rearrangement process must be treated on an equal footing. The theoretical calculations emphasize the fact that in addition to providing fragment energy distributions at a few wavelengths it is necessary to have information about the photodissociation excitation spectrum in the wavelength region under consideration. The ideal experiment would involve predissociation through sharp spectroscopically assignable non-overlapping levels where the product energy distribution can be determined as a function of the initial predissociating level. Experimentalists and theoreticians alike will have to contend with the fact that this ideal limit is not so easily accessible for small triatomic molecules whose potential energy surfaces are also amenable to calculation.

4. The Full Three Dimensional Theory

When applied to the description of the dissociation of linear triatomic molecules, the collinear approximations, discussed in the previous two sections, assumes that all three atoms can be constrained to lie on a nonrotating line. This model, therefore, assumes that the bending and rotational degrees of freedom do not influence the partitioning of energy between fragment vibrational and relative translational degrees of freedom. This model ignores the influence of angular momentum and its conservation constraints on the dissociation, leaving unanswered the interesting question of the nature of fragment rotational distributions and of the dynamical information contained therein. In addition, when large amounts of energy appear in the fragment rotational degrees of freedom, this rotational energy is unavailable for the vibrational and translational degrees of freedom. Therefore, the inclusion of the rotational and bending degrees of freedom can seriously affect the energy partitioning into the vibrational and translational degrees of freedom.

When applied to molecules which are bent in the initial bound electronic state, the simple collinear model must assume that motion of the departing atom is constrained to lie along the initial bond that is being broken. Such motion naturally exerts a torque on the remaining diatomic molecule which can introduce excitation of its rotational degrees of freedom and thereby remove energy that the collinear approximation would partition between the vibrational and translational degrees of freedom of the fragments.

The inclusion of rotational degrees of freedom severely complicates the mathematical description of the photodissociation process. This complication limits the full three dimensional theory to systems containing perhaps only three to four atoms. However, these particular systems are of considerable interest for a number of reasons. Firstly, these systems are the ones for which the calculation of ab initio repulsive potential energy surfaces is the simplest and therefore the case for which these surfaces will be developed initially. Secondly, the small molecules provide the simplest possible spectroscopy. They therefore present the promise of the most detailed state-to-state photodissociation information that experimentally is feasible for polyatomic molecules. Thirdly, the theoretical description of the full three-dimensional scattering process is again the simplest when applied to these smaller systems.

4.1 The Bound State Wavefunction

We limit our discussion to the simplest case of the dissociation processes of triatomic molecules. The theory is briefly sketched for the case of triatomic molecules with a linear equilibrium configuration in the initial bound state. Hence, the initial bound surface nuclear wavefunctions may be represented as a linear superposition of the linear molecule basis functions

$$\psi_i(Q_1) = \left(\frac{2J+1}{8\pi^2}\right)^{\frac{1}{2}} D^{J\,*}_{M,k}(\alpha\beta\gamma)\, \psi_{n_1}(Q_1)\, \psi_{n_2}(Q_2)\psi_\nu{}^k(\delta), \tag{13}$$

where $\psi_{n_1}(Q_1)$ and $\psi_{n_2}(Q_2)$ are harmonic oscillator basis wavefunctions for the two stretches, $\psi_\nu{}^k(\delta^2)$ is the basis harmonic oscillator wavefunction for the doubly degenerate bending mode. k is the bending angular momentum along the triatomic axis, and it is assumed, for simplicity, that no electronic angular momentum is present in the initial electronic state. The wavefunction (13) is adequate for the lowest lying states in the initial electronic manifold, and residual anharmonicities, centrifugal distortions and Coriolis couplings may be treated by using linear combinations of these basis functions in a manner similar to that employed in (4). $\alpha\beta\gamma$ are the Euler angles specifying the orientation of the equilibrium molecule fixed axis relative to space fixed axes, and D is a Wigner rotational function.

4.2 Scattering Equations for Dissociative Surface Wavefunction

A full quantum mechanical description of the photodissociation process requires the consideration of the Schroedinger equation involving the coupled motion on both the initial and final electronic potential energy surfaces [35,51]. The full wavefunction for the system must of course be regular at the origin and it must correspond to purely outgoing flux as Q_1' goes to infinity. (In the case of direct photodissociation there is, of course, incident photon flux on the system.) Rather than writing the coupled two-surface Schroedinger equation for the system, we note that the interaction V_{fi} has been assumed to be weak so that it is possible to obtain the final surface nuclear wavefunction to first order in this coupling operator V_{fi} as [35]

$$\psi_f(\underset{\sim}{Q}') = \int d\bar{\underset{\sim}{Q}}' G_f(\underset{\sim}{Q}', \bar{\underset{\sim}{Q}}')V_{fi}(\bar{\underset{\sim}{Q}})\psi_i(\bar{\underset{\sim}{Q}})\ [J(\bar{\underset{\sim}{Q}}')]^{\frac{1}{2}}. \tag{14}$$

Here $J(\underset{\sim}{Q}')$ is the Jacobian for the transformation between the coordinate systems $\underset{\sim}{Q}$ and $\underset{\sim}{Q}'$ appropriate to the two potential energy surfaces, and G_f is the full Green's function for nuclear motion on the dissociative potential energy surface,

$$[E - H_f(\underset{\sim}{Q}')]\ G_f(\underset{\sim}{Q}', \bar{\underset{\sim}{Q}}') = \delta(\underset{\sim}{Q}' - \bar{\underset{\sim}{Q}}'), \tag{15}$$

where $H_f(Q')$ is the full nuclear Hamiltonian for motion on the dissociative surface. Note that G_f must have the boundary conditions of being regular at the origin and having purely outgoing waves. It is for this reason that no homogeneous solution appears in (14) as no such solution obeying these boundary conditions exist.

We may define a zeroth order distorted wave potential energy in a form similar to that in (11). The potential energy in H_f is equal to that in the distortion potential plus a term $V_I(\underset{\sim}{Q}')$ which accounts for coupling between the rotational, vibrational and translational degrees of freedom on the repulsive surface, i.e. the potential giving rise to final state interactions. Let G_0 be the final surface Green's function appropriate to the zeroth order Hamiltonian $H - V_I$. Then (14) can be rewritten as the integral equation [35,51]

$$\psi_f(\underset{\sim}{Q}') = \int d\bar{\underset{\sim}{Q}}' G_0(\underset{\sim}{Q}', \bar{\underset{\sim}{Q}}')\ [V_{fi}(\bar{\underset{\sim}{Q}})\ \psi_i(\bar{\underset{\sim}{Q}})J^{\frac{1}{2}}(\bar{\underset{\sim}{Q}}') + V_I(\bar{\underset{\sim}{Q}}')\psi_f(\bar{\underset{\sim}{Q}}')], \tag{16}$$

which may in turn be converted into the inhomogeneous driven Schroedinger equation for the final surface nuclear wave function $\psi_f(Q')$,

$$[E - H_f(\underset{\sim}{Q}')]\ \psi_f(\underset{\sim}{Q}') = V_{fi}(\underset{\sim}{Q}')J^{\frac{1}{2}}(\underset{\sim}{Q}')\ \psi_i(\underset{\sim}{Q}). \qquad (17)$$

Equations (14) (16) and (17) represent the final surface nuclear wave function in terms of the propagation of the initial surface wave function $\psi_i(\underset{\sim}{Q})$ multiplied by the coupling operator $V_{fi}(\underset{\sim}{Q}')$ and the Jacobian factor $J^{\frac{1}{2}}(\underset{\sim}{Q}')$ on the repulsive final state potential energy surface. This bears a strong similarity to the approach of HELLER [52] who has introduced a description of direct photodissociation using a time dependent wavepacket propagation on the repulsive potential surface of the product of the initial bound state nuclear wave function times the nuclear coordinate dependent transition dipole function. Our approach is the exact time independent quantum mechanical formulation of this picture. It is applicable to predissociation as well as direct dissociation, and it indicates the Jacobian corrections necessary to account for the transformation between initial and final potential energy surfaces.

SHAPIRO [48,53] has developed a way to treat final state interactions by evaluation of the poles in the T-matrix in the system of coupled channel scattering equations involving both the repulsive <u>and</u> bound nuclear basis functions plus a fictitious entrance channel. In order to avoid the nonseparable Franck-Condon problem, he utilizes the same nuclear coordinate system, the repulsive surface reaction coordinates on both surfaces. However, the method is purely a numerical one and does not admit analytically soluble models. Furthermore, for weak incident radiation in the case of direct photodissociation or for weak bound-continuum coupling in the case of predissociation there is no need to consider a problem of dimensionality equal to the number of repulsive surface basis functions <u>plus</u> the number of bound surface basis functions. Equations (14) through (17) demonstrate that all that is required is the treatment of driven single surface scattering equations. This considerably reduces the dimensionality of the scattering calculation and it also obviates the search for poles in the entrance channel T-matrix of the SHAPIRO approach.

4.3 Repulsive Surface Basis Functions and Axis Switching Transformation

Consider a situation in which the atomic fragment departs at an orientation given by the polar angles θ_{SF} and ϕ_{SF} with respect to the center of mass of the diatomic fragment, where θ_{SF} and ϕ_{SF} are the polar angles in the laboratory coordinate system. Zeroth order basis functions, eigenfunctions of the simple zeroth order Hamiltonian $H_f - V_I$, are given by

$$\psi_{Enlj\mu m}(Q') = Y_{jm}(\beta'\alpha')Y_{l\mu}(\theta_{SF}, \phi_{SF})\ \psi_n(Q_2')\ \psi_{Enlj}(Q_1')/(k_{nj}Q_1'), \qquad (18)$$

where Y_{jm} is a spherical harmonic function, j and l denote the angular momentum of the diatomic molecule and relative angular momentum of the atom with respect to the center of mass of the diatomic molecule, respectively. $\psi_n(Q_2')$ is the vibrational wavefunction for the diatomic fragment, and $\psi_{Enlj}(Q_1')/(k_{nj}Q_1')$ is the distorted wave scattering wave function on the repulsive potential energy surface. The angles $\beta'\alpha'$ give the polar angles of the diatomic molecule with respect to the space fixed axis. Note that because of molecular bending in the initial electronic state, the angles $\alpha\beta\gamma$ specifying the orientation of the equilibrium triatomic axis and $\beta'\alpha'$ specifying the orientation of the axis of the diatomic fragment are in general different [54]. This dynamic axis switching phenomena may appear physically to average out over a bending period but its inclusion is necessary to describe the correct partitioning of the initial rotational angular momentum (J) of the triatomic molecule into the angular momentum of rotation of the diatomic fragment and the orbital angular momentum of the atomic fragment. The effect of the dynamic axis switching is relatively small for very light departing atoms, but it becomes considerable when the departing atom is much heavier than the diatomic fragment [54]. A rotation through the Euler angles $0\phi 0$ takes the axis system fixed in the triatomic molecule into the axis system fixed in the diatomic fragment. Because the initial bound state of the molecule can be described, c.f. (13), in terms of small amplitude bending vibration, we may invoke the small angle approximation in which it is found that [54]

$$\phi = \rho\delta, \tag{19}$$

where the parameter ρ can be represented in terms of the masses and equilibrium bond lengths in the bound initial state

$$\rho = \frac{m_3 r_{23}[(m_1+m_2)r_{23}+m_1 r_{12}]}{m_3(m_1 + m_2)r_{23}^2 + 2\, m_1 m_3 r_{12} r_{23} + m_1(m_2+ m_3)\, r_{12}} . \tag{20}$$

It should be emphasized that the solution of the integral equations (16) or the solution of the driven differential equations (17) by the use of an expansion in the basis functions (18) requires the evaluation of the primitive transition matrix elements,

$$\left\langle \psi_{Enlj\mu m}(\underset{\sim}{Q}') \right| V_{fi}(\underset{\sim}{Q}')J^{\frac{1}{2}}(\underset{\sim}{Q}') \left| \psi_i(\underset{\sim}{Q}) \right\rangle, \tag{21}$$

which are now six-dimensional integrals. Even given the complication of the axis switching process, the evaluation of the rotational integrals over the Euler angles in (21) is straightforward but tedious [54]. Again invoking Condon-like approximations this then leaves us with the evaluation of three-dimensional non-separable integrals involving either the three vibrational coordinates in the initial bound molecular state or the final state scattering coordinates involving the diatomic vibrational coordinate Q_2', the reaction coordinate Q_1' describing the distance between the atom and the diatom and finally the orientation of the vector from the atom to the center of mass of the diatomic molecule with the axis of the diatomic molecule. The transformation between the two coordinate systems $\underset{\sim}{Q}$ and $\underset{\sim}{Q}'$ is a nonlinear one, and this is the reason for the presence of the Jacobian factor in (14), (17) and (21). This nonlinearity also severelv complicates the treatment of the Franck-Condon integrals, but it can be shown that they can likewise be reduced to sums of one-dimensional integrals of the form (8) multiplied by factors arising from the bending and rotational degrees of freedom [4]. These latter factors incorporate all the constraints of angular momentum conservation and they have been evaluated analytically for cases where the equilibrium bound state geometry is linear or bent. As the details are rather involved, they will be presented elsewhere [54, 55].

Given the primitive basis function integrals (21) it can be shown [51] that the transition amplitudes for producing fragments in given internal states can be obtained from the asymptotic limit of the functions $\psi_{Enlj}(Q_1')/(k_{nj}Q_1')$ in the limit of $Q_1'\rightarrow\infty$. The matrix elements (21) provide us with the direct transition amplitudes which ignore the coupling between degrees of freedom on the repulsive surface that occurs during the half-collision. A full description of calculations of these direct transition amplitudes for model molecular systems would be too lengthy to be presented here, and they will therefore be given elsewhere [54,55].

This section is designed to sketch how a full quantum mechanical theory can be developed which includes a proper description of the final state interactions as well as a full description of the sudden Franck-Condon rearrangement process in a true three-dimensional world in which bending vibrations and rotational degrees of freedom are correctly treated. The Franck-Condon integrals are shown to provide the direct transition amplitude, while the solution of the half-collision scattering equations requires the input of these Franck-Condon integrals or something mathematically equivalent.

5. Discussion

Photodissociation is central to the understanding of basic physical processes in the chemistry of photochemical systems and of upper atmosphere and planetary dynamics. The description of these phenomena in polyatomic molecules presents us with a multidimensional relaxation process in which the available energy is partitioned into rotational, vibrational, translational and even electronic energy of the fragments. The theory emphasizes two qualitatively different, yet physically inseparable, steps in the photodissociation process. Firstly, there is a sudden Franck-Condon transition whereby the molecule is transported from a bound potential energy surface to the dissociative potential energy surface. These two potential energy surfaces generally have totally different natural coordinate systems for describing the nuclear motion. This sudden change in nuclear coordinate systems is shown to provide one important source for pumping energy into the internal degrees of freedom of the fragments. There are then inter-fragment interactions which appear on the repulsive potential energy surface which can lead to relaxation of the nascent energy distributions produced in the sudden Franck-Condon rearrangement. This half-collision process can likewise produce excitation or deexcitation in these internal degrees of freedom.

Despite the apparent highly nonseparable character of the multidimensional bound-continuum Franck-Condon integrals, it is shown that the use of harmonic oscillator basis functions enables these integrals to be reduced to simple one-dimensional bound-continuum forms which may be evaluated numerically or which may be approximated analytically by semiclassical methods. For the more general case of a full three-dimensional description of photodissociation, the inclusion of rotational degrees of freedom and the bending degree of freedom severely complicates the mathematics, so the theory has been limited to date only to the case of triatomic molecules. An outline is given of the description of the final state interactions in terms of driven scattering equations where the initial bound state wavefunctions, multiplied by an appropriate coupling operator, is propagated on the final state repulsive potential energy surface. The solution of these driven half-collision equations requires as input the sudden Franck-Condon transition matrix elements, and the treatment of these quantities is briefly discussed for the case of the dissociation of an initially linear triatomic molecule. Work describing the rotational energy distributions produced by the three dimensional theories as well as the fragment angular distributions will be given elsewhere [54,55].

6. Acknowledgments

The work discussed in this talk has been carried out in collaboration with Professor Yehuda Band at the Ben Gurion University of the Negev, Beer-Sheva, Israel and with Dr. Michael Morse of the University of Chicago. This research is supported, in part, by NSF Grant CHE77-24652.

References

1 O. K. Rice: J. Chem. Phys. 1, 373 (1933)
2 R. A. Harris: J. Chem. Phys. 39, 978 (1963)
3 R. Zare: Mol. Photochem. 4, 1 (1972)
4 S. Yang, R. Bersohn: J. Chem. Phys. 61, 4400 (1974)
5 S. Mukamel, J. Jortner: J. Chem. Phys. 61, 5348 (1974)
6 W. H. Miller, T. F. George: J. Chem. Phys. 56, 5637 (1972)
7 W. H. Miller: Adv. Chem. Phys. 25, 63 (1974)
8 J. R. Laing, K. F. Freed: Phys. Rev. Lett. 34, 809 (1975); Chem. Phys. 19, 91 (1977)
9 M. J. Berry: Chem. Phys. Lett. 29, 323 (1974); G. A. West, M. J. Berry, J. Chem. Phys. 61, 4700 (1974)
10 P. J. Robinson, K. A. Holbrook, Unimolecular Reactions (Wiley Interscience, New York, 1972)
11 K. F. Freed: Chem. Phys. Lett. 42, 600 (1976); K. F. Freed, Faraday Discuss. 67 (in press)
12 J. D. Rybrant, B. S. Rabinovitch: J. Chem. Phys. 54, 2275 (1971); J. Phys. Chem. 75, 2164 (1971)
13 D. L. Bunker, S. A. Jayich: Chem. Phys. 13, 129 (1976)
14 For a more extensive review see J. P. Simons: Chem. Soc. Specialist Periodic Rep. Gas Kinetics and Energy Transfer, 2, 58 (1977)
15 K. E. Holdy, L. C. Klotz, K. R. Wilson: J. Chem. Phys. 52, 4588 (1970)
16 M. Dzvonik, S. Yang, R. Bersohn: J. Chem. Phys. 61, 4408 (1974)
17 M. Kawasaki, S. J. Lee, R. Bersohn: J. Chem. Phys. 63, 809 (1975)
18 J. G. Moehlmann, J. D. McDonald: J. Chem. Phys. 62, 3052, 3061 (1975)
19 W. M. Jackson, R. J. Cody: J. Chem. Phys. 61, 4183 (1974)
20 M. Sabety-Dzvonik, R. J. Cody: J. Chem. Phys. 64, 4794 (1976)
21 A. Baronavski, J. R. McDonald: Chem. Phys. Lett. 45, 172 (1977)
22 T. Carrington: J. Chem. Phys. 41, 2012 (1964)
23 A. Mele, H. Okabe: J. Chem. Phys. 51, 4798 (1969)
24 H. Okabe: J. Chem. Phys. 62, 2782 (1975)
25 L. C. Lee, D. L. Judge: Can. J. Phys. 51, 378 (1973); J. Chem. Phys. 63, 2782 (1974)
26 T. G. Slanger, R. L. Sharpless, G. Black, S. V. Filseth: J. Chem. Phys. 61, 5022 (1974)
27 G.M. Lawrence: J. Chem. Phys. 56, 3435 (1972)
28 R. C. Mitchell, J. P. Simons: Discuss. Faraday Soc. 44, 208 (1967); J. P. Simons, P. W. Tasker: Mol. Phys. 26, 1267 (1973); 27, 1691 (1974);
29 M. N. R. Ashfold, J. P. Simons: J. C. S. Faraday Trans. II, 73, 858 (1977) 74, 280, 1263 (1978); Chem. Phys. Lett. 47, 65 (1977); M. N. R. Ashfold, M. T. MacPherson, J. P. Simons, Chem. Phys. Lett. 55, 84 (1978)
30 F. Flouquet: Chem. Phys. 13, 257 (1976); F. Flouquet and J. A. Horsley: J. Chem. Phys. 60, 3767 (1974)
31 G. J. Vazquez, J. F. Gouyet: Chem. Phys. Lett. 57, 385 (1978); (in press)
32 R. L. Jaffe, S. R. Langhoff: J. Chem. Phys. 68, 1638 (1978)
33 H. S. Johnston: Gas Phase Reaction Theory (Ronald Press, New York, 1966)
34 K. F. Freed, S. H. Lin: Chem. Phys. 11, 409 (1975)
35 K. F. Freed, Y. B. Band: Excited States 3, 109 (1978)
36 Y. B. Band, M. D. Morse, K. F. Freed: J. Chem. Phys. 68, 2702 (1978) and references therein
37 R. G. Gilbert, I. G. Ross: Aust. J. Chem. 24, 1541 (1971)
38 Y. B. Band, K. F. Freed: Chem. Phys. Lett. 28, 328 (1974); J. Chem. Phys. 63, 3382 (1975); M. D. Morse, K. F. Freed, Y. B. Band: Chem. Phys. Lett. 49, 399 (1977)
39 P. Fink, C. F. Goodeve: Proc. Roy. Soc. (London) A163, 592 (1937); D. Porret, C. F. Goodeve: Proc. Roy. Soc. (London) A165, 31 (1938); Trans. Faraday Soc. 33, 690 (1937)

40 A Gordus, R. B. Bernstein: J. Chem. Phys. 22, 790 (1954); 30, 973 (1959); H. Friedman, R. B. Bernstein, H. E. Gunning: J. Chem. Phys. 30, 973 (1959)
41 M. Shapiro, R. D. Levine: Chem. Phys. Lett. 5, 499 (1970)
42 S. Mukamel, J. Jortner: J. Chem. Phys. 60, 4760 (1974)
43 R. A. Marcus: J. Chem. Phys. 45, 4493 (1966); J. C. Light, R. Walker: J. Chem.Phys. 65, 4272 (1976)
44 Y. B. Band, K. F. Freed: J. Chem. Phys. 68, 1292 (1978)
45 P. Pechukas, J. C. Light: J. Chem. Phys. 44, 3897 (1966)
46 S. Mukamel, J. Jortner: J. Chem. Phys. 65, 3735 (1976)
47 O. Atabek, J. A. Beswick, R. Lefebvre, S. Mukamel, J. Jortner, Mol. Phys. 31, 1 (1976)
48 J. A. Beswick, M. Shapiro, R. Sharon: J. Chem. Phys. 67, 4045 (1977); O. Atabek, R. Lefebvre: J. Chem. Phys. 67, 4983 (1977)
49 G. A. West; M. J. Berry: private communication
50 R. T. Pack: J. Chem. Phys. 65, 4765 (1976)
51 Y. B. Band, K. F. Freed, D. J. Kouri: to be published
52 E. J. Heller: J. Chem. Phys. 68, 3891 (1978); K. C. Kulander, E. J. Heller: J. Chem. Phys. 69, 2439 (1978)
53 M. Shapiro: J. Chem. Phys. 56, 2582 (1972); Isr. J. Chem. 11, 691 (1973)
54 M. D. Morse: thesis, University of Chicago (1980); M. D. Morse, K. F. Freed: to be published.
55 The theory without dynamic axis switching is given in M. D. Morse, K. F. Freed, Y. B. Band: J. Chem. Phys. 70, 3604, 3620 (1979); Y. B. Band, M. D. Morse, K. F. Freed: Chem. Phys. Lett. 67, 294 (1979)

Determination of Molecular Relaxation Constants by Coherent Transient Spectroscopy

T. Shimizu

Department of Physics, University of Tokyo
Hongo, Bunkyo-ku, Tokyo 113, Japan

ABSTRACT

Procedures to determine rate constants of the individual fundamental relaxation processes among vibrational-rotational energy levels of molecules are described. The experiments were carried out on microwave and infrared transitions of NH_3, OCS, and H_2O by methods of coherent transient spectroscopy with lasers.

1. Introduction

It admits of no doubt that the recently developed method of coherent transient spectroscopy has opened a unique and useful field for investigations of dynamic properties of atoms, molecules and solids. In these several years various transient phenomena due to interactions of the material system with a coherent radiation field have been demonstrated in a viewpoint of providing optical analogs of the magnetic resonance. However, there are problems to be overcome until the method is established as a really spectroscopic one. The spectroscopy is based on accumulations of accurate and reliable data. The method has to provide unambiguous physical constants repeatedly, which are of course consistent with ones obtained by other methods.

Apparent signals obtained in the coherent transient spectroscopy of gaseous molecular system are generally results of complex mixtures of several relaxation processes and are also frequently disturbed by some "spurious effects" such as Doppler dephasing, field inhomogeneity, collisional velocity jump, saturation effect, and so on. For example, in Figs.1 and 2 are shown typical signals of the coherent transients at microwave and infrared frequencies. These are due to heterodyne beat detections of emissions from the macroscopic dipole moments induced in the molecular system by the coherent radiation field. Besides the transition frequencies, the experimental conditions in both cases are almost identical. Decays of the signals should represent relaxation of the induced dipole moments. The signal at infrared frequency, however, shows much faster decay than that at microwave frequency. This is due to a large Doppler broadening of the infrared transition. The infrared signal is a superposition of responses from a group of moving molecules which have individual emission frequencies in reference to the laboratory frame. On the other hand the Doppler shift at microwave frequency is negligibly small in comparison with an inverse of observation time of the transient signal. The microwave signal is free from Doppler dephasing, but is not completely free from other effects. For instance, the microwave field in the wave-guide cell is rather inhomogeneous and each molecule is subjected to a radiation field with different amplitude. This causes another type of dephasing effect in the coherent transient signal.

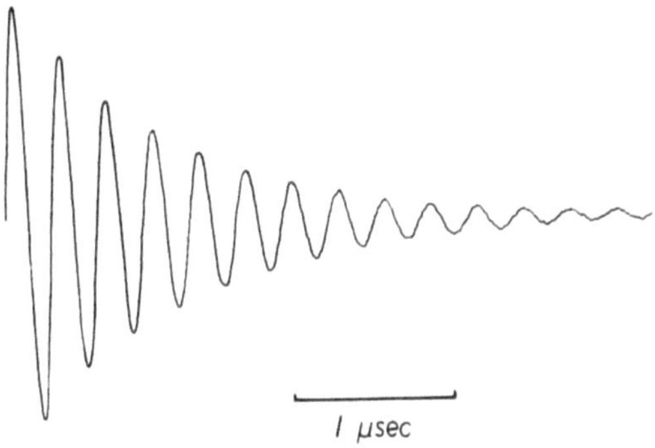

Fig.1. Free induction beat signal observed on the J=3, K=2 microwave inversion transition of $^{15}NH_3$. The pressure is 7.2 mTorr.

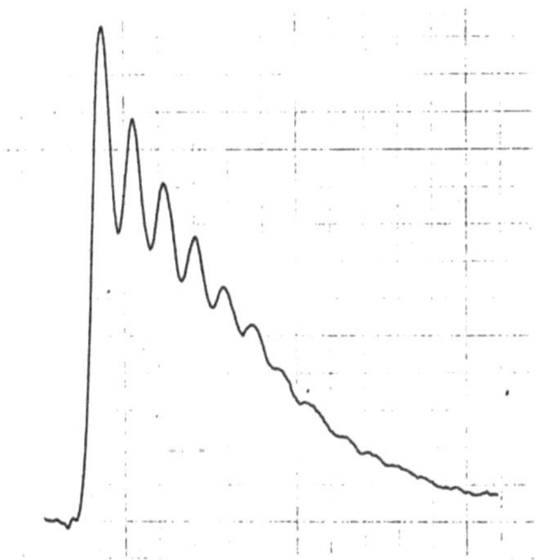

Fig.2. Free induction beat and nutation signal observed on the $\nu_2{}^qQ(J=4,K=4)$ transition of $^{15}NH_3$ by using the P(15) transition of N_2O laser. The pressure is 5 mTorr and the time scale is 0.1 μs/division.

ort we will discuss
emove the "spurious effects" from the observations,
solate an individual relaxation process from others,
ons among the results obtained from various methods including
the frequency domain,
ons on merits and demerits of various switching techniques in
excitation, and
cal interpretations of the obtained rate constants.

)f time-resolved emission at microwave frequency was first pro- ›rved by DICKE and ROMER in 1955 [1]. Only a few works [2] had until Brewer and Shoemaker succeeded in observation of infrared ıal with lasers by using Stark switching method [3]. Activities :roscopy have stimulated microwave spectroscopists and a number of : since then [4]. The microwave and infrared approaches are com- ı each other to study relaxation mechanisms among vibration-rota- ›vels of molecules. We carried out various types of time-resolved observations in both microwave and infrared frequencies and data synthetically to obtain the rate constants of individual elaxation processes.

2. Basic idea of the method

Time development of the components of density matrix of a two-level system due to interaction with the classical coherent radiation field, E $\cos\omega t$, is described on the semiclassical approximation in the rotating frame by

$$\frac{d}{dt}\,\mathrm{Re}\rho_{12} = -(\omega_0-\omega)\mathrm{Im}\rho_{12} \quad , \tag{1}$$

$$\frac{d}{dt}\,\mathrm{Im}\rho_{12} = (\omega_0-\omega)\mathrm{Im}\rho_{12} + \frac{\bar{\mu}E}{\hbar}(\rho_{22}-\rho_{11}) \quad , \tag{2}$$

$$\frac{d}{dt}(\rho_{22}-\rho_{11}) = -\frac{4\bar{\mu}E}{\hbar}\,\mathrm{Im}\rho_{12} \ , \tag{3}$$

where $\omega_0=(E_2-E_1)/\hbar$ and $\bar{\mu}$ are the Bohr angular frequency and the dipole matrix element of the system, respectively. A transient behavior of the molecule can be geometrically represented by motion of the Bloch vector in a three-dimentional space, where the Bloch vector has components of $r_1=2\mathrm{Re}\rho_{12}$, $r_2=2\mathrm{Im}\rho_{12}$, and $r_3=\rho_{22}-\rho_{11}$. The torque vector is $\omega = (2\bar{\mu}E/\hbar,\ 0,\ \omega_0-\omega)$.

The system is also subjected to a random perturbation, $H^{(r)}(t)$, which may destroy coherence of the system produced by radiation and also may change population densities on the relevant levels. If a mean amplitude and a correlation time of the random perturbation are so small that $\langle|H^{(r)}(t)|\rangle\,\tau_c \ll \hbar$, the components of density matrix vary slowly with time as expressed by

$$\frac{d\rho_{12}}{dt} = -[\tfrac{1}{2}(\sum_{k\neq 1} w_{1k}+\sum_{k\neq 2} w_{2k})+\gamma_{12}^{a}]\rho_{12} = -[\tfrac{1}{2}(\Gamma_1+\Gamma_2)+\gamma_{12}^{a}]\rho_{12} \ , \tag{4}$$

$$\frac{d\rho_{nn}}{dt} = -(\sum_{k\neq n} w_{nk})\rho_{nn}+\sum_{k\neq n} w_{kn}\rho_{kk} \ , \tag{5}$$

where

$$w_{nk} = \frac{1}{\hbar^2}\int_{-\infty}^{\infty}\langle H_{nk}^{(r)}(t)H_{nk}^{(r)}(t-\tau)\rangle \exp[-i\omega_{kn}\tau]d\tau \tag{6}$$

is the rate of collision induced transition from the level n to the level k and $\Gamma_i=\sum_{k\neq i} w_{ik}$ represents an overall nonadiabatic relaxation rate from the level i.

$$\gamma_{12}^{a} = \frac{1}{\hbar^2}\int_{0}^{\infty}\langle\{H_{11}^{(r)}(t)-H_{22}^{(r)}(t)\}\{H_{11}^{(r)}(t-\tau)-H_{22}^{(r)}(t-\tau)\}\rangle d\tau \tag{7}$$

consists of only diagonal elements of the perturbation and represents an adiabatic relaxation in the system. It is obvious from Eq.(4) that the relaxation of the off diagonal element can be generally described by a single parameter of

$$T_2^{-1} = \frac{1}{2}(\Gamma_1+\Gamma_2)+\gamma_{12}^{a} \tag{8}$$

On the other hand only when the conditions of $w_{kn}\rho^0_{kk}=w_{nk}\rho^0_{nn}$ (detailed balance), $\rho_{kk}\sim\rho^0_{kk}(k\neq 1,2)$, and $w_{12},w_{21}\ll\Gamma_1,\Gamma_2$ are satisfied, and also the levels 1 and 2 are identical in respect of relaxation, that is $\Gamma_1\sim\Gamma_2$, the decay of population difference $(\rho_{22}-\rho_{11})$ can be described by a single parameter of

$$T_1^{-1} = w_{12}+w_{21}+\frac{1}{2}\sum_{k\neq 1,2}(w_{1k}+w_{2k}) \quad . \tag{9}$$

If the levels 1 and 2 are not identical the population difference may decay in a form of

$$A_1\exp[-\Gamma_1 t]+A_2\exp[-\Gamma_2 t] \quad . \tag{10}$$

When a duration of collision is much shorter than a mean interval between two successive collisions as in spectroscopy of low pressure gas, the coherent and the random perturbations can be manipulated independently. In the most favorable case Eqs.(1),(2), and (3) and Eqs.(4) and (5) are simply combined and an equation of motion of the Bloch vector can be described by

$$\frac{d\underline{r}}{dt} = \underline{\omega}\times\underline{r} - \frac{r_1}{T_2}\underline{1} - \frac{r_2}{T_2}\underline{2} - \frac{r_3-r_3^0}{T_1}\underline{3} \tag{11}$$

where $\underline{1}$, $\underline{2}$, and $\underline{3}$ are the unit vectors in $\underline{r}$-space. Since behaviors of actual molecular systems may deviate more or less from Eq.(11), characteristic features of the individual system should be discussed case by case.

3. Removal of spurious effects

3.1 Doppler dephasing

It is well known that the signal of photon echo is primarily free from the Doppler effect. A polarization vector induced on an individual molecule by a $\pi/2$ pulse of coherent radiation makes a precession in the $\underline{1}$-$\underline{2}$ plane around the $\underline{3}$ axis with a Doppler shifted angular frequency of $\omega_0(v)-\omega$. Due to these inhomogeneously distributed precession speeds the macroscopic dipole moment may decay quite rapidly. However, after an application of a π-pulse, the dephasing process is reversed and the individual polarization vectors rephase to result in the macroscopic dipole again, which gives rise to an echo emission.

If the velocity of molecule is conserved during these processes, the resultant dipole is diminished only by collisional relaxation of the induced polarization. The decay of echo amplitude as a function of the interval between two pulses gives the relaxation time constant of T_2. An observation time should be comparable with or fairly longer than the relaxation time.

3.2 Effect of velocity changing collision

A characteristic time constant of collisional velocity jumping process may be interpreted as a cross relaxation time or a spectral diffusion constant in the inhomogeneously broadened Doppler line. The effect, however, disturbs the transient signals. If the velocity jump occurs frequently during the observation time, the rephasing in the photon echo process cannot be complete and the resultant macroscopic dipole diminishes rather rapidly with an increase of the time interval between two pulses.

The effect of velocity changing collision on the photon echo signal may be appreciable when $t_{obs} \gtrsim (k\Delta u)^{-1}$, where k is the wave number vector of radiation and Δu the average amount of velocity jump in a single collision. For instance, in a case of the infrared transition of NH_3, $k\Delta u$ is typically of an order of 10^5Hz. If the echo experiment is carried out at relatively high pressure, the observation time can be set to be not longer than several microseconds and the echo signal is fairly free from the effect of velocity jump.

The effect of velocity jump may disturb also other types of coherent transient phenomena. The criterion of the observation time range for the free induction decay experiment may be similar with one described above. For the delayed nutation experiment described below, t_{obs} should be less than $\Delta\nu_h(k\Delta u)^{-2}$ where $\Delta\nu_h$ is homogeneous hole produced by the laser radiation in the Doppler broadened line profile. For radiation power density of several hundreds milliwatts per square centimeter, the saturation width is of an order of 10^6Hz and $\Delta\nu_h(k\Delta u)^{-2}$ is roughly a hundred microseconds. Determination of T_1 in several microseconds region may be quite free from the effect of velocity change.

A sophisticated way to escape from the effect may be an application of the multiple pulse echo method [6] (Carr-Purcell echo), where the time interval between successive two pulses is shorter than the characteristic time constant of velocity jump, but an overall observation time is reasonably long compared with the relaxation time. The detailed discussions on the effect may be found in the references [6].

3.3 Inhomogeneity of radiation field

Several types of coherent transient experiments may not be performed satisfactorily at the microwave frequency. Inhomogeneity of microwave field in a wave guide type absorption cell is appreciably large. Molecules are subjected to different fields and then the individual molecules have different Rabi frequencies, $\bar{\mu}E/\hbar$. This causes another type of dephasing effect. The nutation signal may decay with a much shorter time constant than the relaxation time. The $\pi/2$ pulse condition in the photon echo experiment cannot be satisfied simultaneously for an entire molecular system. The echo signal may diminish fast with time because of mismatching in the conditions.

The rotary echo method so gives a signal being free from field inhomogeneity as the photon echo signal is free from the Doppler broadening. The rotary echo is interpreted as the processes of dephasing and rephasing of the polarization vectors in the 2-3 plane. In Fig.3 a series of movements of the polarization vectors in the Bloch space are depicted. In thermal equilibrium all vectors are orientating along the 3-axis. Applied radiation fields make the vectors rotate around the 1-axis inducing polarization on each molecule (a). Since the rotation speeds are different among them, the resultant vector gradually diffuses (b) and the nutation signal from the molecular system decays fast. Then the radiation is switched off for the interval Δt so that $(\omega-\omega_0)\Delta t=\pi$. Each vector makes field free precession around the 3-axis and comes into the opposite quadrant of the 2-3 plane (c). The radiation is applied again and individual vectors start to rotate around the 1-axis (d). The vector rephases at $t=2T+\Delta T$ (e), and then gives a nutation echo (f). The echo amplitude is smaller than that of the first nutation, because each component of the Bloch vector is subjected to T_1 and T_2 relaxations during this process. Fig.4 shows the rotary echo signal observed on the $J=5\leftarrow4$ microwave

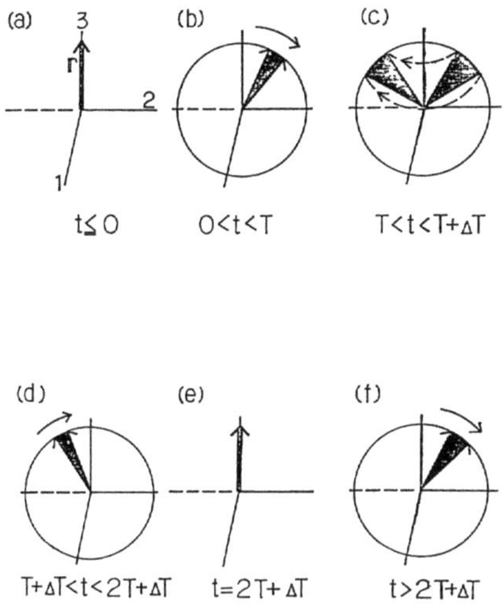

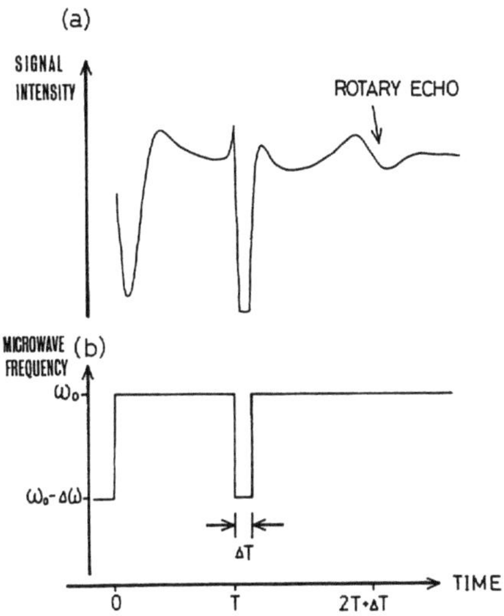

Fig.3 A schematic representations of the rotary echo in the Bloch space. The resonant radiation field is applied from t=0 to t=T, the frequency is switched by $\Delta\omega=\omega=\omega_0$ for an interval of Δt, and then applied again at $T+\Delta t$. Behaviors of the Bloch vectors are described in the text.

Fig.4 The rotary echo signal on the $J=5\leftarrow4$ microwave transition of OCS (a). The voltage pulse applied on the reflector electrode of klystron is also shown (b). The microwave field starts to drive the molecules at t=0. A short voltage pulse switches off the microwave frequency by $\Delta\omega$ for an interval of Δt, so that the phase shift of the microwave field is $\Delta\omega\cdot\Delta T=\pi$.

transition of OCS molecule. If we observe the amplitude of nutation echo as a function of $2T+\Delta T$, the relaxation rate constant of $(T_1^{-1}+T_2^{-1})/2$ of the system is obtained. The rotary echo signal is also free from the dephasing due to the presences of various nutation frequencies of unresolved M components.

3.4 Power dependence of the signal decay constant

Since the free induction and the nutation are due to the motion of polarization, the decay rate constants of signals are primarily determined by the T_2 relaxation. Because of saturation effect, a change in the population also comes into the processes. The effective decay rate constants are functions of T_1, T_2, and the saturation parameter (Rabi frequency), $\bar{\mu}E/\hbar$. This may be written by

$$\frac{1}{T_{eff}} = [(\frac{1}{T_2})^2+(\frac{T_1}{T_2})(\frac{\bar{\mu}E}{\hbar})^2]^{\frac{1}{2}} \tag{12}$$

at microwave frequency, and also by

$$\frac{1}{T_{eff}} = \frac{1}{T_2} + [(\frac{1}{T_2})^2+(\frac{T_1}{T_2})(\frac{\bar{\mu}E}{\hbar})^2]^{\frac{1}{2}} \tag{13}$$

at infrared frequency on the Doppler-limit condition. The latter is apparently faster than the former, because of the Doppler dephasing.

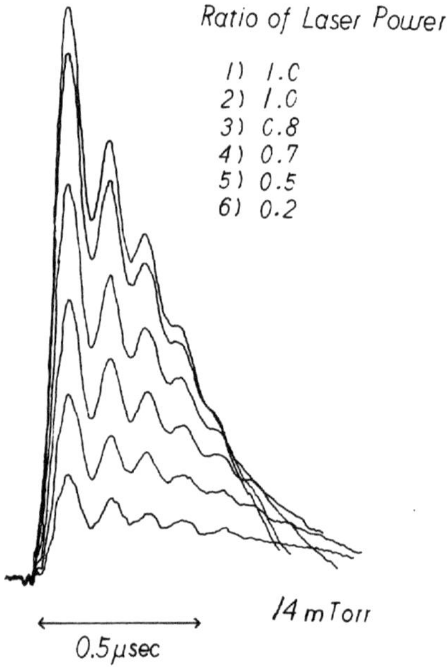

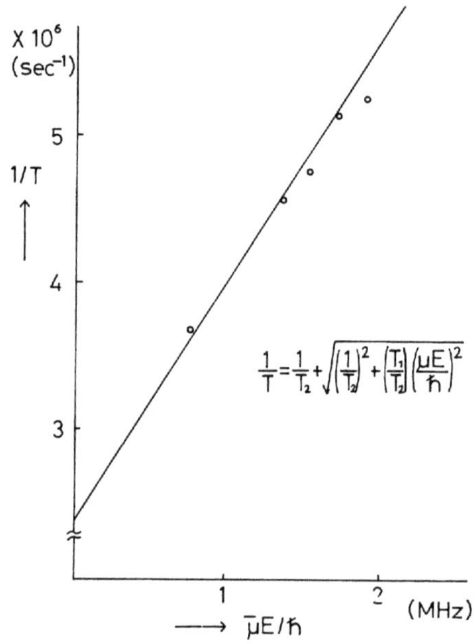

Fig.5 The free induction beat signal observed on the ν_2:asQ(J=5, K=3, M=5) vibration-rotation transition of NH_3 with the P(32) line of CO_2 laser. The relative laser power 1.0 corresponds to 1.9 W/cm^2. The sample pressure is 14 mTorr.

Fig.6 The effective decay rate constant of the free induction beat as a function of the saturation parameter. The observations are fitted with the parameters of $(2\pi p)^{-1}\Gamma_2$=8.1[MHz/Torr], $(2\pi p)^{-1}\Gamma_1$=24.1[MHz/Torr], and then $(2\pi T_1 p)^{-1}$=12.4[MHz/Torr], and $(2\pi T_2 p)^{-1}$=28[MHz/Torr].

The free induction beat signals observed at various power levels of the laser radiation are shown in Fig.5. The experiment was carried out on the $\nu_2$asQ(J=5, K=3, M=5) transition of NH_3 by using the P(32) line of CO_2 laser. An apparent decay constant depends on the radiation power. In Fig.6 the decay rate constants are depicted as a function of the saturation parameter and the graph is satisfactorily fitted by Eq.(13). From this analysis the effect of radiation power can be eliminated.

4. Discrimination of individual relaxation processes

In Fig.6 an extrapolated value of T_{eff}^{-1} at the zero field strength will give the value of T_2 at this observation pressure. It is also possible to determine T_1 from a gradient of the curve. However an accuracy in determination of T_1 from this method may not be good. An alternative method, the delayed nutation, will be described here, as an example.

Irradiation of the molecular system for a sufficiently long time results in saturation in population difference described by

$$r_3^S = r_3^0 \frac{(\omega-\omega_0)^2+1/T_2^2}{(\omega-\omega_0)^2+1/T_2^2+(T_1/T_2)(\bar{\mu}E/\hbar)^2} \quad . \tag{14}$$

When the radiation is switched off at $t=t_1$, the population difference starts to recover to its equilibrium value, r_3^0, in a way of

$$r_3(t) = r_3^0+(r_3^S-r_3^0)\exp[-(t-t_1)/T_1] \quad . \tag{15}$$

If we apply the radiation field at $t=t_2$ again and observe the nutation, the time development of the signal is written by

$$r_2(t-t_2) = r_2(t_2)\cos\Omega t+(1/\Omega)\{(\omega-\omega_0)r_1(t_2)+(\bar{\mu}E/\hbar)r_3(t_2)\}\sin^2\Omega t \ , \tag{16}$$

where $\Omega = [(\omega-\omega_0)^2+(\bar{\mu}E/\hbar)^2]^{1/2}$. For the infrared transition, because of a large inhomogeneous broadening, the values of $\langle r_1(t_2)\rangle$ and $\langle r_2(t_2)\rangle$ averaged over Doppler line profile may be neglected, and then one obtains

$$r_2(t-t_2) \sim (\bar{\mu}E/\hbar\Omega)r_3(t_2)\sin^2\Omega t \quad . \tag{17}$$

The amplitude of the nutation signal is proportional to $r_3(t_2)$, and this recovers to its equilibrium value with the time constant T_1. When T_1 is not well-defined, $r_3(t)$ may be given by

$$r_3(t) = r_3^0 - \frac{r_3^0(\bar{\mu}E/\hbar)^2/2T_2}{(\omega-\omega_0)^2+1/T_2^2+[(\bar{\mu}E/\hbar)^2/T_2](\Gamma_1^{-1}+\Gamma_2^{-1})/2}$$

$$\times\{\frac{1}{\Gamma_1}\exp[-\Gamma_1 t]+\frac{1}{\Gamma_2}\exp[-\Gamma_2 t]\} \tag{18}$$

in place of Eq.(15). When isolation of the two exponential functions in Eq.(18) is good, one may obtain the nonadiabatic relaxation rate constants of Γ_1 and Γ_2 of the ground and excited states independently.

At microwave frequency we may set $\omega=\omega_0$. Eq.(16) becomes

$$r_2(t-t_2) = r_2(t_2)\cos[\bar{\mu}Et/\hbar] + r_3(t_2)\sin[\bar{\mu}Et/\hbar]. \tag{19}$$

T_1 is generally well defined for the microwave transitions, and the amplitude of the nutation signal is proportional to

$$A \sim [\{\frac{(\bar{\mu}E/\hbar)T_2}{1+(\bar{\mu}E/\hbar)^2T_1T_2}\exp[-(t_2-t_1)/T_2]\}^2$$

$$+\{1-\frac{(\bar{\mu}E/\hbar)^2T_1T_2}{1+(\bar{\mu}E/\hbar)^2T_1T_2}\exp[-(t_2-t_1)/T_1]\}^2]^{1/2} . \tag{20}$$

When the radiation field is strong enough so that $\bar{\mu}E/\hbar \gg (T_1T_2)^{-\frac{1}{2}}$, Eq.(20) becomes

$$A \sim 1-\exp[-(t_2-t_1)/T_1] \tag{21}$$

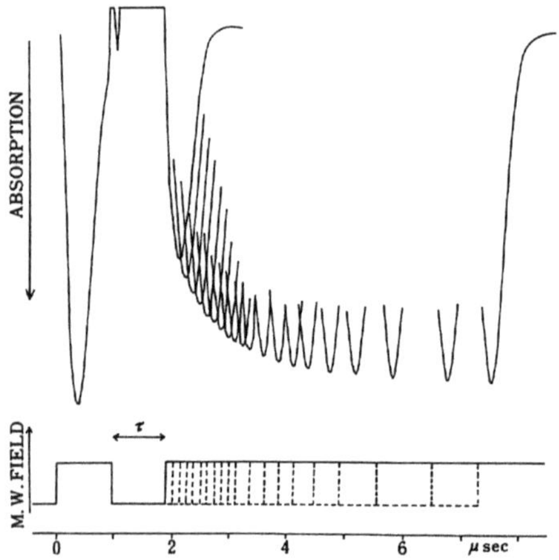

Fig.7. A composite chart of the delayed nutation signals as a function of delayed time. The recovery of amplitudes of the second nutations gives the relaxation time constant T_1. The experiment was carried out on the J=3, K=3 inversion transition of $^{15}NH_3$ by using the power switching technique.

As shown in Fig.7, the recovery of the delayed nutation with the time interval of (t_2-t_1), gives a fairly accurate value of T_1.

Another method of determination of the T_1 relaxation time is the adiabatic rapid passage technique. If one inverts the torque vector $\underline{\omega}$ fairly slowly in the Bloch space, the Bloch vector $\underline{r}$ will follow it adiabatically. The passage can be performed by sweeping the radiation frequency ω across the resonant frequency ω_0 as shown in Fig.8. This causes a change in the sign of the third component, $(\omega_0-\omega)$, of the torque vector. An inversion of the Bloch vector $\underline{r}$ in this passage results in population inversion in the molecular system. After the passage the Bloch vector will recover to its equilibrium position with the time constant of T_1, which may be obtained by monitoring the population difference in the two-level system by applying the second passage.

The adiabatic condition is expressed by

$$\frac{\Delta\omega}{\Delta t} \ll \left(\frac{\bar{\mu}E}{\hbar}\right)^2$$

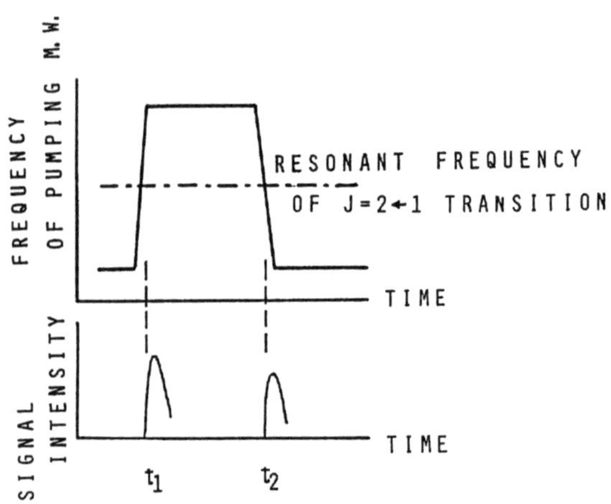

Fig.8. A schematic diagram showing the adiabatic rapid scans of the radiation frequency across the resonant frequency. The transient changes in absorption appeared in the probing transition are also shown.

where $\Delta\omega$ and Δt are the frequency shift in and the duration of the passage, respectively. Δt must be much shorter than the T_1 relaxation time. Both conditions can be satisfied simultaneously by preparing reasonable experimental conditions.

Since this process does not affect the transverse components of the Bloch vector, we cannot observe the molecular responses by transient absorption or emission from the relevant two-level system. We have successfully demonstrated a method of observation of this process by using double resonance technique on a three-level systems of NH_3[7] and OCS[8] molecules. As shown in Fig.9. the passages are carried out across the resonant frequency of the $J=2\leftarrow1$ transition. Population inversion occurs between the J=2 and J=1 levels. A change in population on the J=1 level is detected by measuring an intensity of the $J=1\leftarrow0$ transition. In Fig.10 the recovery of intensities of the transient absorption as a function of the time interval between two successive passages. The recovery time is exactly the T_1 relaxation time constant of the $J=2\leftarrow1$ transition.

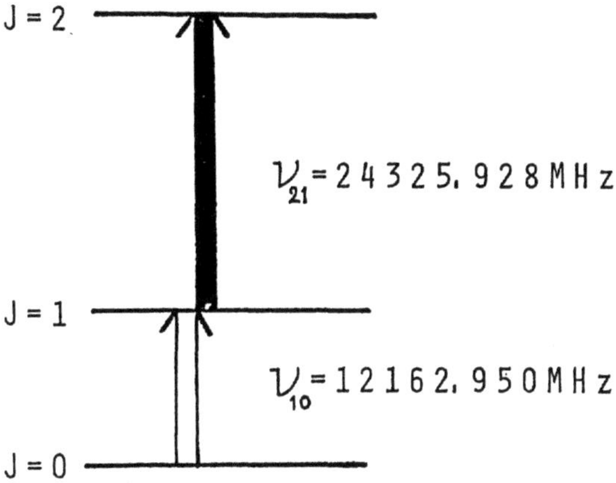

Fig.9. A three-level system of rotational transitions of OCS. The adiabatic rapid scan is applied on the $J=2\leftarrow1$ transition. The effect is monitored on the $J=1\leftarrow0$ transition.

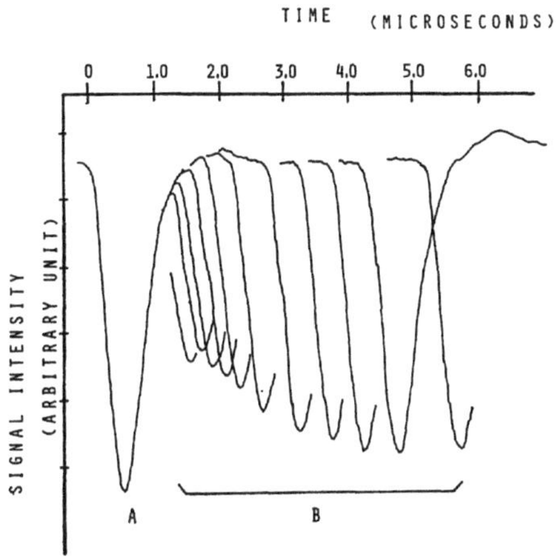

Fig.10. A composite chart of double resonance signals at two successive adiabatic rapid passages. A and B are observed at the first and the second passages, respectively.

5. Switching methods

In the Stark switching the resonance frequencies of molecules are switched by pulsed Stark field to be resonance with the fixed frequencies of radiation fields. Since the first successful demonstration by BREWER and SCHOEMAKER [3] the method has been conveniently used for a variety of transitions, because the experimental set up is rather simple and reasonably good results are obtained. This method may be practically only one available technique at infrared frequency.

However, the method is not applicable for a transition with small Stark coefficient. Degeneracy of an energy level is removed by application of the Stark field and the signal is a complex superposition of responses from various M components. The relaxation process itself might be affected by the high Stark field. An alternative switching technique should be employed on an occasion.

The frequency switching is easily achieved at microwave frequency by applying a pulsed voltage on the reflector electrode of klystron. However, a long term stabilization of the klystron frequency is essential to detect weak signals by integration of repeated responses. We have developed a novel method, in which the mean value of modulated microwave frequency is stabilized being referred to the standard frequency of the phase locked klystron [9]. The gate time of a microwave frequency counter is set to be much longer than the switching period. The mean value is compared with the preset standard in a digital circuit. A discrepancy is D-A converted and the error voltage is fed back to the reflector electrode of the klystron. By this method a long term frequency drift of the X-13 klystron at 11 GHz was kept within 10 kHz for several hours. Since the klystron can be controlled by a low voltage pulse, it is very easy to tailor a series of pulses which are necessary for observation of the transient responses.

The free induction beat and the rotary echo signals have been observed on the $6_{16} \leftarrow 5_{23}$ transition of H_2O, which has a very small Stark coefficient. The former gives the value of T_2 and the latter $(T_1^{-1}+T_2^{-1})/2$. The longitudinal relaxation constant of this transition has been determined for the first time by this method [10].

The power switching may be free from difficulties of the frequency stabilization and of the Stark effect described above. However a large change in the power level at the detector causes gain saturation in an amplifier and it is difficult to detect a weak transient signal superposed on the highly modulated carrier.

A useful method of power switching has been developed. A schematic diagram is shown in Fig.11. The microwave power from the phase-locked klystron is divided into two circuits. The microwave in the main wave guide $E_0\cos\omega t$ is switched by a PIN diode before a sample cell. The microwave in the side arm is attenuated and phase-shifted by π so as to be

$$\frac{1}{2}E_0\cos(\omega t+\pi),$$

and fed to the detector through a magic-T coupler. The resultant microwave power at the detector is proportional to the square of

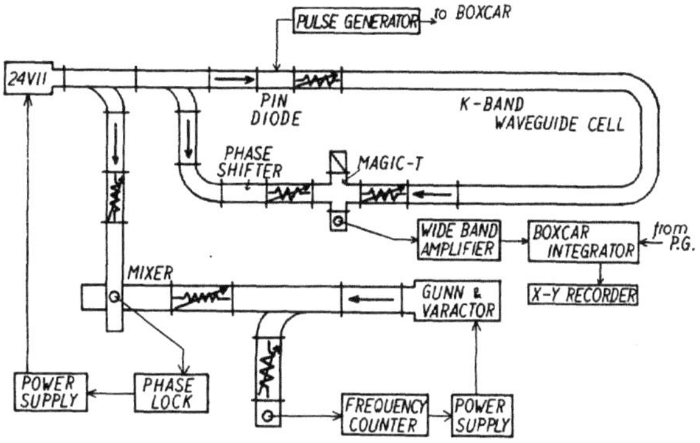

Fig.11 A schematic representation of the experimental setup for microwave power switching method.

$$E_0\cos\omega_0 t + \frac{1}{2}E_0\cos(\omega t+\pi) = \frac{1}{2}E_0\cos\omega t \quad ,$$

when the diode switch is open, and also proportional to the square of

$$0 + \frac{1}{2}E_0\cos(\omega t+\pi) = -\frac{1}{2}E_0\cos\omega t \quad ,$$

when the switch is closed. Therefore the microwave power level at the detector does not change by switching and only the signal from molecules can be amplified without gain saturation. The delayed nutation and the free induction decay signals have been observed on the several tens inversion transitions of NH_3. The experiment was carried out satisfactorily and the field free T_1 and T_2 values were determined.

6. Results and discussions

NH_3 molecule has a plenty of spectral lines in the microwave region. These are the transitions between the inversion doublets, which are well isolated two-level systems. These transitions are good targets of a systematic investigation of vibrational-rotational relaxations. A number of reports have been appeared especially on $^{15}NH_3$, because the transitions of this isotopic molecule are free from complex quadrupole hyperfine structures. Some results are shown in Tables I and II.

Generally T_2 relaxation constants can be determined accurately by well established methods. As seen in Table I agreements among the results obtained by various methods in the time and frequency domains are satisfactory. On the other hand, even when T_1 is well defined, some sophisticated experimental techniques are required to determine it. In Table II the agreements between the results obtained by two experiments in the time domain are fairly well, but noticeable discrepancies with the results from the frequency domain experiment are found.

A stationary-state solutions of Eq.(11),

$$\rho_{22}-\rho_{11} = (\rho_{22}^0-\rho_{11}^0)\frac{(\omega-\omega_0)^2+(1/T_2)^2}{(\omega-\omega_0)^2+[1/T_2^2+(\bar{\mu}E/\hbar)^2T_1/T_2]} \qquad (22)$$

Table I The transverse relaxation rate constants of inversion transitions in the ground vibrational state of $^{15}NH_3$

Rotational state J,K	$(2\pi pT_2)^{-1}$[MHz/Torr]		
	Power switching a)	Stark switching b)	Saturated linewidth c)
1.1	21.5±1.1	22.4	21.6
2.1		15.4	14.6
3.2	18.0±0.9	19.1	17.5
3.3	25.1±1.2	25.9	24.4
4.4	25.8±1.9*	25.0	
5.3			18.4
5.4	23.4±2.4*		22.0
8.7		23.4	23.7

a) Present investigation, reference [10].
b) Reference[11], these figures are recalculated from the data appeared in the literatures.
c) Reference[12].
* These values were obtained on the transitions of $^{14}NH_3$.

Table II The longitudinal relaxation rate constants of inversion transitions in the ground vibrational state of $^{15}NH_3$

Rotational state J,K	$(2\pi pT_1)^{-1}$[MHz/Torr]			
	Power switching a)	Stark switching b)	Saturated linewidth c)	Calculation d)
1.1	30.5±2.3	31.8	28.0	42
2.1		17.5	17.8	24
3.2	30.2±2.1	30.3	21.9	34
3.3	40.9±3.1	28.6(M=1) 38.2(M=2) 48.5(M=3)	34.6	51
4.4	44.7±4.3*	46.2		52
5.3			24.2	34
5.4	33.9±1.9*		31.5	43
8.7		27.1	33.7	50

a) Present investigation, reference [10].
b) Reference[11], these figures are recalculated from the data appeared in the literatures.
c) Reference[12].
d) Present investigation, reference[10].
* These values are obtained on the transitions of $^{14}NH_3$.

represents a spectral line shape and then the saturated linewidth is given by

$$\Delta\omega = [(\frac{1}{T_2})^2 + (\frac{T_1}{T_2})(\frac{\mu E}{\hbar})^2]^{\frac{1}{2}} \quad . \tag{23}$$

The dominant contribution to the linewidth comes from the T_2 process and the rate constant is determined accurately. However one may not expect a high accuracy in determination of T_1 from Eq.(23).

A nonempirical calculation of the rate constant has been carried out by using an impact theory by ANDERSON [13]. A dipole-dipole interaction may be the dominant contribution to the perturbation at NH_3-NH_3 collision. A classical trajectory is assumed. The rate constant of nonadiabatic relaxation from the relevant level, i, is calculated from the sum of probabilities of collision induced transitions to all connecting levels and may be written,

$$P_i = \frac{1}{\hbar^2}\sum_f \left|\int_{-\infty}^{\infty} \langle i|V|f\rangle \exp[\frac{i}{\hbar}(E_f-E_i)t]dt\right|^2 \tag{24}$$

with

$$V = [\underline{\mu_1}\,\underline{\mu_2} - \frac{3(\underline{\mu_1}\,\underline{r})(\underline{\mu_2}\,\underline{r})}{r^2}]r^{-3} \tag{25}$$

An integration in Eq.(24) is expressed by a modified Bessel function. However the dependence on an energy discrepancy, (E_f-E_i), may be roughly estimated from

$$\exp[-\frac{b}{v}\frac{E_f-E_i}{\hbar}] \tag{26}$$

where b is the impact parameter and v the molecular velocity. An inversion splitting in the ground state is less than 1 cm^{-1}. With typical values of the parameters, b=10A and v=500 m/s, the argument in the exponential function of Eq.(26) is evaluated to be of an order 1. Since in the ν_2 vibrational

Table III The transverse relaxation rate constants of the vibration-rotation transitions and the nonadiabatic relaxation rate constants of the rotational levels in the ν_2 excited state of NH_3.

Rotational state J,K	$(2\pi pT_2)^{-1}$[MHz/Torr] Photon echo	$(2\pi p)^{-1}\Gamma_{ex}$[MHz/Torr] Delayed nutation	Calculation
2.1		15±2	31.6
5.3	28±2	8.1±0.7	6.6
5.5		7	6.9
6.3		7.0±0.7	6.8
6.5		7.8±0.8	7.1
8.7	23	6.5*	

* From reference(14)

state a separation of the inversion doublet is about 30 cm^{-1}, the relaxation rate constants in the vibrationally excited state should be much smaller than those in the ground state unless an accidental coincidence between two interacting energy levels occurs. The results of numerical calculations are shown in Table II and III.

In Table III the relaxation rate constants of the vibrational excited state are also shown. The nonadiabatic relaxation rate constants of the rotational levels in the vibrational excited state are obtained isolatedly for the first time [10,14,15]. Although the calculation is based on a crude assumption, agreements between the observations and the calculations in Tables II and III are satisfactorily good.

The ratios of the transverse relaxation time to the longitudinal one are listed in Table IV for several rotational transitions in the ground vibrational state. In the case of inversion transition of NH_3, the two-level system is well isolated from others and each level is almost identical for collisional perturbation. As seen from Eq.(7) γ_{12}^{a} may be negligible in such a case. If the collision induced transition between these two levels is dominant, we obtain

$$T_2^{-1} = \frac{1}{2}(w_{12}+w_{21}) \tag{27}$$

from Eq.(8) and also

$$T_1^{-1} = w_{12}+w_{21} \tag{28}$$

from Eq.(9). From these relations it is found that $T_2/T_1=2$ in an ideal two level case. In another limiting case that $\sum_{k\neq 1,2} w_{ik} \gg w_{12},w_{21}$, the ratio T_2/T_1 is nearly equal to 1.

Table IV The ratios of collisional relaxation time constants, T_2/T_1 (the ground vibrational state)

Molecule	Transition	perturber	T_2/T_1
NH_3	3.3- ← 3.3+	NH_3	1.6
		He	1.1
NH_3	4.4- ← 4.4+	NH_3	1.7
		He	1.1
NH_3	5.4- ← 5.4+	NH_3	1.5
		He	1.0
H_2O	6_{16} ← 5_{23}	H_2O	1.5

As shown in Table IV for NH_3-NH_3 collision T_2/T_1 is appreciably larger than one. This means that the two-level system is almost closed even for collisional perturbation. This is because the electric dipole transition between nearby levels is dominant. This is consistent with the preference rule for collision induced transitions, which is established from the double resonance experiment [16]. The rule is that ΔJ=0, ±1, ΔK=0, and change in the parity of wave functions. This suggests taht the collision induced transition in the inversion doublet is dominant.

In the case of NH_3-He collision T_2/T_1 is nearly equal to 1. A short range interaction potential is valid in this type of collision, and then the transitions to other many levels are simultaneously induced by the collision. The double resonance study suggests that the preference rule for this type of collision may be described as follows: ΔK=0, ±3n, any ΔJ and the parity does not necessarily change. Therefore the monitored levels cannot be an isolated two-level system.

H_2O molecule has only a few transitions in microwave region. The $6_{16} \leftarrow 5_{23}$ transition is important, because a large scale maser action occurs in an interstellar space on this transition. T_1 constant has been determined for the first time in the present experiment using the frequency switching method. The rotational levels 6_{16} and 5_{23} lie accidentally close to each other, but the two-level system is not well isolated in a sense of collisional perturbation. The ratio T_2/T_1 is appreciably smaller than the value of the ideal two-level system.

Acknowledgements

The author would like to express his sincere thanks for collaborations of Drs. S. Kano, T. Amano, T. Kasuga, N. Morita, F. Matsushima, and H. Kuze on the microwave transient spectroscopy, and of Drs. N. Konishi and H. Sasada on the infrared transient spectroscopy.

References

1 R.H. Dicke and R.H. Romer: Rev. Scie. Instr. 26, 915 (1955)
R.H. Romer and R.H. Dicke: Phys. Rev. 99, 532 (1955)
2 R.M. Hill, D.E. Kaplan, G.F. Herrmann and S.K. Ichiki: Phys. Rev. Letters 18, 105 (1967)
3 R.G. Brewer and R.L. Shoemaker: Phys. Rev. Letters 27, 631 (1971)
4 J.C. McGurk, R.G. Schmalz and W.H. Flygare: Adv. Chem. Phys. 25, 1 (1974), R.H. Schwendeman: Ann. Rev. Phys. Chem. 29, 533 (1978)
Lists of references on transient spectroscopy will be found in these review papers.
5 N. Bloembergen: "Nonlinear Optics" Benjamin, N.Y. 1965
6 J. Schmidt, P.R. Berman, and R.G. Brewer: Phys. Rev. Letters 31, 1103 (1973)
7 T. Amano: Chem. Phys. Letters 37, 585 (1976)
8 T. Kasuga, T. Amano, and T. Shimizu: Chem. Phys. Letters 42, 278 (1976)
9 T. Kasuga, H. Kuze, and T. Shimizu: J. Chem. Phys. 69, 5195 (1978)
10 T. Shimizu, N. Morita, T. Kasuga, H. Sasada, F. Matsushima, and N. Konishi: Appl. Phys. 21, 29 (1980)
11 W.E. Hoke, D.R. Bauer, and W.H. Flygare: J. Chem. Phys. 64, 5276 (1976), 67, 3454 (1977)
12 T. Amano and R.H. Schwendeman: J. Chem. Phys. 65, 5133 (1976)
13 P.W. Anderson: Phys. Rev. 76, 647 (1949)
14 S.M. Hamadani, N.A. Kurnit, and A. Javan: Chem. Phys. Letters 49, 277 (1977)
15 M.M.R. Loy: Phys. Rev. Letters 32, 814 (1974)
16 T. Oka: Ad. At. Mol. Phys. 9, 127 (1973)

Index of Contributors

Aihara, M. 156

Burak, I. 206

Freed, K.F. 246

Hanamura, E. 71
Hasegawa, H. 130
Henry, C.H. 19

Ishida, Y. 121

Kanamaru, N. 181
Knox, R.S. 171

Laubereau, A. 101

Miyazaki, H. 71
Mukamel, S. 220,237

Nasu, K. 34

Prior, Y. 206

Rashba, E.I. 88

Shimizu, T. 268
Souma, H. 121

Taira, Y. 121
Takagahara, T. 45
Tanaka, J. 181
Toyozawa, Y. 3
Tsao, J. 206
Tsuchiya, S. 212

Yablonovitch, E. 197,206
Yajima, T. 121

S. Nakajima

The Physics of Elementary Excitations

With contributions by Y. Toyozawa, R. Abe
Translated from the Japanese by S. Nakajima, Y. Toyozawa, R. Abe

1980. 114 figures, 3 tables. Approx. 340 pages
(Springer Series in Solid-State Sciences, Volume 12)
ISBN 3-540-09921-2

Contents:
Families of Elementary Excitations: Crystals and Phonons. Polarization Waves and Dielectric Dispersion. Fermi Liquids. Phase Transitions and Elementary Excitations. – Interaction Between Elementary Excitations: Linear Interactions and Coupled Modes. Renormalization and Damping-Centering Around Electron-Phonon Interaction. Interaction Between Elementary Excitations and Spectral Line Shapes. – Bibliography.

Springer-Verlag
Berlin
Heidelberg
New York

Excitons

Editor: K. Cho

1979. 118 figures, 8 tables. XI, 274 pages
(Topics in Current Physics, Volume 14)
ISBN 3-540-09567-5

Contents:
K. Cho: Introduction. – *K. Cho:* Internal Structure of Exitons. – *P. J. Dean, D. C. Herbert:* Bound Excitons in Semiconductors. – *B. Fischer, J. Lagois:* Surface Exciton Polaritons. – *P. Y. Yu:* Study of Excitons and Exciton-Phonon Interactions by Resonant Raman and Brillouin Spectroscopies.

Picosecond Phenomena

Proceedings of the First International Conference on Picosecond Phenomena
Hilton Head, South Carolina, USA, May 24–26, 1978
Editors: C. V. Shank, E. P. Ippen, S. L. Shapiro

1978. 222 figures, 10 tables. XII, 359 pages
(Springer Series in Chemical Physics, Volume 4)
ISBN 3-540-09054-1

Contents:
Interactions in Liquids and Molecules. – Poster Session. – Sources and Techniques. – Biological Processes. – Poster Session. – Coherent Techniques and Molecules. – Solids. – High-Power Lasers and Plasmas. – Postdeadline Papers.

D. C. Hanna, M. A. Yuratich, D. Cotter

Nonlinear Optics of Free Atoms and Molecules

1979. 89 figures, 10 tables. IX, 351 pages
(Springer Series in Optical Sciences, Volume 17)
ISBN 3-540-09628-0

Contents:
Introduction. – Theory of the Nonlinear Optical Susceptibility. – Propagation of Plane Waves in a Nonlinear Medium. – Sum Frequency and Harmonic Generation. – Stimulated Electronic Raman Scattering. – Raman-Resonant Four-Wave Processes. – Nonlinear Optical Processes in Free Molecules. – Some Miscellaneous Topics. – References. – Subject Index.

FSC
www.fsc.org
MIX
Papier aus verantwortungsvollen Quellen
Paper from responsible sources
FSC® C105338

If you have any concerns about our products,
you can contact us on
ProductSafety@springernature.com

In case Publisher is established outside the EU,
the EU authorized representative is:
Springer Nature Customer Service Center GmbH
Europaplatz 3, 69115 Heidelberg, Germany

Printed by Libri Plureos GmbH
in Hamburg, Germany